World Climatic Systems

World Climatic Systems

John G. Lockwood

School of Geography
University of Leeds

Edward Arnold

First published in Great Britain 1985 by
Edward Arnold (Publishers) Ltd, 41 Bedford Square, London WC1B 3DQ

Edward Arnold (Australia) Pty Ltd, 80 Waverley Road, Caulfield East,
Victoria 3145, Australia

Edward Arnold, 300 North Charles Street, Baltimore, Maryland 21201, U.S.A.

British Library Cataloguing in Publication Data

Lockwood, John G.
 World climatic systems.
 1. Climatology
 I. Title
 551.6   QC981

 ISBN 0-7131-6404-2

Library of Congress Catalog Card Number
84-071799

Text set in 10/11 pt English Times
by Colset Private Limited, Singapore
Printed and bound by Richard Clay (The Chaucer
Press) Ltd, Bungay, Suffolk

Contents

Preface   ix
Acknowledgements   x

Part I   *Aspects of the Climatic System*   1

Chapter 1 *The general atmospheric circulation*   1
    1 The climatic system   1
        1.1 Isolated, closed and open systems   1
            Isolated systems   1
            Open systems   1
        1.2 Cascading systems and the atmosphere   2
        1.3 Physical components   3
        1.4 Further development   4
    2 Driving Forces   4
        2.1 Radiation distribution   5
            Global radiation   5
            Albedo   5
            Radiation balance   8
        2.2 The earth's rotation   8
            Baroclinic instability   11
    3 General Atmospheric Circulation   11
        3.1 The mean meridional circulation   11
        3.2 The subtropical jetstreams   12
    4 Tropical circulation patterns   13
        4.1 The trade winds   13
        4.2 The equatorial trough   14
        4.3 The Walker circulation and the southern oscillation   14
        4.4 Monsoon climates of southern Asia   15
    5 Middle and high latitude circulation patterns   17
        5.1 Atmospheric thermal patterns   18
        5.2 Index cycles and blocking systems   22
        5.3 Polar regions   24
    6 Circulation of energy in the atmosphere   24
        6.1 Energy content   24
        6.2 Energy transport   25
    7 Geological evolution of the general atmospheric
        circulation   30
        7.1 Changes in solar luminosity   30
        7.2 Albedo   30
        7.3 Greenhouse effect   31

7.4 Computational results   32
References   33

Chapter 2 Oceanic Subsystems   35
    1 Ocean circulation patterns   35
    2 The atmosphere–ocean interface   37
    3 Tropical and equatorial oceans   44
    4 Tropical oceanic weather systems   45
        4.1 Subtropical anticyclones   45
        4.2 The trade winds   45
        4.3 The equatorial trough   46
        4.4 Tropical rain-forming disturbances   47
        4.5 Easterly waves   48
        4.6 Tropical storms and cyclones   51
    5 Temperate oceanic weather systems   54
        5.1 Extratropical cyclones   55
        5.2 Precipitation and continentality over temperate
            oceans   56
    6 Polar oceans   59
    7 Atmosphere–ocean interactions   62
        7.1 Ocean surface temperature anomalies and climate   62
        7.2 Influence of the atmosphere on the ocean   63
        7.3 Effect of the ocean on the atmosphere   64
        7.4 The southern oscillation and El Niño   65
    References   68

Chapter 3 Glacial Subsystems   73
    1 Polar climates   75
        1.1 Radiation and temperature   75
        1.2 Inversions   86
        1.3 Surface temperatures   86
        1.4 Snow cover, precipitation and water balance   87
    2 Physics of large ice-sheets   88
    3 Snow and ice on mountains   89
        3.1 Present day snow-balance in Britain   90
    4 The history of ice ages through geological time   95
        4.1 PreCambrian glaciations   95
        4.2 Tertiary glaciations   97
        4.3 The influence of the Antarctic ice-sheet on global
            climate   102
        4.4 Quaternary glaciations   103
        4.5 The 18,000 BP continents   108
    5 Milankovitch theory and ice ages   110
        5.1 The astronomical theory of climatic change   110
    References   117

Chapter 4 Arid Subsystems   122
    1 Characteristics of desert climate   122
        1.1 Arid climates   122

1.2 Deserts   124
2 Radiation and temperature   125
3 Rainfall   127
  3.1 Rainfall variability   129
4 Evapotranspiration and water balance   130
  4.1 Southern Sahara   131
  4.2 North Africa and Middle East   135
5 African rainfall and teleconnections   135
6 The Indian Monsoon and global deserts   136
7 Closed lakes and the climatic history of North Africa   138
  7.1 African monsoons and variations in the earth's orbital
      parameters   142
8 Desertification   143
  8.1 Vegetation, albedo and climate   143
  8.2 Vegetation, soil moisture and climate   146
References   148

Chapter 5 Grassland and Vegetated Subsystems   153
1 Vegetation and water balance   153
2 Microclimates of grasslands   154
  2.1 Radiation   154
  2.2 Albedo   157
  2.3 Atmospheric turbulence   165
  2.4 Convection   169
3 Evapotranspiration   169
  3.1 The Bowen ratio   174
4 Hydrology of grasslands   176
  4.1 Resistances   177
  4.2 The Penman–Monteith evaporation equation   179
  4.3 Water balance   180
  4.4 Potential evapotranspiration   182
  4.5 Soil moisture models and actual evaporation   182
References   186

Chapter 6 Forest Subsystems   189
1 Forest microclimates   191
  1.1 Radiation   191
  1.2 Temperature and humidity   193
  1.3 Vertical exchanges   195
2 Forest hydrology   195
3 Forest clearance   209
  3.1 The multilayer crop model   209
  3.2 Some results of a multilayer crop model   211
  3.3 Experimental studies   221
4 Interception, evaporation and run-off in tropical forests   223
  4.1 Deforestation of the Amazon basin   229
References   232

viii   *Contents*

Part II   Impacts of the Climatic System   238

Chapter 7 Climate and Energy   238
      1   Climate as an energy source   241
      2   The pollution of the atmosphere by energy use   241
         2.1 Carbon dioxide   241
         2.2 Aerosols   249
         2.3 Waste heat   250
         2.4 Urban climates   252
      3   Climatic controls on energy use and the effects of climatic variations   254
         3.1 Long-term climatic changes   254
         3.2 Scenario of future climatic changes   257
      References   257

Chapter 8 Climate and Food   261
      1   Climate and plant growth   261
         1.1 Temperature   265
         1.2 Soil moisture   265
      2   Climate and grain production   266
      3   Drought   271
         3.1 Definitions of agricultural drought   272
         3.2 Teleconnections   278
         3.3 Drought and increasing atmospheric CO_2 278
      References   283

Index   285

Preface

The importance of climatology to various human activities has been growing in recent years. This is seen particularly in the fields of agriculture and energy use. Over the last decade climatic variations, including widespread drought in tropical countries, have caused major crop failures. Tropical deserts appear to be extending and this is sometimes blamed on incorrect agricultural techniques. Man has also upset the global carbon cycle by burning fossil fuels and by deforestation and changing land use. The net result of these activities has been to increase the CO_2 content of the atmosphere and the oceans, and to cause changes in both global temperature distributions and the global hydrological cycle over the next 50 years. The various climatic systems involved in these important areas of human activity are described in *World Climatic Systems*.

World Climatic Systems provides second and third year University and Polytechnic students in geography, environmental science, and related subjects such as agriculture, with a broad picture of the major climatic processes. It also forms a background text for postgraduate students in the climatological sciences. It is complementary to *Causes of Climate*, and assumes some basic knowledge of mathematics and physics, which should not be beyond physical geography students.

Dr L. Musk is thanked for his advice on the original manuscript.

John G. Lockwood
University of Leeds
February, 1984

Acknowledgements

The publishers would like to thank the following for permission to include copyright figures:

Academic Press Inc. for figs 1.1; 1.3; 2.4; 2.5; 5.7; The American Association for the Advancement of Science for figs. 2.11; 2.12; 3.12; 7.9; The American Meteorological Society for figs 1.2; 1.4; 2.2; 2.3; 2.13; 7.4; 7.5; Applied Science Publishers and A.H. Bunting for figs. 8.1; 8.2; 8.3; 8.4; Edward Arnold (Publishers) Ltd for fig. 7.6; Cambridge University Press for figs. 2.10; 8.9; Climatic Change for figs. 8.5; 8.10; 8.11; 8.12; Elsevier Publishing Co. for figs. 3.2; 3.3; 3.4; 3.5; 5.2; Her Majesty's Stationery Office for fig. 1.9 which is crown copyright and is used by permission of the Controller of HMSO; The Institute of Hydrology for figs. 5.9; 5.11; McGraw-Hill Company Ltd for fig. 4.1; Methuen and Co. Ltd for figs. 7.7; 7.8; National Academy Press for figs. 3.10; 7.1; 7.2; Macmillan Journals Limited for figs. 3.9; 3.14; Lawrence Livermore National Laboratory, University of California, the Department of Energy and Macmillan Journals Limited for fig. 4.10; Plenum Press for fig. 7.3; Quaternary Research Center for fig. 3.13; D. Reidel Publishing Co. for figs. 2.1; 4.9; 6.15; The Royal Meteorological Society for figs. 1.11; 1.3; 2.9; 4.8; Tellus for figs. 2.6; 3.6; UNESCO for figs. 4.5; 4.6; 4.7; John Wiley & Sons Inc. for figs. 1.12; 2.7; 2.8; World Meteorological Organization for figs. 4.4; 7.1; 7.2; 8.6; 8.7; 8.8 and Yale University Press Ltd for fig. 3.11. Detailed citations may be found by consulting captions and bibliography.

1 The General Atmospheric Circulation

1 The climatic system

A system may be defined as a structural set of objects and/or attributes, where these objects and attributes consist of components or variables that exhibit discernible relationships with one another and operate together as a complex whole, according to some observed pattern. The concept of the system is very useful in providing a means of understanding complex phenomena, provided that it is clearly understood that systems try to describe what happens in nature, and that nature cannot necessarily be forced into the mould of some particular preconceived system. Systems can be classified in terms of their function and also in terms of their internal complexity.

1.1 Isolated, closed and open systems

A common functional division is into isolated, closed and open systems.

   (a) *Isolated systems* have boundaries which are closed to the import and export of both mass and energy. Such systems are rare in the real world, though they may occur in the laboratory, i.e. a mass of gas within a completely sealed and insulated container.
   (b) *A closed system* is one in which there is no exchange of matter between the system and its environment though there is, in general, an exchange of energy. The planet earth together with its atmosphere may, very nearly, be considered a closed system.
   (c) *An open system* is one in which there is an exchange of both matter and energy between the system and its environment. There are numerous examples of open systems in nature, i.e. precipitating clouds, river catchments, plants, etc.

Isolated systems
Gas within a completely sealed and insulated container provides a good example of an isolated system. Whatever the original temperature gradients within the gas, temperatures will eventually become uniform, and while the system remains isolated nothing can check or hinder this inevitable levelling down of differences. Stated more generally, in an isolated system there is a tendency for the levelling down of existing differentiation within the system, and towards the progressive destruction of the existing order. In such a system there is always a decrease in the amount of free energy available for causing changes and doing work, and eventually the free energy will become zero.

Open systems
Open systems need an energy supply for their maintenance and preservation, and

are in effect maintained by the constant supply and removal of material and energy. Closed systems may be considered as a special case of open systems, there being no exchange of matter with the environment. It has already been noted that most of the systems observed within the natural environment belong to the open group. In particular, the open system has one important property which is not found in the isolated system, that is, it may attain a condition known as steady-state equilibrium. This is the condition of an open system wherein its properties are invariant when considered with reference to a given time-scale, but within which its instantaneous condition may oscillate due to the presence of interacting variables. Stated rather more simply, the general features of the system appear to remain constant over a long period of time, though there may be minor changes in details. Meteorological storms, such as hurricanes or thunderstorms, are good examples of open systems in a steady state, in that their general features remain relatively constant over periods of time ranging from several days in the case of the hurricane to several hours for a thunderstorm.

Open systems in the natural environment can be divided into three general categories, which may be termed decaying, cyclic and haphazardly fluctuating. Some systems always belong to one broad category while others change from one to another over relatively short periods of time.

Decaying systems consume their own substance, which may be energy or matter, or both. A good example is the decay of river-flow in dry weather, when the flow decreases each day but the rate at which the flow decreases also decreases with time and is proportional to the available water stored in the rocks. The rocks in the river catchmnt act as a store which supplies water to the river. In this case the river-flow approximates to a negative-exponential decay curve, and the amount of water stored in the rocks decreases to one-half of its original value in a given constant time interval.

The input of short-wave radiation follows diurnal and annual cycles, and these are imposed on many natural systems to form cyclic systems. Heat balances of land surfaces are largely controlled by the input of solar energy, and therefore show both diurnal and annual cycles. Air temperatures reflect the state of the heat balance of the surface and therefore also show marked diurnal and annual cycles. The variations in many cyclic systems when observed over a period of time appear to approximate to a mathematical curve known as a sine curve, which may be obtained by plotting the sine of an angle against the angle itself.

Haphazardly fluctuating systems change in a random and irregular manner, fluctuations occurring at unpredictable times and by unpredictable amounts. Turbulence in fluids or the occurrence of earthquakes are good examples, since neither can be exactly predicted. On small space- and time-scales most systems exhibit some degree of unpredictability.

1.2 Cascading systems and the atmosphere

From a climatological viewpoint, the atmosphere, oceans and land surfaces may be considered as consisting of a series of open systems of the type known as cascading systems.

Cascading systems are composed of a chain of subsystems, having both magnitude and geographical location, which are dynamically linked by a cascade of mass or energy and in this way, the output of mass or energy from one subsystem becomes the input for the next subsystem. Typically, the subsystems consist of an input into a store, which may contain a regulator controlling the amount of mass or energy remaining in the store or forming the output. The regulator may be a physical property of the store itself or it may be completely external to the store. More complex subsystems may have several inputs and outputs and even several regulators which decide how the mass or energy is divided between the various outputs. Many of the processes taking place in the atmosphere can be interpreted in terms of cascading systems, an example being provided by the cycle of water. Water may be stored in the oceans, the atmosphere (as water vapour), the soil, the deep rocks, rivers, etc., and the transfer of water from one store to another is controlled by various physical regulators. The output from the atmospheric store in the form of rain constitutes the input into the soil, where in turn one of the outputs forms the input into the deep rock storage, and so on until the water arrives back into the ocean where evaporation forms the input into the atmospheric store.

Interception of rainfall by a forest is a good example of a subsystem. The amount of water that can be carried on a leaf surface is limited, and so there is a definite upper limit to the amount of water that can be stored in a tree canopy and thus to the store of the subsystem. The input into the subsystem is rainfall and the outputs are the evaporation of the intercepted water and the gradual drip of water out of the trees onto the soil surface. At the start of the rainfall the tree canopies will be dry and no water will reach the soil, except through holes in the canopy. After some time the canopies will become completely saturated with water, and when this occurs most of the succeeding rainfall will eventually drip onto the soil surface. So the regulators controlling the amount of water reaching the soil will be the physical geometry of the tree canopies and the percentage saturation of the canopies. There is also a loss of water by evaporation from intercepted water in the canopies. This loss is controlled by the prevailing meteorological conditions and thus by a regulator which is outside of the physical bounds of the subsystem.

The climatic system consists of those properties and processes that are responsible for climate. The properties of the climatic system may be broadly classified as: thermal properties, which include the temperature of the air, water, ice and land; kinetic properties, which include the wind and ocean currents, together with the associated vertical motions, and the motion of ice masses; aqueous properties, which include the air's moisture or humidity, the cloudiness and cloud water content, groundwater, lake levels, and the water content of snow and of land- and sea-ice; and static properties, which include the pressure and density of the atmosphere and ocean, the composition of the air, the oceanic salinity, and the geometric boundaries and physical constants of the system. These variables are interconnected by the various physical processes occurring within the system, such as precipitation and evaporation, radiation, and the transfer of energy by advection and turbulence.

1.3 Physical components

It is normal to divide the complete climatic system into five physical components – the atmosphere, hydrosphere, cryosphere, lithosphere and biosphere. These components have quite different physical characteristics, and are linked to each other and to conditions external to the system by a variety of physical processes. The atmosphere is the central component of the climatic system, and displays a spectrum of conditions varying from microclimates to the climate of the entire planet. Because of the ease with which the atmosphere can be heated and set in motion, it may generally be expected to respond to an imposed change more rapidly than the other components of the climatic system. A close second to the atmosphere in terms of its overall importance in the climatic system is the hydro-sphere. The extent and bulk of the world's oceans and the prevalence of surface water on the land ensures a potentially plentiful supply of water for the global hydrological cycle of evaporation, condensation, precipitation and run-off. The cryosphere, like the hydrosphere, consists of a portion closely associated with the sea (sea-ice) and portions associated with the land (snow, glaciers and ice sheets). The importance of the cryosphere to the climatic system lies in the high albedo and low thermal conductivity of snow and ice. The surface lithosphere, in contrast to the atmosphere, hydrosphere and cryosphere, is a relatively passive component of the climatic system. An exception to this is the amount of soil moisture which is closely related to the local surface and ground hydrology. Soil moisture exerts a marked influence on the local surface balance of moisture and heat, through its influence on the surface evaporation rate and on the soil's albedo and thermal con-ductivity. The remaining component of the climatic system, the surface biomass, interacts with the other components on time-scales which are characteristic of the life cycles of the earth's vegetative cover. The trees, plants and ground-cover modify the surface radiation balance and surface heat flux, and play a major role in the seasonal variations of local surface hydrology.

1.4 Further development

World Climatic Systems considers the components of the global climatic system in detail. There are chapters on the general atmospheric circulation, oceanic sub-systems, glacial subsystems, and also two chapters looking at aspects of atmo-spheric interactions with the biosphere. The hydrosphere is a central theme throughout much of the book, while interactions with the lithosphere in the form of soil moisture and albedo are discussed in a number of places. The major com-ponents of the climatic system are considered in Part I of the book. The importance of climatology to various human activities has been growing in recent years. This is seen particularly in the fields of agriculture and energy use. Man has upset the global carbon cycle by burning fossil fuels and by deforestation and changing land use. The net result of these activities has been to increase the CO_2 content of the atmosphere and the oceans, and to cause future changes in both global temperature distributions and the global hydrological cycle, including the incidence of drought. Some of these applied aspects of the climatic system are considered in Part II on Impacts of the Climatic System.

2   Driving forces

The global atmosphere circulation consists of the observed wind systems with their annual and seasonal variations, and is the principal factor determining the distri-bution of climatic zones. The two major causes of the global wind circulation are inequalities in radiation distribution over the earth's surface and the earth's rota-tion. The global radiation distribution drives the global circulation while the earth's rotation determines its shape. Basically the mean surface circulation con-sists of easterly winds with equatorial components in the tropics and westerly winds with poleward components in middle latitudes, the corresponding meridional flows aloft being reversed. Weak surface easterlies are found in the polar regions and extensive areas of calms in the equatorial and subtropical regions. Strong upper westerly winds are found poleward of about 25°N and S.

2.1 Radiation distribution
The planet earth receives heat from the sun in the form of short-wave radiation, but it also radiates an equal amount of heat to space in the form of long-wave radiation. This balance of heat gained equalling heat lost only applies to the planet as a whole over several annual periods; it does not apply to any specific area for a short period of time. The equatorial region absorbs more heat than it loses, while the polar regions radiate more heat than they receive. The distribution of radiation over the earth's surface is reviewed in Lockwood (1979). Nevertheless, the equatorial belt does not become warmer during the year, nor do the poles become colder, because heat flows from the warm to the cold regions, thus maintaining the observed temperatures. An exchange of heat is brought about by the motion of the atmosphere and upper layers of the oceans, thus forming the general circulation of the atmosphere and oceans.

Global radiation
Global radiation is the sum of all short-wave radiation received, both directly from the sun and indirectly from the sky, on a horizontal surface. Generalized isolines of average annual global radiation are shown in Figure 1.1, which conveys a very general picture of the distribution of global radiation. The actual distribution of global radiation reflects closely astronomical factors and the distribution of cloud. Thus the areas receiving most global radiation are found in the subtropics where there are unusually clear skies because of the prevailing anticyclonic conditions.

Albedo
Radiation reflected directly back to space from the earth constitutes a loss of available energy to the earth–atmosphere system. The distribution of albedo values over the earth's surface must therefore be considered together with the global radiation. The annual albedo of the earth–atmosphere system is shown in Figure 1.2. The albedo map clearly reveals the land–sea distribution and the general atmospheric circulation as it is represented by the mean cloud patterns over both hemispheres. The high-reaching convective clouds associated with the intertropical convergence zones and partly with the Asian monsoon appear as a belt of relatively high albedos of more than 25–30 per cent. Similarly, low persistent stratus clouds along the western coastal areas of North and South America and Africa appear with albedos between 25 and 35 per cent. The albedo of both polar regions is considerably higher than 50 per cent because of the associated permanent snow and ice-fields. Regions of major gain of radiative energy are the oceanic areas in the subtropics of both hemispheres. The planetary albedo shown in Figure 1.2 is made up of three main components. These are the light reflected from the actual land and sea surfaces of the earth, the light reflected by clouds, and the light scattered upwards by the atmosphere. Estimates of the albedo of the ground and also locations of persistent cloud fields and of ice and snow can be made when travelling or otherwise changing cloud-fields are removed by displaying only the lowest observed satellite albedo value in each area. This approach is based on the simple assumption that the albedo of the earth–atmosphere system is higher over each area in the presence of clouds than for a cloud-free atmosphere. A map produced by Raschke *et al.* (1973) in this manner shows the much lower albedos observed over the oceans as compared with those over the continents. Similarly over Africa the high albedos of the desert surfaces are clearly to be seen, with a relatively sharp boundary where the southern Sahara grades into regions with rather more vegetation. Minimum albedos greater than 40 per cent belong to ice-fields at their smallest extent during July over the Arctic and during January over the Antarctic, respectively.

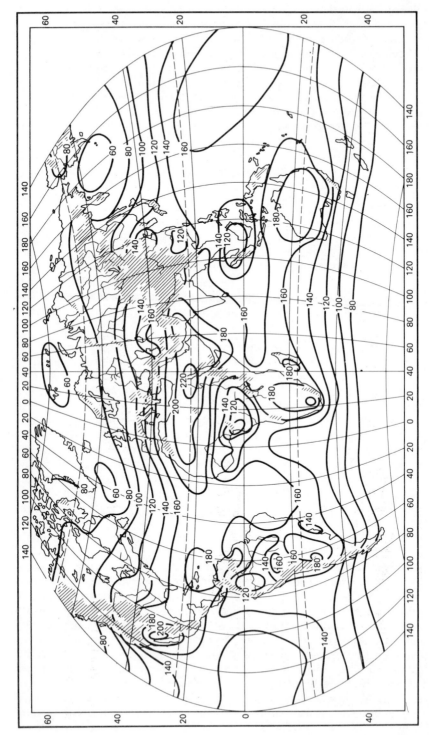

Figure 1.1 Average annual global radiation (Kcal cm^{-2} yr^{-1}) (1 cal cm^{-2} min^{-1} = 698 W m^{-2}) (after Budyko, 1974).

Nimbus 3 1969-1970

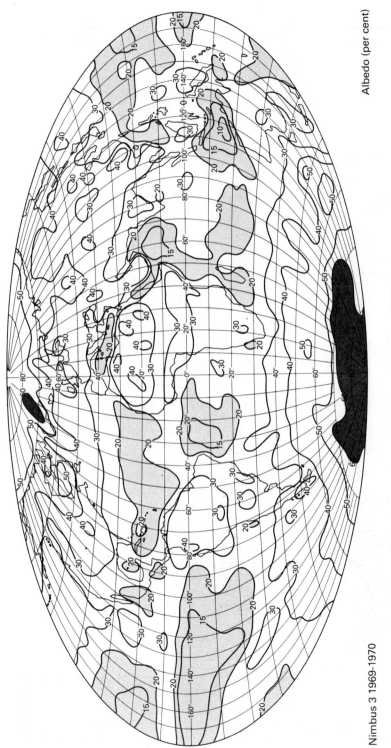

Albedo (per cent)

Figure 1.2 Annual albedo of the earth-atmosphere system (after Raschke *et al.*, 1973).

Outgoing long-wave radiation
The outgoing long-wave radiation distribution reflects the temperatures of the emitting surfaces. Thus high clouds are cold and have a low emission, while low-level surfaces are warm and have a high rate of emission. Over Africa, low rates of emission are observed from the high convective clouds of the intertropical convergence zone while over the Sahara high rates are observed since the atmosphere is clear and the warm sands of the desert surface are visible.

Outgoing radiation from the whole globe corresponds to an effective temperature of 255 K, and the planetary albedo is found to be about 30 per cent.

Radiation balance
Figure 1.3 shows the geographical distribution of annual net radiation at the earth's surface only, for the atmosphere is excluded. This figure reveals that the annual means of the net radiation balance over the greater part of the earth's surface are positive, and thus signifies that the absorbed short-wave radiation is greater than the long-wave outgoing radiation. This pattern is the result of the greater transparency of the atmosphere for short-wave radiation in comparison with long-wave radiation, and the excess of energy at the earth's surface is transferred to the atmosphere by turbulent heat exchange and by evaporation.

A satellite estimate of the annual radiation balance of the earth–atmosphere system is shown in Figure 1.4. Positive values, mostly found between 40°N and 40°S, imply a heat gain by the surface–atmosphere system, while negative values elsewhere imply a general heat loss to space. The net radiation integrated over the whole globe must come to zero, because there is, over an annual period, almost an exact balance between solar energy absorbed and infrared radiation emitted to space. Significant changes in the earth's radiation budget occur within latitudinal zones, especially in the tropics. Thus the areas of the major gains of radiative energy are the oceans in the subtropics of both hemispheres, while the African and Arabian deserts at the same latitude actually have a radiative deficit.

Seasonal changes of radiation balance are dominated by seasonal variations in solar declination. The radiative balance of the surface–atmosphere system is positive during the whole year only in the narrow equatorial zone between the latitudes 10°N to 10°S, for elsewhere the sign of the radiation balance changes twice a year. For about 3 summer months in a year the radiation balance of the whole of each hemisphere is positive, but in late summer, zones of negative balance arise near the poles and then gradually spread towards the equator, reaching latitude 30 after 5 months; a similar process of retreat begins in the spring.

If the earth's surface were homogeneous and the planet were not rotating, the imposition of a latitudinal heating gradient would result in a single circulation cell in each hemisphere with an upward limb at the equator and downward limbs at the poles. These cells are energetically direct cells because they are the result of a transformation of potential to kinetic energy. They are often referred to as Hadley cells in tribute to the eighteenth-century scientist who first deduced their existence.

2.2 The earth's rotation

When the earth rotates, the situation is altered in a number of ways, since the rotation generates east–west motions in the atmosphere. To an observer on the rotating earth it appears that a force is acting on moving air particles which causes them to be deflected from their original path, and this apparent force per unit mass is termed the 'Coriolis force'. The Coriolis force turns the wind to the right relative to

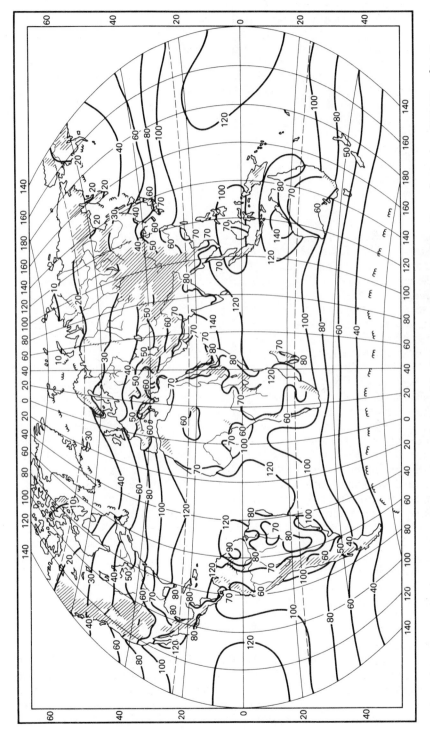

Figure 1.3 Radiation balance of the earth's surface (Kcal cm^{-2} yr^{-1}) for an annual period (1 cal cm^{-2} min^{-1} = 698 W m^{-2}) (after Budyko, 1974).

Nimbus 3 1969-1970

Radiation balance (cal cm^{-2} min^{-1})

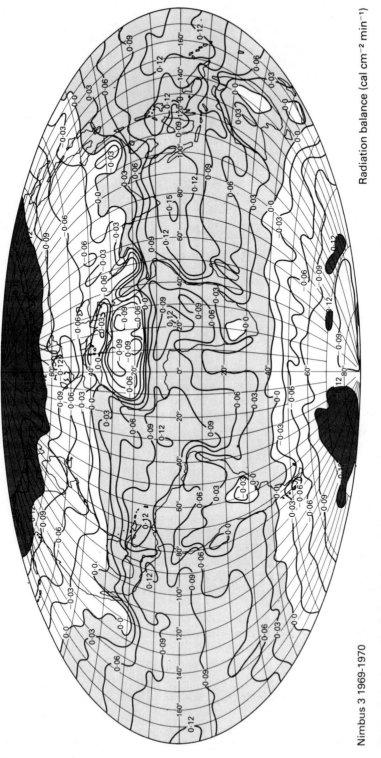

Figure 1.4 Annual radiation balance of the earth–atmosphere system (cal cm^{-2} min^{-1}) (1 cal cm^{-2} min^{-1} = 698 W m^{-2}) (after Raschke *et al.*, 1973).

the earth in the northern hemisphere and to the left in the southern hemisphere. Therefore, in the northern hemisphere the upper poleward current of the Hadley cells assumes a strong eastward component (westerly wind) and the lower equatorial current assumes a westward component (easterly wind). For a given heating gradient, the turning of the wind tends to reduce the efficiency with which the single cell can transport heat poleward. For a sufficiently large rotation rate, a balance cannot be maintained and the meridional temperature gradient continues to increase, while remaining circularly symmetric about the pole. There comes a time, however, when small meridional displacements of a particular east–west size can become dynamically unstable and grow in size, the process being known as baroclinic instability.

Baroclinic instability
The growth of these wave disturbances under baroclinic instability is characterized by the ascent of the warmer, and the descent of the colder, air masses, thus causing a decrease of potential energy and an associated release of kinetic energy. This ascent and descent of air masses takes place in a manner described by the term slantwise convection. In the normal atmosphere, equal potential temperature (isentropic) surfaces slope upward from lower to higher latitudes, and in slantwise convection the trajectories of individual air particles are tilted at an angle to the horizontal that is comparable with, but less than, the slope of the isentropic surfaces (Figure 1.5). Thus while an air parcel may be prevented from rising vertically because of a subadiabatic lapse rate, poleward travel brings it to an environment denser than itself, thus enabling it to rise. Air at higher levels and higher latitudes may similarly descend by equatorward movement. Further details may be found in Lorenz (1967).

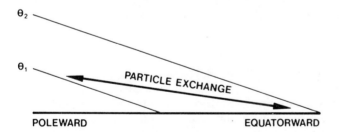

θ_2

θ_1

PARTICLE EXCHANGE

POLEWARD   EQUATORWARD

Figure 1.5 Height–latitude cross-section of the troposphere showing potential temperature θ increasing equatorward and upwards. The system is stable for vertical convection, but energy can be released if particles are exchanged by slantwise convection along the thick solid arrow.

3 General atmospheric circulation

3.1 The mean meridional circulation
It can be shown theoretically that disturbances with the dimensions of 5 to 8 waves about the pole will grow fastest under baroclinic instability. At the surface anti-cyclones develop which grow and ultimately move equatorwards and find a final residence in the subtropics. Similarly, surface low pressure systems develop and move poleward. Thus the mean meridional circulation is no longer simple. The single equator-to-pole direct cell is contracted equatorward, so that the poleward

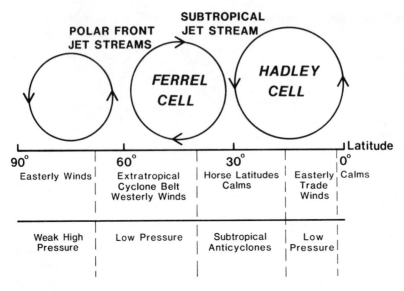

Figure 1.6   A schematic representation of the mean meridional circulation of the northern hemisphere.

descending limb coincides with the axis of the surface subtropical anticyclones (Figure 1.6). In the tropics baroclinic instability is virtually non-existent because of the weak Coriolis force, so that the direct mean meridional circulation (the Hadley cell) is still the most efficient means for effecting a poleward heat transfer. In mid-latitudes, the unstable wave motion is the most efficient means for the poleward heat transfer. A weak energetically indirect mean meridional circulation exists in middle latitudes that is known as the Ferrel cell.

3.2 The subtropical jetstreams

A schematic representation of the mean meridional circulation in the northern hemisphere during winter is shown in Figure 1.6. The simple Hadley cell circulation is clearly seen south of 30°N. Eastward angular momentum is transported from the equatorial latitudes to the middle latitudes by nearly horizontal eddies, 1,000 km or more across, moving in the upper troposphere and lower stratosphere. Such a transport leads to an accumulation of eastward momentum between 30° and 40° latitude, where a strong meandering current of air, generally known as the sub-tropical westerly jet stream, develops. The cores of the subtropical westerly jet stream in both hemispheres and in both the seasons occur at an altitude of about 12 km. The air subsiding from the jet streams forms the subtropical anticyclones. More momentum than is necessary to maintain the subtropical jet streams against dissipation through internal friction is transported to these zones of strong winds. The excess is transported downwards to maintain the eastward flowing surface winds of the middle latitudes against the ground friction. The supply of eastward momentum to the earth's surface in middle latitudes tends to speed up the earth's rotation. Counteracting such a continuous speeding up of the earth's rotation, air flows from the subtropical anticyclones towards the equatorial regions, forming the so-called trade winds. The trade winds, with a strong component directed towards the west, retard the earth's rotation, and in turn pick up momentum.

4 Tropical circulation patterns

The mean winds in the tropics are directed predominantly toward the equator within latitudes 30°N and S. Such motions can exist only if some force accelerates the air in that direction, and the only force in the atmosphere known to produce accelerations of large-scale windfields is the pressure force. Therefore high pressure will exist in the subtropics and low pressure in the equatorial zone, the resulting pressure gradients driving the mean winds. The average latitude of the equatorial low pressure trough is about 5°S in the northern winter and 15°N in the southern winter. The annual mean latitude is about 5°N, so the meteorological southern hemisphere is larger than the northern one.

4.1 The trade winds

Between the equatorial low pressure trough and the subtropical high pressure belt lie two belts of tropical easterlies, each with an equatorial meridional flow. These wind systems are known as the northeast trade winds in the northern hemisphere and the southeast trade winds in the southern. There tends to be a certain monotony about the weather of the trade winds since the extreme steadiness of the trades reflects the permanence of the subtropical anticyclones, which are inclined to be the most intense in winter, making the trade winds strongest in winter and weakest in summer. All tropical oceans have extensive areas of trade winds except the northern Indian Ocean. Each trade wind region contains definite centres of high resultant wind speed that reach 6–8 m s^{-1}. In contrast, wind speeds are low in the subtropics and in the equatorial trough zone. According to Riehl (1979) when wind speed is averaged without regard to direction, an almost uniform value of 7 m s^{-1} appears everywhere except within the equatorial trough.

During the summer of 1856 an expedition under the direction of C. Piazzi-Smyth visited the island of Tenerife in the Canary Islands to make astronomical observations from the top of the Peak of Tenerife. On two of the journeys up and down the 3,000 m mountain, Piazzi-Smyth carefully measured the temperature, moisture content, wind direction and speed of the local trade winds. He found that an inversion was often present and that it was not located at the top of the northeast trade regime, but was situated in the middle of the current; thus it could not be explained as a boundary between two air-streams from different directions. He also noticed that the top of the cloud layer corresponded to the base of the inversion. The observations by Piazzi-Smyth have been confirmed many times and the trade-wind inversion is now known to be of great importance in the meteorology of the tropics. Broad-scale subsidence in the subtropical anticyclones is the main cause of the very dry air above the trade-winds inversion. The subsiding air normally meets a surface stream of relatively cool maritime air flowing towards the equator. The inversion forms at the meeting point of these two air-streams, both of which flow in the same direction, and the height of the inversion base is a measure of the depth to which the upper current has been able to penetrate downward.

Subsidence is most marked at the eastern ends of the subtropical anticyclonic cells, that is to say along the desert cold-water coasts of the western edges of America and Africa, for it is here that the trade-wind inversion is at its lowest. Normally as the trade winds approach the equator, the trade inversion increases in altitude and conditions become less arid. Over the oceans the intense tropical radiation evaporates water which is carried aloft by thermals and eventually distributed throughout the layer below the trade inversion. The result is that the layer below the inversion becomes more moist as the trade wind nears the equator and the continual convection in the cool layer forces the trade inversion to rise in height.

4.2 The equatorial trough

Trade winds from the northern and southern hemispheres meet in the equatorial trough, which is a shallow trough of low pressure, generally situated on or near the equator. Its position is clear cut in January, apart from the central South Pacific, and in July the trough can be located with ease over Africa, the Atlantic and the Pacific to about 150°E. The equatorial trough shows a marked tendency to meander with longitude in January from 17°S to 8°N, in July from 2°N to 27°N. This is largely derived from wide oscillations in the monsoon regions, that is, the whole of southern Asia plus North Africa in the northern summer, and southern Africa and Australia in the southern summer. The excursions into the southern hemisphere are smaller than those into the northern, because the more constant westerly circulation of the southern hemisphere middle latitudes constrains the equatorial trough to near the equator. Over southern Asia and the Indian Ocean mean seasonal positions are around 15°N and 5°S, a difference of 20° latitude. In contrast the trough is quasi-stationary over the Pacific and Atlantic where seasonal displacement is restricted to 5° latitude or less.

Winds in the equatorial trough are generally calm or light easterly. Over some areas, such as the Indian Ocean and southeast Asia, the structure of the trough is very complex and westerlies may be found.

4.3 The Walker circulation and the southern oscillation

Dietrich and Kalle (1957) have mapped the difference between the sea-surface temperature and its average global value along the part of each latitude circle situated over the oceans. Cold-water regions being defined as ocean areas with sea-surface temperatures below the global average. By far the most extensive cold-water area is the South Pacific cold water, for it stretches westwards from the coast of South America by about 85° longitude; in contrast the South Atlantic cold water continues westward from the coast of Africa by only about 40° longitude.

Bjerknes (1969) considers that when the Pacific cold ocean water along the equator is well developed, the air above will be too cold to take part in the ascending motion of the Hadley cell circulation. Instead, the equatorial air flows westward between the Hadley cell circulation of the two hemispheres to the warm west Pacific, where, having been heated and supplied with moisture from the warmer waters, the equatorial air can take part in large-scale, moist adiabatic ascent. In Indonesia huge cloud clusters with a diameter of more than 600 m develop each day, giving an area-averaged rainfall amount around 2,200 mm per year, equivalent to a release of latent heat of about 170 W m^{-2}, which is much more than the net radiation near the surface. This thermally driven circulation (Figure 1.7) between an equatorial heat centre – the 'maritime continent' of Indonesia – and a cooling area in the eastern Pacific is often known as a Walker circulation. As shown in Figure 1.7, the equatorial Atlantic is analogous to the equatorial Pacific in that the warmest part is in the west, at the coast of Brazil, but west–east contrasts of water temperature are much smaller than in the Pacific. However, in January a thermally driven Walker circulation may operate from the Gulf of Guinea to the Andes, with the axes of the circulation near the mouth of the Amazon.

The strength of the Walker circulation varies with the surface water temperature of the Pacific ocean and in turn forms part of a wider tropical fluctuation known as the 'southern oscillation'. According to Berlage (1966), 'the southern oscillation is

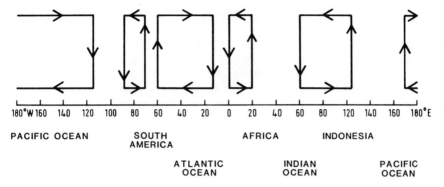

Figure 1.7 Schematic representation of the Walker circulation near the equator during the northern winter.

a fluctuation of the intensity of the intertropical general atmospheric and hydrospheric circulation. This fluctuation, dominated by an exchange of air between the southeast Pacific subtropical high and the Indonesian equatorial low, is generated spontaneously. The period varies between roughly 1 and 5 years and amounts to 30 months on the average.' In general terms, when pressure is high in the Pacific Ocean, it tends to be low in the Indian Ocean from Africa to Australia; these conditions are associated with low temperature in both these areas, and rainfall varies in the opposite direction to pressure (Walker and Bliss, 1932).

4.4 Monsoon climates of southern Asia

Originally the term 'monsoon' was applied to the surface winds of southern Asia which reverse between winter and summer, but the word is now used for many different types of phenomena, e.g. the stratospheric monsoon, the European monsoon, etc. The characteristics of the monsoon climate are to be found mainly in the Indian subcontinent, where over much of the region the annual changes may conveniently be divided as follows:

1 The season of the northeast monsoon
   (a) January and February, winter season
   (b) March to May, hot-weather season
2 The season of the southwest monsoon
   (a) June to September, season of general rains
   (b) October to December, post-monsoon season.

Halley proposed the first explanation of the Asiatic monsoon in a memoir presented to the Royal Society in 1686. His theory is general and can be applied to all continental regions, not only Asia. Over the cold continents during winter the intense cooling leads to the establishment of thermal high-pressure systems, while pressure remains relatively low in the warmer lighter air over the oceans. The surface air flow is therefore from the highs over the land towards the lows over the ocean. During summer, the land is warmer than the sea and a reverse current circulates from the relatively cool sea towards the heated land. The average sea-level pressure charts for the two extreme seasons show a reversal of pressure over Asia and its maritime surroundings. In winter the Siberian anticyclone accumulates sub-zero masses of air ($-40°C$ to $-60°C$) under a pressure of between 1,040 and 1,060 mbar, while in summer the torrid heat (about $40°C$) reduces the pressure to 950 mbar over northwest India. Modern explanations of the Indian monsoon are much more complex (see Lockwood, 1974; and Lighthill and Pearce, 1981), but follow the basic idea of Halley.

According to Miller and Keshavamurthy (1968), the southwest monsoon current in the lower 5 km near India consists of two main branches, which are the Bay of Bengal branch, influencing the weather over the northeast part of India and Burma, and the Arabian Sea branch, dominating the weather over the west, central, and northwest parts of India. The low-level flow across the equator during the southwest monsoon is not evenly distributed between latitudes 40°E and 80°E as was previously thought, but has been found by Findlater (1969, 1972) to take the

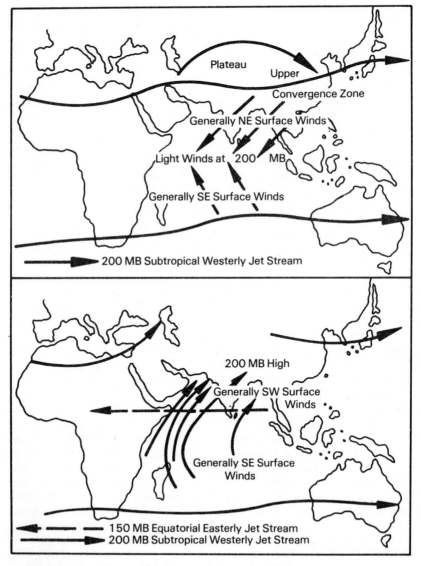

Figure 1.8   Schematic illustration of the major features of the Asian monsoon. Upper: northern winter. Lower: northern summer.

form of low-level high-speed southerly currents which are concentrated between about 39°E and 55°E. A particularly important feature of this flow is the strong southerly current with a mean wind speed of about 14 m s^{-1} observed at the equator over eastern Africa from April to October. The strongest flow occurs near the 1.5 km level, but it often increases to more than 25 m s^{-1} and occasionally to more than 45 m s^{-1} at heights between 1.2 and 2.4 km. According to Findlater, this high-speed current flows intermittently during the southwest monsoon from the vicinity of Mauritius through Madagascar, Kenya, eastern Ethiopia, Somalia, and then across the Indian Ocean towards India.

When upper winds are taken into account, it is found that the Asian monsoon is a fairly complex system. During the northern winter season (Figure 1.8), the subtropical westerly jet stream lies over southern Asia, with its core located at about 12 km altitude. It divides in the region of the Tibetan Plateau, with one branch flowing to the north of the plateau, and the other to the south. The two branches merge to the east of the plateau and form an immense upper convergence zone over China. In May and June the subtropical jet stream over northern India slowly weakens and disintegrates, causing the main westerly flow to move north into central Asia. While this is occurring, an easterly jet stream, mainly at about 14 km, builds up over the equatorial Indian Ocean and expands westward into Africa. The formation of the equatorial easterly jet stream is connected with the formation of an upper-level high-pressure system over Tibet. In October the reverse process occurs; the equatorial easterly jet stream and the Tibetan high disintegrate, while the subtropical westerly jet stream reforms over northern India.

5 Middle and high latitude circulation patterns

The greatest atmospheric variability occurs in middle latitudes, from approximately 40° to 70°N and S, where large areas of the earth's surface are affected by a succession of cyclones (depressions) and anticyclones or ridges. This is a region of strong thermal gradients with vigorous westerlies in the upper air, culminating in the polar-front jet stream near the base of the stratosphere. The zone of westerlies is permanently unstable and gives rise to a continuous stream of eddies near the surface, the cyclonic eddies being thrown poleward and the anticyclonic ones equatorward.

Compared to the Hadley cells the middle-latitude atmosphere is highly disturbed and the suggested meridional circulation in Figure 1.6 is largely schematic. Daily weather charts for any large region in middle and high latitudes reveal well defined dynamic systems that normally move from west to east with a speed that is considerably smaller than the speed of the upper air currents. The structure of these systems varies considerably with altitude, for near the earth's surface they are of relatively small dimensions (1,000 to 3,000 km) and show a maximum of complexity, while in the middle and upper troposphere the systems are relatively large and have a great simplicity. At the surface the predominant features are closed cyclonic and anticyclonic systems of irregular shape, while higher up smooth wave-shaped patterns are the general rule. The dimensions of these upper waves are much larger than those of the surface cyclones and anticyclones, and only rarely is there a one-to-one correspondence. In typical cases there are four or five major waves around the hemisphere, and superimposed upon these are smaller waves which travel through the slowly moving train of larger waves. The major waves are called long waves or Rossby waves, after Rossby (1939, 1940, 1945) who first investigated their principal properties.

5.1 Atmospheric thermal patterns

The basic thermal pattern of the lower half of the atmosphere in the temperate latitudes is partly controlled by the prevailing mean surface temperatures. The mean
thermal pattern may be usefully investigated by using the concept of thickness
lines. The thickness (or the depth) of an isobaric layer is given by

$$z - z_0 = \frac{RT_m}{g} \ln \left(\frac{p_0}{p} \right) \tag{1.1}$$

where z and z_0 are the heights above sea-level of the top and bottom of the isobaric
layer; p and p_0 are the pressures at the top and bottom of the isobaric layer; R is the
gas constant for dry air; and T_m is the mean temperature of the layer. Clearly the
thickness of an atmospheric layer bounded by two fixed pressure surfaces, 1,000
and 500 mbar for example, is directly proportional to the mean temperature of the
layer. Thus low thickness values correspond to cold air and high thickness values to
warm air.

The geostrophic wind velocity at 500 mbar is the vector sum of the 1,000 mbar
geostrophic wind and the theoretical wind vector blowing parallel to the 1,000–500
mbar thickness lines, with a velocity proportional to their gradient. This theoretical
wind is known as the thermal wind and the magnitude (V_T) of the thermal wind is
given by the expression

$$V_T = \left| \frac{g}{f} \frac{\Delta z'}{\Delta n} \right| \tag{1.2}$$

where g is the gravitational acceleration, f is the coriolis parameter; and $\Delta z'/\Delta n$ is
the horizontal gradient of the thickness of the layer.

Inasmuch as the thickness is but another expression for the mean temperature of
the isobaric layer, the thermal wind will blow along the mean isotherms with lower
temperature to the left in the northern hemisphere. Thus in temperate latitudes the
thermal wind will be westerly, and also since horizontal temperature gradients are
relatively steep, it will be strong. The upper winds are the vector sum of a rather
weak surface wind field and a vigorous thermal wind field. This implies that in the
lower middle-latitude troposphere, winds will become increasingly westerly with
altitude and that the upper wind field is strongly controlled by the thermal wind. A
parallel therefore exists between the mean topography of the 500 mbar level and the
mean 1,000–500 mbar thickness patterns.

The mean January thickness patterns for the northern hemisphere show two
dominant troughs near the eastern extremities of the two continental landmasses,
while ridges lie over the eastern parts of the oceans. A third weak trough extends
from north Siberia to the eastern Mediterranean. Climatologically, the positions of
the main troughs may be associated with cold air over the winter landmasses, and
the ridges with relatively warm sea surfaces. As seen from Figure 1.9, the mean
January 500 mbar wind field is similar to the thickness field.

In July, the mean ridge in the thickness pattern found in January over the Pacific
has moved about 25° west, and now lies over the warm North American continent,
while there is a definite trough over the east Pacific. Patterns elsewhere are less
marked, but a weak trough does appear over Europe, and may perhaps be connected with the coolness of the North Sea, the Baltic, the Mediterranean and the
Black Sea.

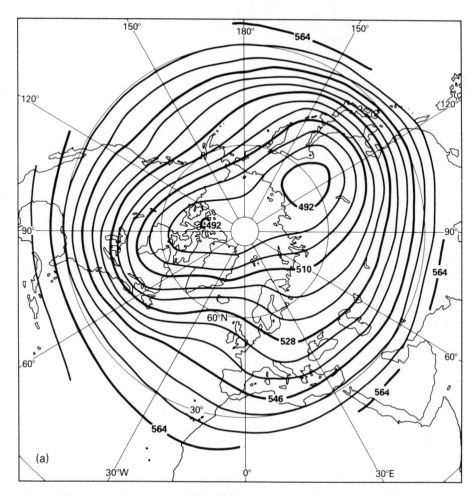

Figure 1.9 (a) Monthly mean 1,000–500 mbar thickness for January 1951–66

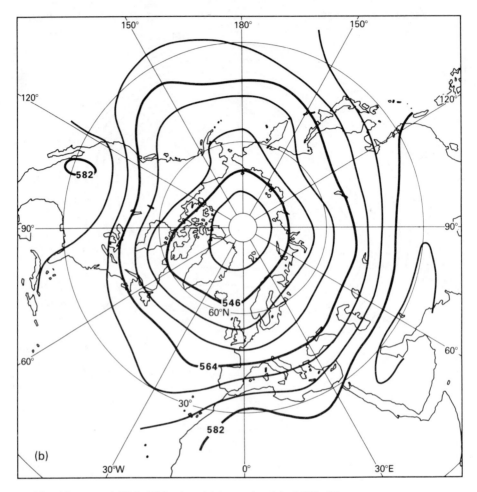

(b) Monthly mean 1,000–500 mbar thickness for July 1951–66

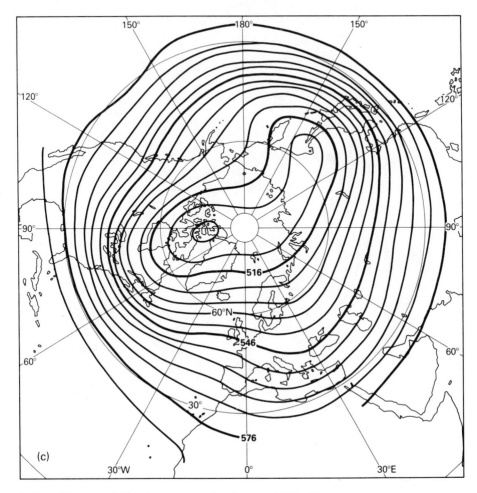

(c) Monthly mean 500 mbar contours for January 1951–66

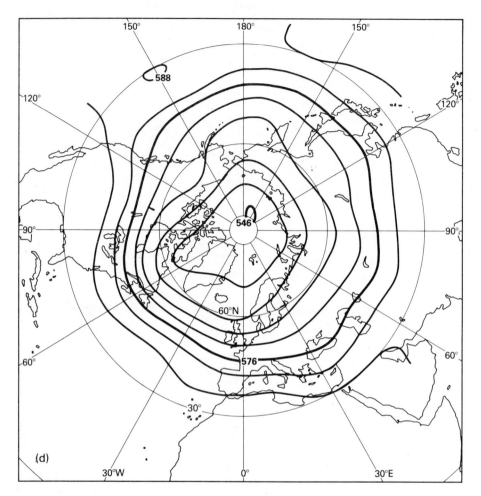

(d) Monthly mean 500 mbar contours for July 1951–66. Isopleths are at intervals of 6 geopotential decametres (after Moffitt and Ratcliffe, 1972)

5.2 Index cycles and blocking systems

Long-term mean circulation charts for the middle latitudes show only a faint resemblance to individual synoptic charts. The long-term mean charts are composed of averages of circulation patterns for a great variety of scales and types and do not reveal the traits of individual synoptic situations. The strength of the zonal circulation is conveniently measured by the mean pressure difference between two latitude circles, which is called the zonal index, for sea-level conditions 35° and 55°N are often used.

A typical zonal index cycle can be divided into four stages (Figure 1.10) with the following characteristics:

1 An initial high zonal index with strong sea-level westerlies and a long-wavelength pattern in the upper atmosphere;
2 An initial lowering of the sea-level zonal index with an associated shortening of the wavelength pattern aloft;

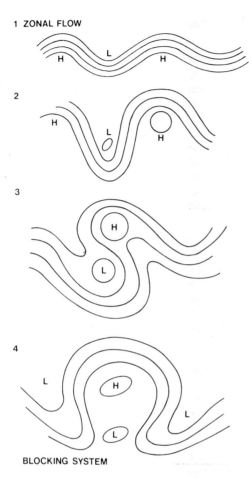

1 ZONAL FLOW

2

3

4

BLOCKING SYSTEM

Figure 1.10 Schematic representation of the consecutive stages of the breakdown of the zonal flow in the middle troposphere during an index cycle. The development takes place during a period of 10 days and the scale is some thousands of kilometres.

3 The lowest sea-level zonal index characterized by a complete break-up of the sea-level zonal westerlies with closed cellular centres, and with a corresponding breakdown of the wave pattern aloft;
4 Finally, an initial increase of the sea-level zonal index with a gradual increase of the westerlies and the development of an open wave pattern aloft.

Since the high index situation is characterized by strong latitudinal temperature gradients in middle latitudes, and by little air-mass exchange, the cyclones and anticyclones of the mid-latitude belt drift eastward, often with considerable speed. Under conditions of low zonal index, a warm cut-off high forms in the 1,000–500 mbar thickness pattern, and sea-level cyclones have a tendency to be steered around these highs, either to the north of the cut-off or to the south along the base of the pattern. Because the normal westerly current is blocked and reversed, this last arrangement is often called a 'blocking pattern' or simply a 'block'. Anticyclonic development is associated with the warm ridge and sometimes cyclonic development with the cold trough of the blocking pattern, and so an extensive blocking anticyclone often exists in middle and high latitudes, while a non-frontal low exists to the south. Blocking patterns once formed are relatively stable and usually persist for a period of 12–16 days.

5.3 Polar regions

Both polar regions are located in regions of general atmospheric subsidence, though the climate is not particularly anticyclonic and the winds are not necessarily easterly. The moisture content of the air is low because of the intense cold, and horizontal thermal gradients are normally weak, with the result that energy sources do not exist for major atmospheric disturbances, which are rarely observed. Vowinckel and Orvig (1970) suggest that the Arctic atmosphere can be defined as the hemispheric cap of fairly low kinetic energy circulation lying north of the main course of the planetary westerlies, which places it roughly north of 70°N. The situation over the south polar regions is more complex and the boundary of the Antarctic is not so clear.

6 Circulation of energy in the atmosphere

A major function of the atmospheric circulation is the transport of heat, water vapour, angular momentum, etc. If S is the amount of the quantity being studied in a unit volume of air, then the flux of this quantity across a surface of unit area is given by the product VS, when V is the velocity of the air normal to the surface. Now let an over-bar represent the mean value in time or in space, or in both, then at any instance $V = \overline{V} + V'$ and $S = \overline{S} + S'$ where the prime denotes the instantaneous deviation from the mean. Therefore

$$VS = (\overline{V} + V')(\overline{S} + S') = \overline{V}\,\overline{S} + \overline{V}S' + V'\overline{S} + V'S'$$

and the mean flux is given by

$$\overline{VS} = \overline{V}\,\overline{S} + \overline{\overline{V}S'} + \overline{V'\overline{S}} + \overline{V'S'}$$

The mean value is defined such that

$$\overline{S'} = 0 = \overline{V'}$$

and hence

$$\overline{\overline{V}S'} = \overline{V}\,\overline{S'} = 0 \text{ and } \overline{V'\overline{S}} = \overline{V'}\,\overline{S} = 0$$

The mean flux is therefore given by

$$\overline{VS} = \overline{V}\,\overline{S} + \overline{V'S'} \tag{1.3}$$

This equation can be expressed in the form:

total flux = advective flux + eddy flux.

The advective flux is due to the mean motion of the atmosphere while the eddy flux depends on the correlation between V and S as they fluctuate about their mean values. This means that both the mean circulation of the atmosphere and the large-scale eddies have to be studied when considering the transport of water vapour, angular momentum, etc.

6.1 Energy content

The total energy content of one gram of moist air may be expressed by the following equation:

$$Q = \frac{AV^2}{2} + Bgz + C_pT + Lq \tag{1.4}$$

where Q is the total energy content, V is the wind speed, g is the acceleration due to gravity, z is the height above mean sea level, C_p is the specific heat of air at constant pressure, T is the temperature in degrees Kelvin, L is the latent heat of condensation of water vapour, q is the specific humidity, and A and B are conversion constants appropriate to the units employed.

If the vertical motion of the parcel of air is frictionless and adiabatic and if it is assumed that there is no mixing with the surrounding atmosphere, then there is a direct conversion between the potential energy (gz) and the sensible heat content ($C_p T$). It is, therefore, convenient to consider these two quantities together under the heading of total potential energy (Margules, 1903). The total potential energy is, typically, 100 to 1,000 times greater than the kinetic energy ($V^2/2$) of the atmosphere; for example, in a powerful jet stream the kinetic energy content will be equivalent to 0.6 to 0.9 cal g^{-1}, but the total potential energy at jet-stream level may be betwen 75 and 80 cal g^{-1}. Normally the kinetic energy content is about equal to the probable error in estimating the total potential energy and may, therefore, be neglected in estimations of the total energy content (Q). The total potential energy content is normally about 10 times greater than the latent heat content (Lq); in a very warm, moist atmosphere the latent heat content may be between 10 and 12 cal g^{-1} while the total potential energy will be about 70 cal g^{-1}.

If condensation is taking place in a rising air parcel, there is a conversion of latent heat into total potential energy. If it is assumed that there is no mixing with the surrounding atmosphere and the energy losses due to radiation and the fall-out of condensed water are ignored, the energy content of the parcel will remain constant during ascent. The graph of Q for such a parcel will be a vertical straight line. The amount of temperature fall attributable mainly to radiation cooling in the atmosphere depends on several factors, but falls of between 1 and 2°C per day are probably normal, and this is equal to a cooling rate of between 0.25 and 0.5 cal g^{-1} per day. Air subsiding very rapidly under conditions of no mixing will maintain an approximately constant Q content, but if the air is subsiding slowly, as in an anticyclone, it may take 15 to 20 days to sink from the 300 mbar level to the 600 mbar level, and under these conditions the air may have cooled gradually during the descent by 5 to 10 cal g^{-1}. Profiles of Q through a slowly sinking air-mass will, therefore, show a decrease in Q values from the top to the bottom. If no energy is supplied to the atmosphere by the earth's surface and if descent takes place almost to the surface, there will be a minimum in the value of Q at the surface. However, in anticyclones there is normally a low-level inversion and below the inversion active convection may take place. Moreover if the anticyclone is over a warm surface, energy can be transported upward by convection and the layer containing the minimum Q values will be lifted from the surface into the lower atmosphere.

6.2 Energy transport

In recent years much attention has been given to the problem of energy transport by the atmosphere and oceans. In the past, oceanic energy transport was assessed as being small, that is, not more than 10 per cent of the total energy transfer. Recent computations indicate, however, that the oceans may carry a much larger fraction of the total energy transfer, reaching perhaps 25 per cent in the subtropics and middle latitudes. Comment was made earlier that the total flux of an atmospheric quantity can be divided into two components, which are the advective flux and the eddy flux. Figure 1.11 shows the eddy sensible heat flux as computed by Mintz (1951) for January 1949, and the eddy latent heat flux as computed by Starr,

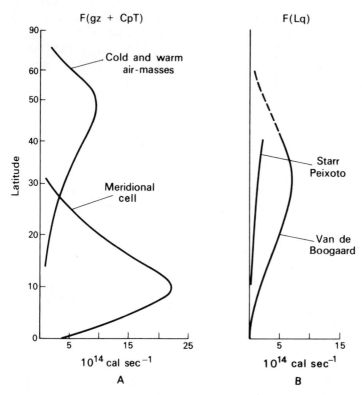

Figure 1.11   A: poleward flux of sensible heat by eddies in January 1949 (Mintz, 1951) and poleward transport of potential energy plus sensible heat by the mean meridional circulation in winter.
B: poleward flux of latent heat by eddies after Starr, Peixoto and Livadas (1957) and after Van de Boogaard (1964) (after Riehl, 1969).

Peixoto, and Livadas (1957) and Van de Boogaard (1964); this last computation applies only to a single day, 12 December 1967, hence the flux may be unrepresentative. This diagram suggests that the eddy fluxes are largest in the middle latitudes and decrease toward the tropics, where they may approach zero. The large eddy fluxes in the middle latitudes reflect the importance of air-mass exchanges, because cold, dry air normally flows from the poles and warm, moist air from the tropics. Within the tropics, where the atmosphere tends to be uniform over large areas and the flow is relatively undisturbed, it appears that the mean meridional circulation of the Hadley cell is the prime mechanism of net heat exchange, and this suggests that the mean advective flux is relatively more important within the tropics than in the middle latitudes.

There is within the tropics general low-level flow towards the equator, and since there is a large exchange of energy between the earth's surface and the atmosphere, obviously there must be a general advection of energy towards the equator. This energy is mainly in the form of latent heat (Lq) carried in the lower layers of the trade winds rather than as total potential energy. The Hadley cell circulation implies that energy must be carried vertically aloft in the equatorial zone and then exported at high altitudes to the subtropics. Figure 1.12 illustrates the mean height

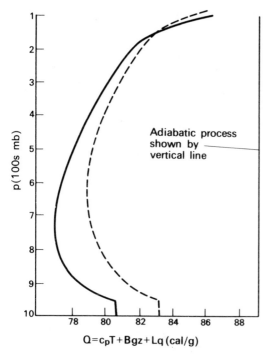

$$Q = c_p T + Bgz + Lq \, (cal/g)$$

Figure 1.12 Vertical profiles of total energy content. Solid profile is for 20° latitude from equatorial trough. Dashed profile is for equatorial trough itself (after Riehl and Malkus, 1958).

distribution of total energy content (Q) in both the equatorial trough zone and at 20° latitude away from it. The diagram shows that in both cases the Q values are high near the surface, and then decrease to a minimum near 700 mbar, after which they increase again. In the equatorial zone large quantities of energy are exported aloft and the mean meriodional circulation demands ascent, but a uniform, gradually rising circulation would produce cooling above the level of minimum Q. Furthermore, the upper troposphere is continually cooling because of radiation losses and this energy has to be replaced from below to maintain the shape of mean Q curve. It would appear therefore that the picture of a Hadley cell with general ascent in the equatorial trough is incorrect.

Riehl and Malkus (1958) have tried to solve the obvious dilemma concerning the equatorial limb of the Hadley cell by their 'hot tower convection hypothesis'. According to these authors the rising portion of the meridional cell is concentrated in the restricted regions of the towering cloud banks within tropical storms. In ordinary cumulus clouds there is much mixing with the surrounding atmosphere, but within the general area of equatorial disturbances there are imbedded giant cumulonimbus whose central cores are protected from dilution by the large cross-sections of the towers. Normal cumulus clouds rarely penetrate far into the atmo-sphere but the giant cumulonimbus provide a means whereby high energy content air from near the ocean surface can be pumped to great heights to balance the heat losses and provide for the calculated energy exports. In these clouds water vapour is condensed and precipitated as rainfall and the latent heat released is converted into total potential energy which is transported to the high troposphere. According to Malkus (1962) only about 1,500–5,000 active giant cumulonimbus clouds are needed within the 10° latitude width equator belt to provide for its high-level poleward energy export. It thus appears that the rising limb of the Hadley cell is restricted to a few very active synoptic disturbances within the equatorial trough.

If ascent near the equator is limited to synoptic disturbances, the descent in the subtropics is general and widespread. In the subtropics the sinking air masses carry total potential energy gained in the equatorial trough and feed it into the middle latitude circulations. The meridional transfer of total potential energy has a double maximum in both hemispheres, one in the subtropics between 15 and 25° and the other in high middle latitudes between 50 and 60°. The first is located on the poleward side of the tropical rain belt and the second on the poleward side of the most intense cyclonic activity along the polar front. There is little direct heating of the atmosphere by the transfer of sensible heat from the surface, for instead, most atmospheric warming between 70°N and 70°S is due to the release of latent heat by condensation and much of this occurs between 10°N and 10°S (Figure 1.13). Poleward of 70°N and 70°S most warming occurs by the advection of warmer air from lower latitudes.

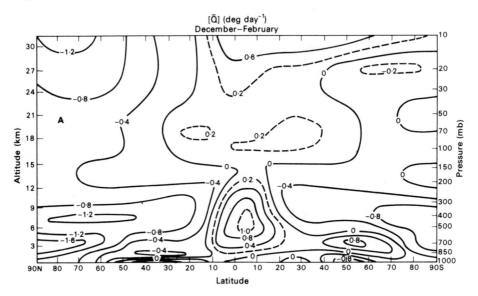

Figure 1.13   Diabatic heating of the atmosphere.
A: total diabatic heating for December–February;
B: June–August;
C, D, E: components of the diabatic heating of the atmosphere for December–February
Units: °C day^{-1} (after Newell *et al.*, 1969).

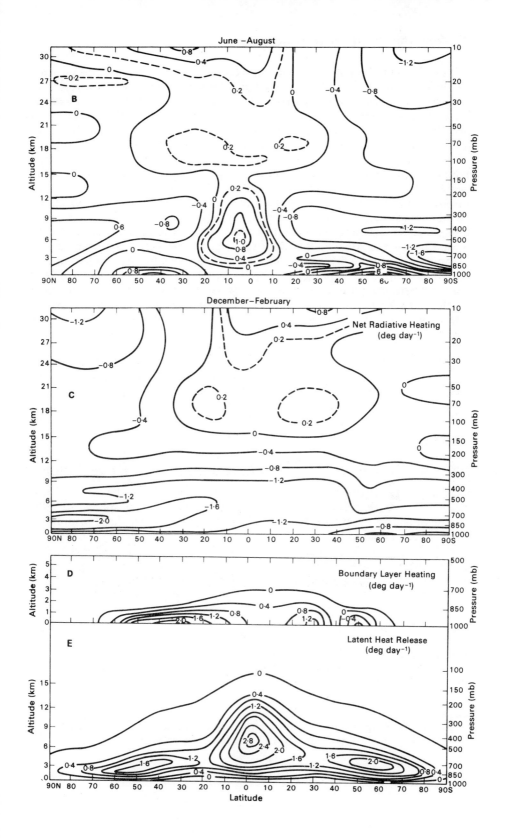

The observed radiation distribution in the tropics raises the interesting question as to whether the Hadley cell is strictly a direct cell, that is to say one in which the potential energy represented by the juxtaposition of relatively dense and light air masses is converted into kinetic energy as the lighter air rises and the denser air sinks. In the normal direct cell the rising air is found over the heat source, but in the tropics the greatest inputs of solar energy are not found near the equator but in the subtropical deserts where there is massive subsidence. It is only when the energy exchanges within the tropics are studied in detail that the nature of the Hadley cell becomes clear. Values of total diabatic (non-adiabatic) heating for the extreme seasons are illustrated in Figure 1.13 where it is seen that the largest positive rates are observed in the middle troposphere between 10°N and 10°S. Figure 1.13 further indicates the magnitudes of the components of the diabatic heating during the northern winter season. It is observed that the net radiative heating of the troposphere is negative, indicating that it is losing heat by long-wave radiation. The transfer of sensible heat into the planetary layer is small except over the hot deserts of the northern hemisphere, leaving the release of latent heat as the main form of atmospheric heating. During the northern winter the release of latent heat occurs mainly in three localities, one of which is found over the equator and the other two in middle latitudes. The maximum in diabatic heating observed over equatorial regions is therefore mainly the result of latent heat released by condensation in the equatorial trough; since it occurs in the rising limb this confirms that the Hadley cell is a direct cell. As described earlier, the large influx of solar radiation into the subtropics does not heat the atmosphere directly but instead is used to evaporate water from the tropical oceans; this is advected by the trade winds into the equatorial trough where it is condensed and the latent heat released to the atmosphere. The Hadley cell is, therefore, only driven indirectly by solar radiation.

7 Geological evolution of the general atmospheric circulation

According to Hart (1978) it has become generally accepted that the earth lost any primordial atmosphere it may have had, and that the present atmosphere is derived from materials degassed from the interior. Recently Hart (1978) has attempted to follow, by a computer simulation, the evolution of the earth's atmosphere over the last 4–5 billion years, with particular emphasis on changes in the chemical composition and mean surface temperature. Many factors have influenced the geological evolution of the Earth's atmosphere, and some are now considered in more detail.

7.1 Changes in solar luminosity
Theoretical analysis of stellar interiors indicate that most main sequence stars gradually increase luminosity with time. Hart (1978) assumed that the sun was 25 per cent less luminous 4–5 billion years ago than it is now. He also assumed that the increase in luminosity of the Sun occured at a uniform rate, and at a nearly uniform effective temperature.

7.2 Albedo
Variations in the earth's albedo play an important part in determining its effective temperature. Hart (1978) estimated the albedo of the earth by the formula:

$$\text{Albedo} = (f_{cloud} \cdot A_{cloud}) + (f_{ice} \cdot A_{ice}) + (f_{sea} \cdot A_{sea}) + (f_{veg} \cdot A_{veg}) + (f_{rock} \cdot A_{rock}) \tag{1.5}$$

Table 1.1   A breakdown of the present albedo of the earth by type of cover (*after Hart, 1978*)

Type of cover	A_i	f_i (present)	$f_i \times A_i$ (present)
Clouds	0.52	0.472	0.245
Rocks	0.15	0.021	0.003
Water	0.04	0.353	0.014
Ice	0.70	0.032	0.022
Vegetation	0.13	0.122	0.016
		1.000	0.300

where f_{cloud} represents the fraction of the earth's surface covered by clouds at a given epoch and f_{ice} represents the fraction by ice (without supervening clouds), etc. The constants A_{cloud}, A_{ice}, etc., represent the mean albedo of clouds, ice, etc. Table 1.1 lists the values used for the A_i, and estimates of the current values of the f_i. It is apparent from the last column that the main contribution to the present albedo comes from cloud cover.

Hart states that the estimate given in Table 1.1 for f_{cloud} today is consistent with satellite photographs. To estimate f_{cloud} at earlier epochs, Hart assumed that f_{cloud} would be proportional to the total mass of water vapour in the atmosphere.

7.3 Greenhouse effect

From the solar luminosity and the earth's albedo it is possible to compute T_{eff}, the effective temperature of the earth. This is because the present effective temperature of the earth represents an energy balance between the visible and near-infrared sunlight that falls on the planet and the middle-infrared thermal emission leaving it. The simplest case is in the absence of an atmosphere, when this equilibrium can be written:

$$0.25\,(1 - \alpha)\,S = \epsilon\sigma\,T_{eff}^4 \tag{1.6}$$

where S is the solar constant, α is the albedo of the earth, ϵ is the mean emissivity of the earth in the middle infrared, T_{eff} is the effective equilibrium temperature of an earth with no atmosphere, and σ is the Stefan–Boltzmann constant. Only that part of the earth that faces the sun is illuminated, while the whole earth loses infrared radiation to space. The factor 0.25 is the ratio of the area πR^2 that intercepts sunlight to the area $4\pi R^2$ that emits thermal infrared radiation to space. When the best estimates of these parameters are used, a value for T_{eff} of about 254 K is obtained, and this is far less than the observed mean surface temperature of the earth of 286 to 288 K. The difference is due to the greenhouse effect, in which visible and near infrared sunlight penetrates through the earth's atmosphere relatively unimpeded, but thermal emission by the ground surface is absorbed by atmospheric constituents that have strong absorption bands in the middle infrared. Thus because of the greenhouse effect $(\Delta T)_{green}$, the mean surface temperature of the earth is given by

$$(T_{surf}) = T_{eff} + (\Delta T)_{green} \tag{1.7}$$

where $(\Delta T)_{green}$ has to be adjusted to include the effects of convective heat transport.

Table 1.2   Evolution of earth's atmosphere according to best-fitting computer run (*after Hart, 1978*)

Time (billions of years BP	Surface pressure (atm)	Mass of atmospheric components							T_{eff} (°K)	(T_{surf}) (°K)
		N_2 (10^{21} g)	O_2 (10^{21} g)	CO_2 (10^{20} g)	CH_4 (10^{21} g)	NH_3 (10^{17} g)	Ar (10^{19} g)	H_2O (10^{19} g)		
4.25	1.25	0.75	0	23.2	3.37	0.74	0.04	4.18	217	305
4.00	1.37	0.56	0	13.4	5.13	1.79	0.16	7.68	218	314
3.75	1.39	0.32	0	8.28	6.00	2.23	0.34	9.67	219	317
3.50	1.32	0.31	0	5.27	5.93	2.29	0.57	8.27	219	315
3.25	1.22	0.52	0	3.54	5.39	2.15	0.86	5.75	220	310
3.00	1.10	0.87	0	2.48	4.56	1.90	1.18	3.77	221	304
2.75	0.97	1.31	0	1.81	3.52	1.62	1.54	2.61	224	299
2.50	0.73	1.57	0	1.37	2.22	1.46	1.93	2.30	230	297
2.25	0.62	2.14	0	1.03	0.93	1.09	2.35	1.86	238	294
2.00	0.75	3.76	0.06	0.646	0	0	2.79	0.77	250	281
1.75	0.76	3.81	0.06	0.365	0	0	3.25	0.69	251	279
1.50	0.77	3.84	0.08	0.215	0	0	3.72	0.69	252	279
1.25	0.77	3.87	0.08	0.117	0	0	4.21	0.71	253	280
1.00	0.78	3.89	0.09	0.049	0	0	4.70	0.72	254	280
0.75	0.78	3.90	0.10	0.024	0	0	5.21	0.79	255	281
0.50	0.79	3.91	0.10	0.024	0	0	5.72	0.90	256	283
0.40	0.85	3.91	0.43	0.024	0	0	5.93	1.02	256	285
0.30	0.90	3.92	0.69	0.024	0	0	6.14	1.10	256	286
0.20	0:94	3.92	0.86	0.024	0	0	6.35	1.16	255	287
0.10	0.97	3.92	1.03	0.024	0	0	6.55	1.21	255	288
0	1.00	3.92	1.21	0.024	0	0	6.76	1.27	255	288

7.4 Computational results

As a starting point for the computer simulations, Hart (1978) assumed that the earth had no atmosphere 4–5 billion years ago, and had an albedo of 0.15 at that time. The simulation was then carried forward, step by step, using intervals of 2.5 $\times$ 10^6 yr. The program calculated the mass of each element added to the atmosphere–ocean system (or lost from it) during each time interval. Some of the results of the 'best-fitting' computer run are presented in Table 1.2. In this simulation, as in all the others which came at all close to the present composition of the atmosphere, most of the water vapour condensed quite promptly to form oceans. The early atmosphere was therefore dominated by CO_2, which was the next most abundant component of the volcanic gases. As CO_2 was removed from the atmosphere–ocean system, an atmosphere developed whose main constituents were methane (or perhaps higher hydrocarbons, or other reduced carbon compounds) and nitrogen. Oxygen was released from photolysis of water vapour, and after the first 800 million by photosynthesis. The oxygen so released gradually destroyed the CH_4 and NH_3. By roughly 2.0 billion years ago all but trace amounts of reduced gases had been removed from the atmosphere, and at that point the atmosphere consisted primarily (about 96 per cent) of N_2. Roughly 420 million years ago the amounts of O_2 and O_3 in the atmosphere became great enough to reduce the intensity of ultraviolet radiation at the surface to a level tolerable by living organisms. Soon thereafter plant life spread to the continents and there was a great increase in the biomass. The resulting increase in photosynthetic activity (and in the rate of burial of carbon) caused a rapid increase in the amount of free oxygen in the atmosphere.

Hart (1978) considers that the most significant results of the computer simulation are those which concern the long-term variations of the earth's effective temperature and mean surface temperature. The greenhouse effect was very large during the first 2–5 billion years, due to the combined effect of reducing gases (such as NH_3 and CH_4) and of large quantities of H_2O and CO_2. The high surface temperature caused a large amount of water to evaporate, which resulted in a cloud-covered earth. That in turn caused a high albedo and therefore the effective temperature was low throughout the first half of earth's history. After CH_4, NH_3, and all other reduced gases were oxidized away the situation changed drastically because the greenhouse effect decreased in intensity, and the mean surface temperature fell. By the time free oxygen appeared in the atmosphere about 2 billion years ago, the mean surface temperature had fallen to about 280 K, and ice caps of considerable size would be able to form. Hart comments that the reason the drop in temperature was not even larger is primarily due to the thermostatic effect of the variations in cloud cover. As the surface temperature falls, the quantity of water vapour in the atmosphere drops, the cloud cover and therefore the albedo decrease, and the effective temperature goes up. However, if the mean surface temperature ever became much lower than 0°C this thermostatic mechanism could break down. With large ice caps and little cloud cover, any further drop in temperature will raise the albedo because of the growth in ice caps, and cause runaway glaciation. Hart reports that runaway glaciation occurred in many of his computer runs. In every such case the computation was continued – often for an additional 2 billion years, or even longer – with CO_2 (from volcanoes) accumulating in the atmosphere, and with the greenhouse effect computed. Hart found that not in a single such case was runaway glaciation reversed.

References

BARRY, R. G. and CHORLEY, R. J. 1976: *Atmosphere, Weather and Climate*. London: Methuen.

BERLAGE, H. P. 1966: The southern oscillation and world weather. *Mededeelingen en Verhandelingen* 88. The Hague: Korunklijk Nederlands Meteorologisch Instituut.

BJERKNES, J. 1969: Atmospheric teleconnections from the equatorial Pacific. *Monthly Weather Review* 97, 163–72.

BUDYKO, M. I. 1974: *Climate and Life*. New York: Academic Press.

CROWE, P. R. 1971: *Concepts in Climatology*. London: Longman.

DIETRICH, G. and KALLE, K. 1957: *Allgemeine Meereskunde* Berlin: Gebrüder Borntraeger.

FINDLATER, J. 1969A: A major low-level air current near the Indian Ocean during the northern summer. *Quarterly Journal of the Royal Meteorological Society* 95, 362–80.

—— 1972: Aerial explorations of the low-level cross-equatorial current over eastern Africa. *Quarterly Journal of the Royal Meteorological Society* 98, 274–89.

HADLEY, G. 1735: Concerning the cause of the general trade-winds. *Philosophical Transactions of the Royal Society*, London, 29, 58–62.

HALLEY, E. 1686: An historical account of trade-winds and monsoons observable in the seas between and near the tropicks with an attempt to assign the physical causes of the said winds. *Philosophical Transactions of the Royal Society*, London. 26, 153–68.

HART, M. H. 1978: The evolution of the atmosphere of the earth. *Icarus* 33, 23–39.

LIGHTHILL, J. and PEARCE, R. P. 1981: *Monsoon Dynamics*. Cambridge: Cambridge University Press.

LOCKWOOD, J. G. 1974: *World Climatology*. London: Edward Arnold.

—— 1979: *Causes of Climate*. London: Edward Arnold.

LORENZ, E. N. 1967: *The Nature and Theory of the General Circulation of the Atmosphere*. Geneva: World Meteorological Organization.

MALKUS, J. S. 1962: Large-scale interactions. In Hill, M. N., *The Sea* Vol. 1. New York: John Wiley, 88–294.

MARGULES, M. 1903: *Über die Energie der Stürme.* Sonderabdruck aus den Jahrbüchern der Zentralanstalt für Meteorologie, Vienna. English translation in C. Abbe (ed.) *The Mechanics of the Earth's Atmosphere.* Smithsonian Institute, Miscellaneous Collection 51 No. 4. 1910.

MILLER, F. R. and KESHAVAMURTHY, R. N. 1968: *Structure of an Arabian Sea Summer Monsoon System.* Honolulu: East–West Centre Press.

MINTZ, Y. 1951: The geotrophic poleward flux of angular momentum in the month of January 1949. *Tellus* 3, 195–200.

MOFFITT, B. J. and RATCLIFFE, R. A. S. 1972: Northern hemisphere monthly mean 500 millibar and 1000–500 millibar thickness charts and some derived statistics (1951–66). *Geophysical Memoirs* 117. London: Meteorological Office.

MONTEITH, J. L. 1981: Evaporation and surface temperature, *Quarterly Journal of the Royal Meteorological Society* 107, 1–27.

NEWELL, R. E., VINCENT, D. G., DOPPLICK, T. G., FERRUZZA, D. and KIDSON, J. W. 1969: The energy balance of the global atmosphere. In Corby, G. A. (ed.), *The Global Circulation of the Atmosphere.* London: Royal Meteorological Society, 42–90.

PIAZZI-SMYTH, C. 1858: An astronomical experiment on the Peak of Teneriffe. *Philosophical Transactions of the Royal Society, London,* 148, 465–534.

RASCHKE, K., VONDER HAAR, T. H., BANDEEN, W. R. and PASTERNAK, M. 1973: The annual radiation balance of the earth–atmosphere system during 1969–70 from Nimbus 3 measurements. *Journal of the Atmospheric Sciences* 30, 341–64.

RIEHL, H. 1969: On the role of the tropics in the general circulation of the atmosphere. *Weather* 24, 288–308.

—— 1979: *Climate and Weather in the Tropics.* London: Academic Press.

RIEHL, H. and MALKUS, J. S. 1958: On the heat balance in the equatorial trough zone. *Geophysica* 6, 503–38.

ROSSBY, C. G. 1939: Relation between variations in the intensity of the zonal circulation of the atmosphere and the displacements of the semi-permanent centres of action. *Journal of Marine Research* 2, 38–55.

—— 1940: Planetary flow patterns in the atmosphere. *Quarterly Journal of the Royal Meteorological Society* 66, 68–87.

—— 1945: On the propagation and energy in certain types of oceanic and atmospheric waves. *Journal of Meteorology* 2, 187–204.

STARR, V. P., PEIXOTO, J. P. and LIVADAS, G. C. 1957: On the meridional flux of water vapour in the northern hemisphere. *Geofisica Pura e Applicata* 39, 174–85.

VAN DE BOOGAARD, H. M. E. 1964: A preliminary investigation of the daily meridional transfer of atmospheric water vapour between the equator and 40°N. *Tellus* 16, 43–54.

VOWINCKEL, E. and ORVIG, S. 1970: The climate of the north polar basin. In Orvig, S. (ed.) *Climates of the Polar Regions.* Amsterdam: Elsevier, 129–252.

WALKER, G. T. and BLISS, E. W. 1932: World Weather V, *Memoir of the Royal Meteorological Society* 4, No. 36, London.

WETHERALD, R. T. and MANABE, S. 1975: The effects of changing the solar constant on the climate of a general circulation model. *Journal of Atmospheric Sciences* 32, 2044–59.

2 Oceanic Subsystems

1 Ocean circulation patterns

The surface layers of the oceans absorb most of the solar radiation reaching the earth's surface, and in turn much of this heat is transferred to the atmosphere. The larger part of this ocean–atmosphere heat exchange occurs as evaporation, which represents a latent heat transfer from the ocean to the air where it is subsequently released at the site of condensation and precipitation. The latent heat flux is effected by the turbulent motions in the lower atmosphere and is dependent upon the low-level vertical moisture gradient and the surface wind speed. Depending upon the low-level vertical temperature gradient the turbulent motions also effect a transfer of sensible heat between the ocean surface and the overlying air. The transfer of momentum between the atmosphere and ocean consists primarily of a turbulent frictional flux from the more rapidly moving atmosphere to the underlying surface. For the atmosphere this surface frictional drag represents the principal mechanism for the dissipation of kinetic energy but for the oceans it represents the major driving force for the system of large-scale currents which are in turn regulated by frictional processes within the oceans. Ocean currents may also be produced by variations of temperature and salinity, which are in turn tied to atmosphere processes. The major physical elements and feedback processes of the coupled atmosphere–ocean system have been summarized by Gates (1981) in Figure 2.1.

The mean global wind system was described in the previous chapter. As the global wind system acts on the ocean, it will produce an east-to-west motion near the poles, a west-to-east motion in the region of the westerlies, and a strong east-to-west motion in the trade-wind belts. If the earth were completely covered by water, these ocean currents would simply circle the globe along parallels of latitudes. The boundaries of the oceans, however, deflect the currents, resulting in a series of circular currents, or gyres. The largest currents, the subtropical gyres, are formed by water being pushed to the west by the trade winds, to be returned eastward in the region of the westerlies. Poleward of those gyres are the subpolar gyres, between the polar easterlies and the westerlies. Finally, on either side of the equator, narrow equatorial gyres are formed by the partial return to the east, in the doldrums, of water that has been pushed to the west by the trade winds. At the boundaries between the gyres, there will be some mixing of the surface water between one gyre and its neighbour. Superimposed on the circulation pattern is a latitudinal temperature gradient due to the variation in the solar energy received, and a latitudinal variation in salinity due to variations in evaporation and precipitation.

The surface wind systems of the world are thus largely responsible for driving the circulation of the upper ocean. Surface currents in the ocean are blown along by

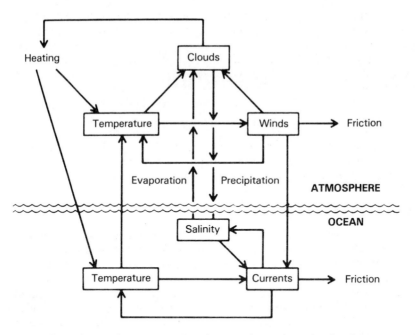

Figure 2.1   The major components of atmosphere-ocean feedback processes (after Gates, 1981).

these winds, and, in the absence of the earth's rotation, would move in the same direction as the winds. This rotation is of considerable influence, even to ocean currents, and the surface currents tend to be driven to the right of the wind direction. Ekman (1905) showed that the net effect of winds driving ocean surface waters on a rotating earth was to produce a net transport of water (later termed 'Ekman Drift') directed 90° to the right (left) of the wind in the northern (southern) hemisphere. Under a surface high, the current has a component directed inward toward the centre and the waters begin to pile up, tending to raise the sea level under the high. Surface water cannot converge indefinitely, instead it must move downward, then outward at some lower level, then upward again far from the high centre to provide mass balance. Downwelling is thus established in the upper ocean under a surface high. This downwelling is persistent, encompasses thousands of square kilometers of ocean area, and is quite weak, only a meter or so in a month. In contrast, a cyclonic low pressure system such as a hurricane will produce strong surface divergence in the upper ocean. In this case upwelling rates of several metres per day may exist. One region of the open ocean where persistent, vigorous upwellings are present is within a few degrees of the equator. On the equatorial sides of the oceanic subtropical anticyclones there exist regions of fairly steady easterly trade winds. Precisely on the equator, the direct influence of the earth's rotation is absent and winds blowing westward tend to drive westward surface currents. Just to the north of the equator the earth's rotation deflects the ocean currents to the right of the wind, while just to the south the deflection is to the left. The net result is a divergence of surface water away from the equator, with replacement water coming from a few hundred metres below the sea surface.

Subtropical oceanic highs are generally situated near the west coasts of continents, and this is not simply coincidental since land–sea contrasts of temperature and surface topography influence the wind systems. Thus there is generally some time during the year in mid-latitudes when the predominant surface winds blow equatorwards along continental west coasts. The Ekman theory predicts an offshore turning of the coastal surface currents under these conditions to the right (left) of the wind in the northern (southern) hemisphere, and a net offshore transport of surface water. To replace the diverging surface water, there must be compensating subsurface currents and upward vertical motion, or coastal upwelling. Recent theoretical and observational studies suggest that not all coastal upwelling is locally wind-driven. For example, continental shelf waves and internal Kelvin waves has been shown to influence coastal upwelling circulations.

1   The atmosphere–ocean interface

At a depth of several hundred metres a sharp temperature gradient, the main thermocline, separates the waters of the upper ocean, in which temperatures are regulated by advection of water masses and by surface diabatic processes, from the cold deep water. The thermocline is a stable region, through which vertical motions are strongly resisted. Accordingly, deep water is thermally insulated from the atmosphere, except in high latitudes, where the main thermocline is essentially absent. The broad pattern of the thermal structure of a high-latitude ocean such as the eastern North Atlantic has been described by Perry and Walker (1977). The normal vertical temperature distribution consists of an isothermal mixed layer beneath which is a layer in which temperature decreases rapidly with depth, that is a thermocline. Seasonal variations occur in the typical vertical temperature distribution. At the end of winter the sea is typically isothermal to a depth of more than 150 m, as a result of convective stirring caused by winter storms. In spring and summer a thermocline is generated by surface heating. In early summer the thermocline may be destroyed by a gale, but later in the summer the vertical temperature gradient is such that even the strongest winds cannot provide the increased mechanical efford needed to overcome the density discontinuity. Once this stage is reached the thermocline intensifies rapidly, since incoming heat is distributed within the limited layer above the seasonal thermocline. The depth of the mixed layer and the detailed thermal structure vary from year to year, depending upon prevailing meteorological and oceanographic conditions. Temperatures fall within the thermocline to low values, and at about 2,000 m or more the water is close to 3°C in all oceans.

Because of the relative uniformity of the abyssal layers most of the horizontal variation of temperature is confined to the warm layer above the permanent thermocline. Wells (1982) has made a comparison of the temperature distribution at the surface in winter (when the seasonal thermocline is weak) and the distribution at 400 m. The temperature at 400 m has no detectable seasonal variation and reflects the slowly changing structure of the main ocean circulation. According to Wells (1982), the most striking feature at 400 m is the warm pool of water with a maximum temperature of 18°C centred at 35°N on the southeast edge of the Gulf Stream system, where the permanent thermocline extends to its deepest level of about 1,000 m. The permanent thermocline not only rises towards the eastern side of the Atlantic but more significantly towards the equator where it reaches its minimum depth of 200 m.

The heat budget equation for the oceanic water body can in approximate form be written (Hastenrath, 1977):

$$SW + LW = Qs + Qe + (Qu + Qt) \tag{2.1}$$

The left-hand terms are net short-wave and long-wave radiation at the ocean surface, and the right-hand terms denote sensible and latent heat flux at the sea–air interface, heat export and storage within the ocean, respectively.

Some indication of the short-wave radiation being absorbed by the ocean surface is given by Figure 2.2. Solar radiation is either absorbed by the sea or reflected from its surface. The degree of reflection is primarily a function of solar elevation and secondarily of sea state. For direct radiation falling upon a smooth sea, albedo varies from about 3 per cent when the sun is vertically overhead to almost 100 per cent when it is close to the horizon, whereas for diffuse radiation it is typically about 8 to 10 per cent. It is seen from Figure 2.2 that the largest amounts of short-wave radiation are absorbed over the tropical oceans with totals falling slightly towards the cloudy equatorial trough and more substantially towards the poles. Not only does the sea absorb and reflect solar energy, it also emits radiation of a wavelength appropriate to its temperature and absorbs long-wave radiation transmitted downwards from clouds and the atmosphere. Long-wave losses from the ocean–atmosphere system are shown in Figure 2.3. Again the largest oceanic losses tend to be in the subtropics with values falling towards both the equatorial trough and the poles. This is because of the clear skies over the subtropical anticyclonic oceans. The distribution of radiation balance at the ocean surface is shown in Figure 1.3. Of considerable importance is the fact that at all ocean surfaces the annual sum of the surface radiation balance is positive. This indicates that there is over the year a mean downward flow of radiant energy into the ocean surface. In Figure 1.3, the abrupt variation at the continental coasts defined by a break in the isolines is very noticeable. The distribution of the radiation balance over the surface of the oceans exhibits, in general, a zonal character. Some deviations from zonality are caused by warm and cold currents. The largest balance, a net income of more than 140×10^3 cal cm^{-2} yr^{-1}, is located in the northwestern part of the Arabian Sea. In this region there is a modest positive balance in winter (when a trade-wind regime prevails) and an exceptionally large positive balance in summer. At the latter season upwelling due to persistent southwesterly winds causes water off the Arabian coast to be comparatively cold and, consequently, values of outgoing long-wave radiation to be relatively small. Cloud amounts are also small allowing large amounts of short-wave radiation to reach the surface.

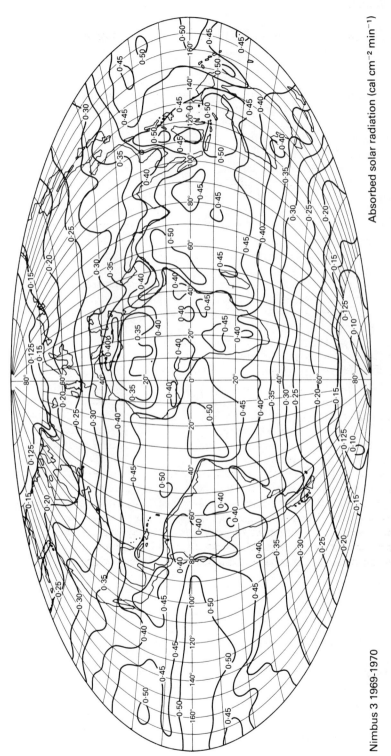

Nimbus 3 1969-1970

Absorbed solar radiation (cal cm^{-2} min^{-1})

Figure 2.2 Solar radiation absorbed annually in the earth-atmosphere system (after Raschke *et al.*, 1973).

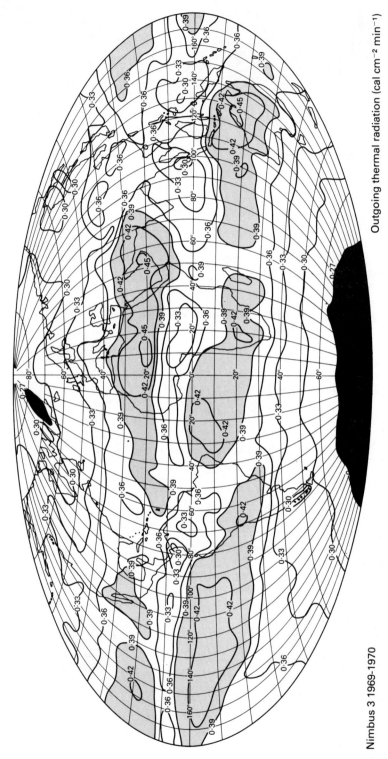

Nimbus 3 1969-1970   Outgoing thermal radiation (cal cm⁻² min⁻¹)

Figure 2.3 Long-wave radiation emitted annually to space (after Raschke *et al.*, 1973).

Over the greater part of the ocean surface, the value of the turbulent sensible-heat flux (Figure 2.4) is small compared to the principle components of the heat balance and usually does not exceed 10–20 per cent of their values. Positive values indicate a flow from the ocean to the atmosphere. The flux reaches large absolute values in those regions where the water is, on the average, much warmer than the air, i.e. in the regions of powerful warm currents (the Gulf Stream), and in high latitudes where the sea remains free of ice. Under such conditions, the turbulent heat flux can exceed 30–40 Kcal cm^{-2} yr^{-1}. Over the oceans the turbulent sensible-heat flux on average increases in absolute values from low to high latitudes, while the opposite profile is observed over the continents. In December, the sensible-heat flux reaches large values in the northwestern portions of the Atlantic and Pacific Oceans, as a result of the activity of warm currents and also the development of heat exchange between the cold continental surface and the warmer ocean. In all the oceanic regions, the sensible-heat flux is small in absolute magnitude, and in some regions, including the zones of the Peru and Benguela cold currents, it turns out to be directed from the atmosphere to the ocean. In June, in the northern portion of the Atlantic Ocean, near the coasts of North America, Europe, and Africa, there exist extensive zones where the sensible-heat flux is directed from the atmosphere to the ocean surface. A similar picture is also found over a considerable portion of the Pacific Ocean in the northern hemisphere. Nevertheless, the sensible-heat flux is small and this indicates that the oceans of the northern hemisphere lose much more heat during the cold season than they gain from the atmosphere during the warm season.

The distribution of the annual values of latent-heat flux is presented in Figure 2.5. The latent-heat flux in extratropical zones diminishes, on the average, with an increase of latitude, but this average regularity is disturbed by large azonal changes. A decrease is also observed near the equator, in comparison with the regions of oceanic high atmospheric pressure. The principal reason for azonal changes in the latent-heat flux over the oceans is the distribution of warm and cold ocean currents. All the major warm currents increase the flux noticeably, and the cold ones decrease it. These changes are clearly apparent in the regions of the warm currents of the Gulf Stream, Kuro Shio, and others, and also in regions of the cold currents, such as the Canary Current, the Benguela, the California, the Peru, the Labrador Current, and so on. Besides ocean currents, azonal variations of latent-heat flux are influenced by the conditions of atmospheric circulation that determine the regime of wind speed and humidity deficit above the ocean. For example, ocean evaporation usually increases in the cold season in comparison with the warm season. The cause of this lies in the increase in the difference between water and air temperature in the cold season, which causes the difference in the water-vapour concentration at the water surface and in the air to increase. Also, in many oceanic regions, mean wind speeds in the cold season are higher than in the warm season, and this also increases evaporation during the cold season. In December the greatest values of evaporation are observed over the oceans in the regions of the northern hemisphere warm currents such as the Gulf Stream and Kuro Shio. In June evaporation is characterized by reduced values in extratropical latitudes of the oceans in the northern hemisphere. These values are, on the average, less than those in the corresponding latitudes on land.

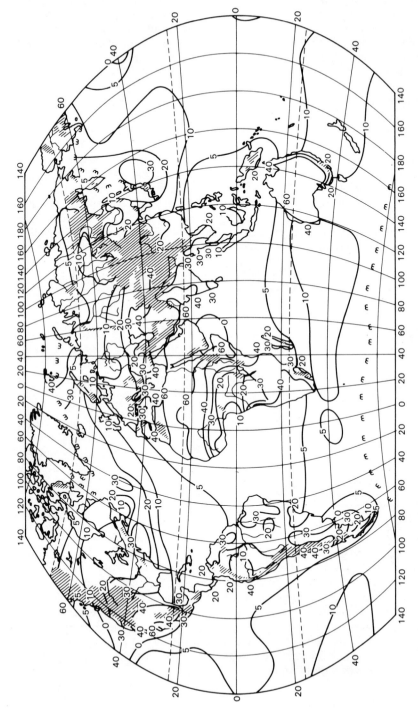

Figure 2.4 The sensible heat flux between the earth's surface and the atmosphere (Kcal cm^{-2} yr^{-1}). (After Budyko, 1978).

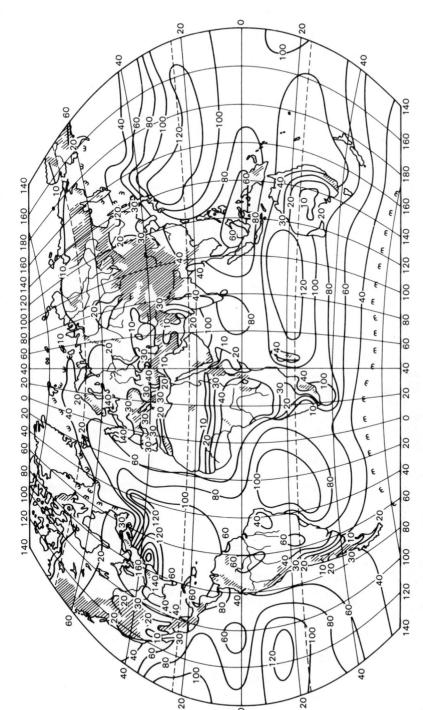

Figure 2.5 The expenditure of heat for evaporation (Kcal cm^{-2} yr^{-1}) (after Budyko, 1978).

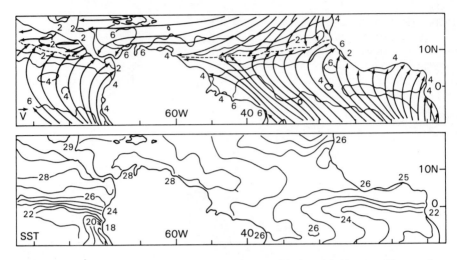

Figure 2.6 Surface wind field and sea temperature during July/August. Top, surface wind field, isotachs in m s^{-1}; bottom, sea surface temperature, in degrees C (after Hastenrath, 1977).

3 Tropical and equatorial oceans

The vast tropical seas have traditionally been thought of as horizontally uniform in thermal pattern and hydrospheric heat budget. Recent work by researchers such as Hastenrath (1977) and Hastenrath and Lamb (1978) has shown that complex tropical oceanic patterns exist. For example, this is illustrated by Figure 2.6 which shows the July/August surface wind field and sea surface temperature for the tropical Atlantic and part of the tropical Pacific. Over both the eastern Pacific and Atlantic, the southern hemisphere trades cross the equator, recurve to south-westerly, and meet the northeast trades along an extended discontinuity in the northern hemisphere. The discontinuity reaches its northernmost position in the northern summer, and recedes equatorward towards winter. Negative vorticity occupies a broad band extending to either side of the equator and the curl of the wind stress shows a similar pattern. The sea surface temperature pattern is charac-terized by a band of cold water to the south of the eastern Atlantic and Pacific equator, and steep meridional temperature gradients across the equator into the northern hemisphere.

According to Hastenrath various mechanisms may cooperate in maintaining the sea surface pattern depicted in Figure 2.6. First, lateral advection may be a factor both in the downstream portion of the Canary Current and in the cold water region of the southeastern equatorial Atlantic. Secondly, wind induced large-scale upwelling may play a role in the cold tongue immediately to the north of the equator. Yoshida and Mao (1957) state that for steady-state ocean conditions away from the immediate vicinity of coasts and the equator, vertical motion near the bottom of the mixed layer may be estimated from:

$$w = \frac{1}{\rho f}\left(\frac{\partial \tau_y}{\partial x} - \frac{\partial \tau_x}{\partial y}\right) \qquad (2.2)$$

where ρ denotes the density of sea water,
     f the Coriolis parameter,
and τ_x and τ_y are the zonal and meridional components of the surface wind stress.

Negative wind stress curl calls for upwelling in the southern and for downwelling in the northern hemisphere with largest values near the equator. A third mechanism, also involving wind-induced vertical motion in the ocean, has been discussed in qualitative terms by Cromwell (1953). He showed that oceanic Ekman transport associated with a southeasterly wind could result in divergence and upwelling in a band extending from the southern hemisphere to north of the equator, and an oceanic convergence band at some low latitudes in the northern hemisphere.

Within a few degrees of the equator, because of the smallness of the Coriolis parameter, the wind tends to generate a surface flow in the direction of the surface wind stress. Therefore the easterly winds drive a flow towards the west Pacific, where it downwells, returns along the equator at around 200 m and upwells at the east Pacific coast. This circulation results in a tilt of the thermocline which is proportional to the strength of the wind. If this equilibrium is disturbed, for example by a reduction in the easterly wind, then the thermocline will try to reduce its tilt and a surge of warm water will move eastwards along the equator.

4 Tropical oceanic weather systems

4.1 Subtropical anticyclones

The two belts of subtropical high pressure at about 30°N and S contain several quasi-permanent anticyclonic cells separated from each other by cols. In the northern hemisphere, the most notable ones are the two oceanic highs, in the Pacific and the Atlantic respectively, and the North African high, which fails to show up at sea-level but emerges clearly at the 3 km level. In the southern hemisphere major anticyclones are found over the Pacific, Atlantic and Indian Oceans.

Subtropical anticyclones show great permanence, for they tend to be located at fixed positions on the globe and undergo only slight seasonal variations, amounting on average to about 5° of latitude. The high-pressure belt is nearest to the equator in winter, but there is a slight asymmetry with regard to the geographical equator, for the southern ridge is situated about 5° latitude closer to it in the mean than the northern one. At the subtropical ridge lines, pressure is practically equal in both hemispheres, varying on average from 1,015 mbar in summer to 1,020 mbar in winter.

4.2 The trade winds

The trade winds occur on the equatorial sides of the subtropical anticyclones, and are found in the greater part of the tropics. In general the trade winds blow from ENE in the northern hemisphere and from ESE in the southern, and are noted for extreme constancy in both speed and direction. Indeed, in no other climatic regime on earth do the winds blow so steadily, for this steadiness reflects the permanence of the subtropical anticyclones; normally interruptions in the flow occur only with the formation of a major atmospheric disturbance. As winter is the season when the subtropical highs tend to be most intense, the trade winds are strongest in winter and weakest in summer.

Air-streams flowing towards the equator tend to subside and diverge, and this can be considered in terms of the concept of potential vorticity:

$$\text{Potential vorticity} = \frac{\zeta + f}{\Delta p} = \text{constant} \qquad (2.3)$$

where Δp is the pressure difference between two adjacent isentropic surfaces differing in potential temperature by the constant value $\Delta\theta$, ζ is the relative vorticity, and f the Coriolis parameter.

It follows from the above equation that since f decreases towards the equator, the depth Δp will also decrease in an air-stream flowing towards the equator, assuming that the relative vorticity ζ remains constant. From these considerations it might be expected that the trade-wind inversion should decrease in height downwind towards the equator, but over the vast ocean areas trade-cumulus convection 'diffuses' energy and water vapour gained from the oceans to higher levels, and causes the trade-wind inversion to rise, even though the mean vertical motion along trajectories is downward. The trade-wind inversion is not therefore a material surface, and subsiding air above the inversion is slowly incorporated through the inversion into the moist layer below.

4.3 The equatorial trough

Carried along by the flow of the trade winds, the trade-wind inversion enters the equatorial trough, but the height of the inversion increases towards the equator and it often vanishes within the trough. The result is that while convection is restricted within the trade winds, towering cumulus clouds develop in the equatorial trough and give rise to heavy showers, thus often creating a relatively wet cloudy zone as compared with the trade winds.

Charney (1967) has developed a theory to explain the formation of convergence zones and their relationship to the equator. He believes that organized cumulus convection is controlled through frictionally induced convergence of moisture in the planetary boundary layer, and that the vertical pumping of mass, and therefore of moisture, out of this layer is proportional to the vorticity of the surface geostrophic wind. Since the air at low levels holds the bulk of the moisture and has the greatest potential buoyancy for moist adiabatic ascent in a conditionally unstable tropical atmosphere, cumulus convection and the release of latent heat are largely determined by this vertical pumping and its associated boundary-layer convergence. Thus a local increase of vorticity in a zonally symmetric flow will give rise to the following sequence of events:

1 increased vertical flux of moisture;
2 increased cumulus convection with release of latent heat;
3 a temperature rise;
4 an accelerative pressure-density solenoidal field;
5 increased low-level convergence;
6 a bringing of high angular momentum air from the equatorward side of the disturbance into juxtaposition with low angular momentum air from the poleward side;
7 a still greater increase of positive vorticity.

This, suggests Charney, is essentially the instability mechanism which would produce a convergence zone if one did not exist, or maintain one if it already existed.

He considers that the character of equatorial flow depends on a parameter η which may be defined as the ratio of the heating of an air column by condensation of the moisture pumped from the frictional boundary layer to the cooling by adiabatic expansion in the rising air:

$$\eta = \frac{Lq}{C_p\, T\, \Delta \ln \theta} \tag{2.4}$$

when q is the specific humidity of the air at the top of the boundary layer; T is a mean mid-atmosphere temperature; $\Delta \ln \theta$ is a characteristic vertical increment of the logarithm of the mean potential temperature, θ; and the other symbols are as defined previously. When η is below a certain critical value, which varies with the other parameters but is roughly between 1 and 4, condensation has only a small effect; but when this threshold value is exceeded, the effect becomes very large and decisively influences the entire circulation. The change occurs when the heating due to condensation in the convergence zones overcomes the cooling due to adiabatic expansion. In this sense the instability is analogous to the conditional instability associated with small-scale cumulus convection, but here it pertains to large-scale flow patterns, and for this reason Charney and Eliassen (1964) have called it 'conditional instability of the second kind'.

Charney's theory suggests that convergence zones can exist anywhere between 15°N and 15°S. It is also likely that a small change in sea-surface temperature (SST) at a given latitude will influence the position and structure of convergence zones, which is interesting in the light of their unequal distribution within the equatorial world.

Work by Pike (1971) suggests that, at least over the sea, the maximum upward motion and precipitation rate associated with the intertropical convergence zone (ITCZ) occurs at or very near the latitude of maximum sea-surface temperature (SST). Hubert *et al.* (1969), using a year of satellite data, have concluded that a single pronounced convergence zone is much more common than a pronounced latitudinal double maximum and that it is usually hemispherically very asymmetric when off the equator. This agrees with some results from an interacting atmosphere and ocean model (Pike, 1971) which suggests that a single convergence zone is comparatively stable, on or off the equator.

Understanding of the mechanism controlling the surface circulation over equatorial oceans has been hampered by the coarse spatial resolution of the climatic mean data hitherto available. Ramage (1974) has summarized, and taken exception to, the widely accepted view of a coincidence of the zones of maximum SST, convergence, rainfall–cloudiness, pressure troughs and kinematic discontinuity, commonly referred to as the ITCZ. Hastenrath and Lamb (1978) have shown the confluence axis to be embedded in a flat low-pressure trough. Over the eastern Pacific, the SST maximum stays to the north of the confluence throughout the year, whereas they are in close proximity in the Atlantic. The convergence maximum lies to the south of the confluence axis and SST maximum throughout the year in the eastern Pacific, and during most of the year in the Atlantic. In the Indian Ocean, the SST maximum is generally situated north of the convergence maximum.

4.4 Tropical rain-forming disturbances

Even in the equatorial trough, rainfall is not particularly frequent. Because in the trade winds proper the sky will contain only small cumulus clouds, it may be rainless for many days. As dry weather, is normal within the tropics, it is necessary to consider under what special circumstances rainfall will occur, which entails a study of tropical synoptic disturbances. Sikdar and Suomi (1971) divide tropical circulations into three broad scales of motion:

1 planetary scale (the equatorial trough and trade-wind regime, subtropical highs and jet streams, monsoons, etc.);
2 synoptic or large-wave scale (easterly waves, tropical cyclones, waves in the upper troposphere, etc.);
3 meso-convective scale (cumulus clouds which can form cloud clusters).

Much of the rain in the tropical world falls from convective clouds, that is, from large cumulus or cumulonimbus. In the most severe storms these clouds are packed side by side to give continuous rain and cloud, but more usually there are wide gaps between the clouds and this leads to a rather random and discontinuous rainfall distribution. Some places will have heavy rainfall, while others nearby will have little or no rain, thus forming a complete contrast to the widespread cloud and rainfall of middle latitudes. Tropical rainfall on many occasions appears to be random, but is nevertheless often caused by some kind of organized disturbance, which may range from a tropical cyclone to a cloud cluster.

Cumuli are the most common tropical clouds and form the outstanding element of the cloud scenery. They are created by convection initiated by solar radiation reacting with the surface and therefore show marked diurnal variations. Their distribution is closely correlated with surface features, though some of the relationships are rather complex. Over most of the tropics the vertical development of cumulus clouds is restricted by the trade-wind inversion, and they do not normally evolve into cumulonimbus. Trade-wind cumulus can give rise to light showers but they do not cause heavy rainfall, which is associated with strong surface convergence and with the absence of the trade-wind inversion, thus allowing a deep unstable moist layer to form.

Related to surface convergences are surface low-pressure systems, ranging from central pressure anomalies of one millibar to those of hurricane strength. Due to mechanisms not yet fully understood, the low-level convergence in these systems becomes concentrated in narrow zones which cover 1–10 per cent of the area of the disturbance, and in which all ascent and precipitation take place in cumulonimbus cloud arrays.

Tropical synoptic disturbances form within a uniform air mass, and horizontal temperature gradients within the tropics are generally small. Therefore, differences in air-mass temperature and density cannot be an important source of energy for these disturbances, instead condensation of water vapour releasing latent heat appears to be the main source of energy. To maintain the supply of water vapour it is necessary for a flow of moist air from over a wide area to enter the synoptic system. This inflow can be organized in a variety of forms but once it ceases, the storm decays very rapidly. This suggests that major tropical disturbances such as tropical cyclones will be restricted to the oceans.

Near the equator, in the equatorial trough, intense rainfall is often associated with minor convergence zones in which pressure differences are very small, making them extremely difficult to study. In the subtropics where systems are more clearly marked and consequently more easily observed, there are two particularly well known pressure systems – the easterly wave and the tropical storm or cyclone.

4.5 Easterly waves
Two types of wave disturbance in the tropical easterlies have been the subject of analysis – the easterly wave of the Caribbean and the equatorial wave of the

Pacific. Historically, the Caribbean wave was examined first, and the model was then applied elsewhere. Observations indicate that waves form in the easterly flow over the Caribbean, and that they move westwards at a speed of 5 to 7 m s⁻¹. The structure of such an easterly wave is shown in Figure 2.7. Waves are typically about 15° latitude across, and the easterly winds flow through them from east to west. The wave structure in the easterlies is found to be weakest at sea level, to increase in intensity up to about 4,000 m, and then above this level to again become weaker. About 300 km ahead of the wave trough, the trade-wind inversion reaches its lowest altitude and exceptionally fine weather prevails, indicating intense sub-

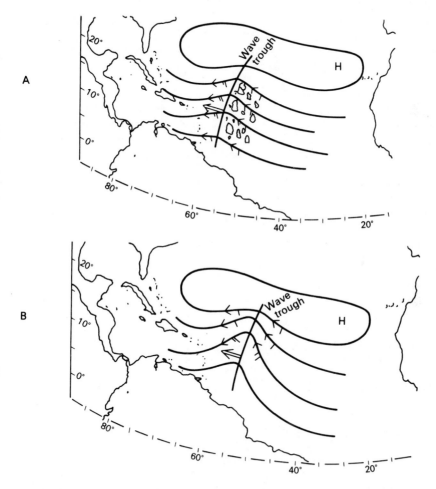

Figure 2.7 Schematic illustration of major features of 'easterly wave' type of tropical disturbance.
A: Surface streamlines of weak-to-moderate amplitude wave, which typically moves westward in direction of heavy arrow at a speed slightly slower than prevailing trade wind. The barbs on the wind arrows denote speed, each small one representing 2.5 m s⁻¹. The area of cloudiness and rain is commonly found to the rear of the trough.
B: 500 mbar streamline pattern in typical moderate easterly wave. The wave amplitude is greater than at the surface.

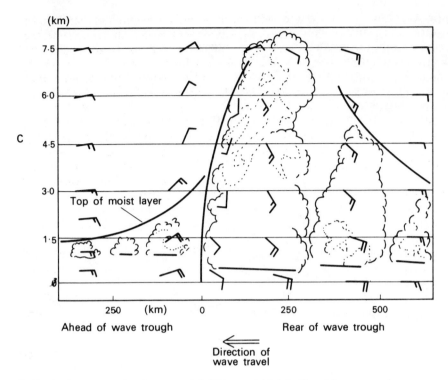

C: Vertical cross-section from west (left) to east (right). Cloud forms are shown schematically and not to scale. *Horizontal* winds are denoted by barbed lines, each short barb representing 2.5 m s^{-1}. Winds flow with barbed lines, east being to the right and west to the left.

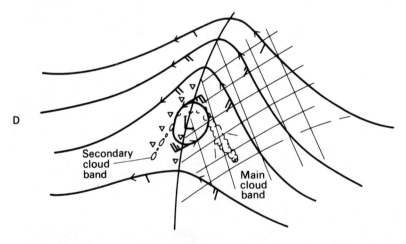

D: Schematic picture of a very deep easterly wave, showing streamlines and major cloud bands. There is a closed central vortex where the surface pressure may be as low as 1,000 mbar. The rain area is indicated by hatching and will contain many subsidiary cloud bands (after Malkus, 1962).

sidence. A rapid rise in the altitude of the trade-wind inversion, and therefore an increase in the depth of the surface moist layer, takes place near the trough line; the moist layer attains a maximum depth of well above 6 km in the zone of ascent behind the trough. Here also are found large squall lines, intense rain, and rows of cumulonimbus clouds, but this region is not completely covered by cumulonimbus because being often organized into lines or rows, there are wide zones of subsidence between individual clouds.

The distribution of weather within a wave in the tropical easterlies has been explained by Riehl (1954) in terms of the concept of potential vorticity mentioned earlier. To the east of the wave surface, air is streaming polewards and therefore entering regions with increasing values of the parameter f. As the air enters the trough, the local vorticity (ζ) will become increasingly positive. The potential vorticity equation (2.3) states that under these conditions the depth of the air-stream must increase to compensate for the change in vorticity. Similar reasoning suggests that ahead of the wave where the surface air is streaming equatorward and away from the trough axis, the vorticity will be decreasing and on that account the vertical depth of the air-stream will decrease. Thus dynamic considerations imply that ascent and the convergence should take place behind the trough line and descent and divergence ahead.

Easterly waves can also be explained in terms of the vorticity equation, which while neglecting the less important terms, can be written for pressure co-ordinates as

$$\frac{d}{dt}(\zeta + f) = -(\zeta + f)\mathrm{div}_p V \tag{2.5}$$

In the lower layers of the atmosphere air flows westward through the wave, and the above equation states that there will be low-level convergence behind the wave and low-level divergence ahead. At higher levels in the troposphere the winds have less strong easterly components and the wave moves westward relative to the air-stream. Under these conditions the vorticity equation states that there will be high-level convergence ahead of the wave and high-level divergence behind, and in this way the observed vertical circulations with their associated weather patterns are explained.

Easterly waves are well known in the Caribbean and similar systems have been described in the west central Pacific. The surges of the trade wind in the northern China Sea are similar to easterly waves, but there is no record of a persistent travelling wave in the China Sea. Though the origins of waves are often obscure, it appears that many over the Caribbean have their sources over Africa.

4.6 Tropical storms and cyclones
Storms with closed circulation systems which can occur anywhere within the tropics outside of the zone 5°N to 5°S, vary from slowly circulating masses of air with scattered cumulonimbus clouds to violent and severe storms. Differences between the various tropical low-pressure systems are as follows:
1 A tropical depression is a system with low pressure enclosed within a few isobars, and it either lacks a marked circulation or has winds below 17 m s^{-1}.
2 A tropical storm is a system with several closed isobars and a wind circulation from 17 to 32 m s^{-1}.
3 A tropical cyclone is a storm of tropical origin with a small diameter (some hundreds of kilometres), minimum surface pressure less than 900 mbar, very

violent winds, and torrential rain sometimes accompanied by thunderstorms. It usually contains a central region, known as the 'eye' of the storm, with a diameter of the order of some tens of kilometres, where there are light winds and more or less lightly clouded sky.

Tropical cyclones are given a variety of regional names. In the southwest Pacific and Bay of Bengal the name 'tropical cyclone' is in use; they are known as 'typhoons' in the China Sea, as 'willy-willies' over western Austrialia, and as 'hurricanes' in the West Indies and in the south Indian Ocean.

Plentiful moisture is a requirement of tropical cyclones, and for this reason their incidence is limited to regions where the highest sea-surface temperatures are found, that is, to the western regions of the tropical oceans in the late summer. Not all incipient tropical low-pressure systems become tropical cyclones; many remain as weak closed circulations with moderate rain and light winds. Indeed, though weak tropical disturbances are relatively common, tropical cyclones are rare, for even in an active year only very seldom will their total number in the northern hemisphere exceed 50, whereas about 20 depressions occur almost daily outside the tropics in winter.

As the structure in Figure 2.8 illustrates, ascent within tropical cyclones takes place in cumulonimbus clouds arranged in spirals converging on the central eye. The spirals, which are sometimes hundreds of kilometres long, are at most a few kilometres wide, and the distance between them is about 50 to 80 km near the edge, decreasing towards the central eye. Hence often only a small fraction of the tropical cyclone, not more than 10 per cent, contains the ascent which gives rise to the bulk of the condensation and rainfall.

Close to the centre, where the clouds form a ring around the eye, the strongest winds and heaviest rainfalls combine to produce the storm's full fury, whereas inside the eye, the winds decrease quickly and the heavy rains cease. Indeed modern research has shown that the eye is actually a region of subsiding warm air at the centre of the storm. This warm core seems to be essential to the formation of a tropical cyclone, and its appearance is one of the first signs that a storm is going to turn into a cyclone.

Despite the strong winds, the rate of movement of a tropical cyclone is only about 5 to 8 m s^{-1}. In the northern hemisphere the storm will drift towards the northwest at first, but when the centre passes latitude 20°N, the speed of travel increases, the direction changes to a more northeasterly path, and the system starts to fill. The point at which the direction of motion changes from westerly to easterly is known as the 'point of recurvature'. The recurvature of cyclone tracks in the southern hemisphere is in the opposite sense, storms moving first towards the southwest and later towards the southeast.

As rainfall rates near the centre of a tropical cyclone can exceed 500 mm per day, a continuous inflow of water vapour into the storm is necessary. In the course of its life, a vast quantity of air is drawn into the circulation and funnelled upward. Given a storm with average inflow and a duration of one week, then the air below the trade-wind inversion contained in a square whose sides are 12–13° latitude long will be raised to 200 mbar or higher in the storm. Thus it is clear that the storm will be consuming water vapour previously evaporated over vast areas of the tropical oceans.

Air spiralling toward the storm centre should decrease in temperature if adiabatic expansion occurs during the pressure reduction, which implies that a dense surface fog should be found in the inner regions. However, observations

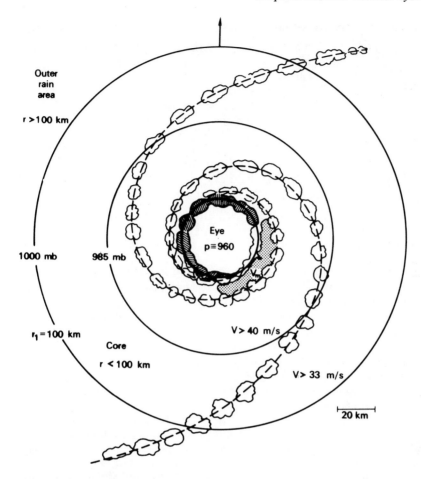

Figure 2.8 Schematic diagram of core region of typical tropical cyclone of moderate strenth. It consists of an 'inner rain area' and an 'eye' with major spiral bands of cumulonimbus clouds along trajectories and forming the eye wall. Stippled region, marked V_m, is region of maximum wind speed, and some typical surface pressures and wind speeds are indicated. Direction of storm movement shown by arrow at top (after Malkus, 1962).

indicate that surface air temperatures within tropical cyclones remain constant or decrease only slightly, and fog is never reported. This suggests that there must be a source of heat and moisture within the storm itself. Greatly agitated by the high winds within the storm, large amounts of water are thrown into the air from the ocean surface in the form of spray. As the air moves towards lower pressure and begins to expand adiabatically, the temperature difference between ocean and air increases; but since the surface of contact is increased to many times the horizontal area of the storm, rapid transfer of sensible and latent heat from the ocean to air is made possible. On the outskirts of the storm, the turmoil of the ocean surface is less and this particular process of heat transfer is not operative. It is now obvious that a fair proportion of the water vapour being used by the storm will have originated from the ocean surface within the storm itself.

5 Temperate oceanic weather systems

It is clear that vertical motion accompanies cyclonic and anticyclonic development and this.vertical motion must be accompanied by surface and upper level convergence or divergence. It is difficult to measure atmospheric vertical motions directly, but they can be estimated if the net difference between the divergence and convergence at lower and upper levels of the storm are known.

In synoptic meteorology the term 'development' is applied to the intensification of both cyclonic and anticyclonic circulations. Sutcliffe (1947) in his 'development theorem' demonstrated, for the flow of the atmosphere at any instant, that areas favourable to cyclogenesis and anticylogenesis can be determined from a knowledge of the vorticity and the vertical shear of the horizontal wind. He found that the horizontal divergence ($\text{div}_p\, V_1$) at the top of an atmospheric layer relative to the bottom ($\text{div}_p\, V_0$), is, with various approximations, given by

$$\text{div}_p\, V_1 - \text{div}_p\, V_0 = -\frac{1}{f}\left(V'\frac{\partial f}{\partial s} + 2V'\frac{\partial \zeta_0}{\partial s} + V'\frac{\partial \zeta'}{\partial s} \right) \tag{2.6}$$

where div_p is the divergence in an isobaric surface, and V_1 and V_0 the wind vectors at the upper and lower pressure levels, f the Coriolis parameter, V' the thermal wind vector between the selected levels, and ζ' its vorticity, ζ_0 the vorticity at the lower level, and where $\partial/\partial s$ denotes differentiation in the direction of the thermal wind.

Sutcliffe has also established that the left-hand side of this equation may be interpreted in terms of vertical motion, and in particular in terms of cyclonic or anticyclonic development, since it is positive in cyclonic development areas (where there is net divergence and therefore a lowering of surface pressure), and negative, in anticyclonic development areas (where there is net convergence and therefore an increase in surface pressure).

Each of the three terms which are contained on the right-hand side of the above equation, has some influence on the development of surface pressure systems. The equation is normally applied to the 1,000–500 mbar layer. The first term represents the latitude effect and can usually be neglected. The second term is the thermal-steering term, and controls the direction of movement of individual depressions and anticyclones. A surface cyclone is a region of maximum relative vorticity, and at the maximum the rate of change of ζ_0 along the thickness lines is zero, which indicates that the second term vanishes. However, ahead of a cyclone, the thermal wind is directed toward decreasing values of ζ_0 and there is relative divergence, but behind the cyclone the thermal-steering term leads to relative convergence. Since similar arguments apply to anticyclones, pressure systems tend to be propagated along the thickness lines at a speed proportional to the thermal wind; this is particularly true in the early stages when the systems are shallow.

The third term is known as the thermal-vorticity effect and is determined entirely by the topography of the thickness chart and the distribution of the thermal winds. It signifies broadly, cyclonic development where the thermal vorticity ζ' decreases in the direction of the thermal wind, and anticylonic development where it increases. Thermal vorticity may be considered as composed of curvature $\left(\dfrac{V'}{r}\right)$ and shear terms $\left(\dfrac{\partial V'}{\partial n}\right)$:

$$\zeta' = \frac{V'}{r} - \frac{\partial V'}{\partial n} \tag{2.7}$$

This equation implies that cyclonic development will be associated with marked ridges and troughs, and also with the entrances and exits to thermal jets. Given a strong thermal wind, cyclones are likely to form downwind of thermal troughs and anticyclones downwind of thermal ridges. Similarly, cyclones tend to form on the equatorward side of the entrances to thermal jets, and on the poleward side of the exits.

The climatological relationships between thickness patterns and surface features now becomes clear. Cyclones will tend to form downwind of the mean thermal troughs in winter over the east coasts of North America and Asia, move northeastwards with the thermal wind, and decay in the thermal ridges over the western regions of the oceans.

Zonal index cycles were discussed in Chapter 1. Since the high index situation is characterized by strong latitudinal temperature gradients in middle latitudes, and by little air-mass exchange, the cyclones and anticyclones of the mid-latitude belt drift eastward, often with considerable speed. Under conditions of low zonal index, a warm cutoff high forms in the middle and upper troposphere, and sealevel cyclones have a tendency to be steered around these highs, either to the north of the cutoff high or to the south along the base of the pattern. Because the normal westerly current is blocked and reversed, this last arrangement is often called a 'blocking pattern' or simply a 'block'. Blocking patterns can have a variety of shapes, but a common one has the shape of the Greek capital Ω and the pattern has been called an Omega-block. Anticyclonic development is associated with the warm ridge and sometimes cyclonic development with the cold trough of the blocking pattern, and so an extensive blocking anticyclone often exists in middle and high latitudes, while a non-frontal low exists to the south.

Clearly, very different weather types will be allied to high index and blocking situations. In order to isolate typical blocking cases, Rex (1950) has introduced the following criteria which apply to the 500 mbar level:

1 the basic westerly current must split into two branches;
2 each branch current must transport an appreciable mass;
3 the double-jet system must extend over at least 45° of longitude;
4 a sharp transition from zonal type flow upstream to meridional type downstream must be observed across the current split; and
5 the pattern must persist with recognizable continuity for at least 10 days.

Blocking patterns once formed are relatively stable and usually persist for a period of 12-16 days. In the northern hemisphere they are most frequently initiated in two relatively narrow longitudinal zones, one (Atlantic) centred aloft at 10°W and the other (Pacific) at 150°W. The number of cases occurring over the Atlantic exceeds those over the Pacific, probably by a factor of two to one.

5.1 Extratropical cyclones

The formation and movement of extratropical cyclones are closely linked to the upper flow patterns discussed earlier. In the northern hemisphere they typically form downward of the large thermal troughs in the thickness patterns over the eastern coasts of Asia and North America. During the early stages the cyclones move rapidly northeast with the thermal wind, but after about 24-30 hours they start to occlude and slow down. They frequently become stationary over the eastern parts of the oceans, just upwind of the major thermal troughs over the western regions of North America and Europe. Once stationary and fully occluded, the cyclones decay slowly over several days. It is the frequency with which they become stationary near Iceland or the Aleutian Islands which gives rise on mean pressure charts to the Iceland and Aleutian lows. From this discussion it will have become apparent that extratropical cyclones, or frontal depressions as they are sometimes called, are primarily maritime storms which are best developed over the oceans. In the interior of continents, particularly to the east of the Rocky Mountains and in central and eastern Eurasia, the areas of continuous cloud cover and precipitation may be small, and in many cases the precipitation area may be broken or absent. Indeed, the weather distribution within a cyclone depends very much on the location and the time of year and it is not possible to generalize with safety. For example, cold fronts are often active with severe thunderstorms over the eastern USA, but they rarely give any significant weather in western Europe.

5.2 Precipitation and continentality over temperate oceans

Precipitation amounts over land are greatly influenced by relief, while those measured over the ocean are presumably free from orographic influence. Unfortunately, it is difficult to measure precipitation over the open ocean directly from ships. Tucker (1961) has carefully estimated from meteorological 'present weather' observations the distribution of precipitation over the North Atlantic for the 5-year period 1953–1957, and his results are reproduced in Figure 2.9. Tucker's average annual precipitation map suggests that precipitation increases from about 400 mm near the Azores anticyclone to over 1,200 mm in the seas between Greenland and Iceland. Precipitation amounts also increase towards the east, reaching a maximum just off the coast of western Europe.

It might be expected that the maximum precipitation over the North Atlantic would occur in the summer, since this is the season when the air is warmest and will therefore hold the most moisture. The seasonal charts in Figure 2.9 clearly show that this is not so, and that the maximum precipitation occurs in winter over large areas of the ocean. Estimates of mean monthly precipitation at the various Atlantic Ocean weather ships are given in Table 2.1 and all except Ship K report maximum monthly precipitation values during the winter months of December, January, and February. Similarly, minimum monthly precipitation values occur either in July and August or April and May. These seasonal changes indicate that precipitation over the North Atlantic is very much controlled by the incidence of synoptic systems.

The degree of influence of land on climate is expressed in the concept of continentality, which is usually applied to the temperature characteristics of a given place. Annual temperature curves can conveniently be represented by three characteristics, which are the mean, the amplitude and the phase; it is sometimes convenient to include a fourth parameter which is indicative of the degree of symmetry of the temperature curve. All these parameters are influenced by changes in the distribution of land and sea, but not always in the same manner, for example the amplitude of the curve may be increased while the mean remains unchanged.

Most climatologists take the mean annual temperature range as the best measure of continentality, and this is used in the index suggested by Conrad (1946):

$$K = 1 - 7A/\sin(\phi + 10) - 14 \qquad (2.8)$$

where K is the index of continentality; A is the average annual temperature range; and ϕ is the latitude. K should be zero for a completely oceanic earth and 100 for the corresponding continental planet. In practice the formula yields values of nearly zero for Thorshavn (62° 2′N, 6° 44′W) and nearly 100 at Verkhoyansk (67° 33′N, 133° 24′E) in eastern Siberia.

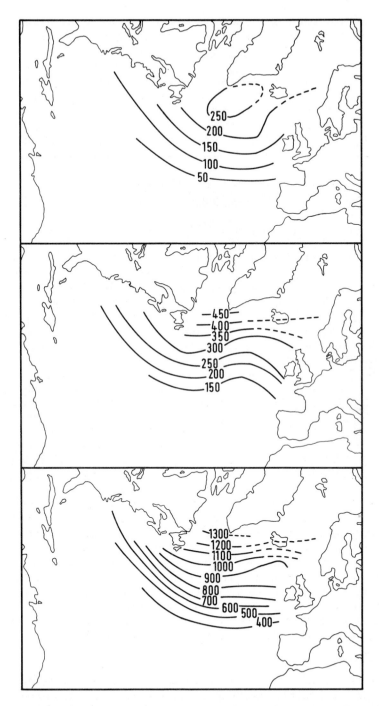

Figure 2.9 Precipitation over the North Atlantic Ocean during the period 1953–57 (mm). Top: average summer; middle: average winter; foot: average annual (after Tucker, 1961).

Table 2.1 Mean monthly and mean annual precipitation (mm) over Atlantic Ocean weather ships during the 5-year period January 1953–December 1957 (*after Tucker, 1961*)

Ocean weather ship	Jan	Feb	Mar	Apr	May	Jun	Jul	Aug	Sept	Oct	Nov	Dec	Annual
I	97	86	50	42	55	65	61	72	92	92	78	103	893
J	78	55	49	43	55	64	56	77	69	68	64	92	770
K	35	47	52	24	35	26	20	29	28	27	28	34	385
M	142	146	88	92	78	73	59	52	102	121	116	125	1,194

An examination of the lag of temperature behind solar radiation has been made by Prescott and Collins (1951) using the technique of harmonic analysis. This lag is an index of continentality in that coastal stations are late in phase and continental stations relatively early. They discovered the largest lags along the western margins of the continents, with substantial differences between west and east coasts. Outside of the tropics considerable areas with lags of 25 days or less occur in Argentina, South Africa, and central Asia.

An extremely continental climate is therefore characterized by very large changes in mean temperature between winter and summer, with the temperature following very closely the annual march of solar radiation. Thus the temperature changes in spring and autumn are relatively rapid. Oceanic climates have small annual ranges with a large lag of temperature behind solar radiation, and there is also some evidence that the mean annual temperatures of land areas in high latitudes are below those in oceanic locations at a similar latitude. Land or sea-ice to the west appears to influence the temperature regime far more than a similar amount of land or ice to the east, and this probably reflects the mean westerly winds in middle latitudes.

6 Polar oceans

At the present time ice surrounds both poles of the globe. In the northern hemisphere, permanent ice cover takes up more than two-thirds of the surface of the Arctic Ocean, and also covers Greenland and several other islands. Arctic sea-ice is an enormous lens, in the centre of which mean ice thickness reaches 3–4 m, decreasing towards its periphery. As this ice consists of a great many individual ice fields, it is constantly moving under the action of air and sea currents, which leads to fluctuations in the boundaries of the ice cover. In summer months, the ice thickness diminishes as a result of thawing, which occurs mainly at the upper surface of the ice. In the cold season, the ice becomes thicker as a result of the freezing of sea water at the lower surface of the ice.

Antarctica is buried beneath an ice-sheet whose surface area is about 13.5 × 10^6 km² and average thickness approximately 2 km. Slowly the ice-sheet creeps outwards from the central plateau towards the Southern Ocean, and more quickly great glaciers advance through portals in the mountain ranges. Spread over the continental shelf is an ice-shelf, consisting of a mass of old sea-ice and superincumbent compressed snow, attaining in places a height of nearly 100 m above sea-level. The ice-shelf is constantly moving seawards, driven by the force of the inland ice. Surrounding Antarctica and its ice-shelf is an area of pack-ice which covers about 25.5 × 10^6 km² at its maximum extent and exhibits a seasonal variation of about 75 per cent of the maximum. Embedded within the Antarctic circumpolar current are icebergs, some of which take as long as 10 years to melt. Bergs may be encountered anywhere south of latitude 55°S, and it is not uncommon to find them as far north as 40°S in the South Atlantic Ocean and the southwestern Indian Ocean.

In Figure 2.10, Kukla (1978) shows the extent of snow and ice at four characteristic intervals during 1974. According to Kukla (1978) the snow and ice conditions during that year were close to the average for the 9 years of available data. In January, the snow cover in the northern hemisphere, after repeated outbreaks of polar air into the middle latitudes, approached its maximum seasonal extent. At the same time, pack-ice in the southern hemisphere shrank close to its seasonal minimum. In April, pack-ice in the northern hemisphere reached its maximum extent, whilst in the southern hemisphere the pack-ice started its northward advance in the zone of almost complete darkness. Where not hindered by winds, large sections of open ocean quickly froze over and wide strips of fast ice formed along the Antarctic coast. Large oceanic segments around 65°S and 20°W and in the Ross Sea, which are affected by upwelling, developed only discontinuous or thin ice cover, a condition which persisted for most of the winter. By the beginning of September, snow and ice in the northern hemisphere had shrunk to its minimum extent, while the Antarctic approached its maximum.

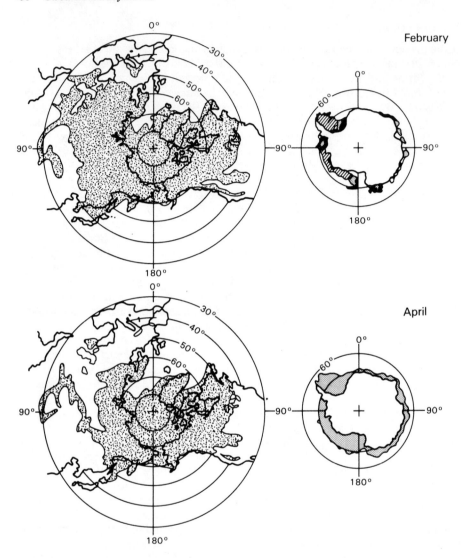

Figure 2.10   Extent of snow and ice at four characteristic intervals of a year. From the National Environmental Satellite Service of the National Oceanographic and Atmospheric Administration weekly snow and ice boundary charts and the Fleet Weather Facility weekly ice charts. Occurrence of snow in South America and Oceania not included (after Kukla, 1978).

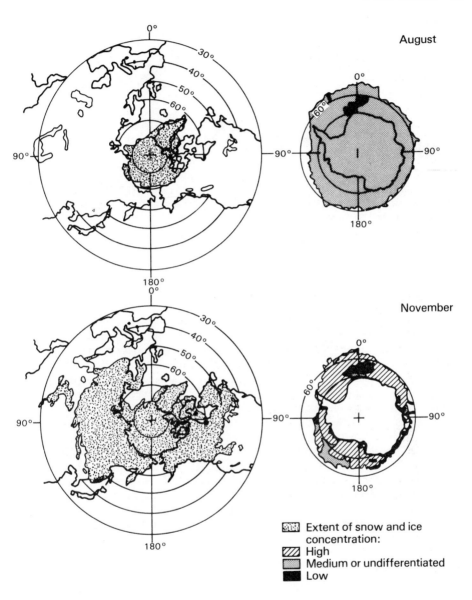

August

November

Extent of snow and ice concentration:
High
Medium or undifferentiated
Low

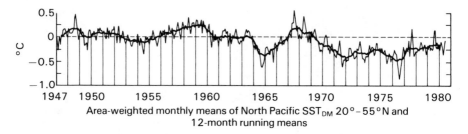

Area-weighted monthly means of North Pacific SST$_{DM}$ 20°–55°N and
12-month running means

Figure 2.11   Area-weighted sea surface temperature average departures from the 1947–66 mean (dashed line) for the North Pacific in the region 15° to 60°N and 130°E to 110°W. Light lines connect monthly anomalies: the dark curve shows the 12-month centred running means (after Namias and Cayan, 1981). Copyright 1981 by the AAAS.

7   Atmospheric–ocean interactions

7.1 Ocean surface temperature anomalies and climate

Since 1965 hundreds of studies of air and sea interactions have appeared in the scientific literature, and have promoted an increased awareness of the importance of the oceans in short-term climate variability. Although it is small-scale processes that govern the details of air–sea interactions, it is the large-scale phenomena that ultimately drive the temporally and spatially averaged exchanges of heat, momentum and water vapour. Namias and Cayan (1981) have provided an outstanding example of the low-frequency nature of the upper ocean's temperature field, and this is shown in Figure 2.11. Figure 2.11 shows that North Pacific areal average sea surface temperature (SST) anomalies for the period 1947 to 1980, both monthly values and their 12-month running means are shown. The running means exhibit a large signal, approaching 0.5°C in amplitude, with a period of several years, a fact quite impressive as they represent oscillations over the bulk of the North Pacific.

There is a mismatch in life expectancy between large-scale atmospheric patterns and ocean surface temperature patterns, and this has also been illustrated by Namias and Cayan (1981). Figure 2.12 shows the persistence of North Pacific monthly average sea surface temperature patterns in contrast to sea-level pressure (SLP) and contours of the 700 mbar surface. The curves represent autocorrelations of monthly patterns at time lags of 1 to 12 months. While the 'memory' of the 700 mbar height and sea-level pressure fades after 1 to 2 months, the SST anomalies persist for 6 months or more. This great SST persistence results from the large thermal capacity of the upper ocean and its relatively sluggish velocity structure. The SST anomalies represent large amounts of stored heat, since they often penetrate to depths greater than 200 m. Thus long-lived, large ocean SST anomaly patterns provide stabilizing boundary influences on the more turbulent overlying atmospheric flow.

The tropical ocean–atmosphere system deserves special mention because of its unusually strong low-frequency, large spatial scale behaviour (Philander, 1979). For much of the low latitudes, the year-to-year variability is larger than the regular

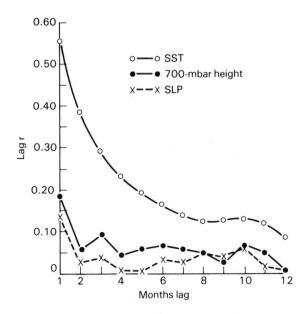

Figure 2.12   Autocorrelations of standardized values for monthly mean patterns of SST, 700-mbar height, and sea-level pressure (SLP) from a 5° grid of points covering the North Pacific (north of 20°N), based on the 20-year period 1947 to 1966; *r*, correlation coefficient (after Namias and Cayan, 1981). Copyright 1981 by the AAAS.

seasonal variations of typical atmosphere and oceanic variables. A prominent example of low-frequency, coupled phenomena is El Niño, which affects the eastern tropical Pacific every 3 to 7 years with unusually warm surface water and an altered trade-wind circulation.

7.2 Influence of the atmosphere on the ocean

Air–sea coupling is easiest to demonstrate in the way that the atmosphere drives the ocean. An obvious example of the influence of the atmosphere on the ocean is that of the wind-driven currents described in the introduction to this chapter.

Numerical investigations to identify the physical processes responsible for the generation, evolution and dissipation of oceanic thermal anomalies (OTA) have been made by Huang (1979). Two simulated cases were studied, one demonstrating the generation and evolution of OTAs under the anomalous atmospheric wind forcing of winter 1949–50, and the other portraying the evolution and dissipation of the OTAs under the climatological atmospheric conditions in winter 1971–72. Results of case 1 show that OTAs are generated by the anomalous wind through wind-induced advections of normal thermal disturbances. Vertical advection plays an important role in OTA generation and their later development, especially in the tropic and subtropic regions during the winter when the thermocline is relative shallow and the Ekman pumping is relatively strong. As a result of the horizontal convergence of the surface currents, downwellings were found under the anomalous anticyclone in the subarctic and upwellings in the anomalous cyclone in the subtropics. The simulated OTAs evolved similarly to the observed SST anomaly for winter 1949–50, with an overall pattern correlation coefficient of

0.72. The study of case 2 indicates that under the assumed boundary conditions, the existing OTAs decrease exponentially in time. It was also shown that sufficiently deep anomalies can persist for a long period of time without much dissipation (at least 120 days) and still maintain more than half of their original strengths in a recognizable pattern under normal atmospheric forcing.

7.3   Effect of the ocean on the atmosphere

Namias and Cayan (1981) comment that despite the apparent logic of the thesis that the sea influences the atmosphere, this thesis is difficult to prove, particularly the influence of SST anomalies on the contemporary or future atmospheric state. Attempts to demonstrate this quantitatively through dynamical models or empirical techniques have not yet yielded entirely conclusive results.

The results of general circulation model testing of atmospheric sensitivity to prescribed mid-latitudinal SST anomalies have not been conclusive according to Namias and Cayan (1981). In most cases the model showed a response localized in the atmosphere directly overlying the anomaly, but significant downstream effects were not unequivocally shown. Rowntree (1979) found from numerical model experiments that warm anomalies of the order of 1 K in middle latitudes will be associated with negative surface pressure anomalies of up to 1 mbar near the warm anomaly. He also commented that numerical experiments with anomalies in middle latitudes show that the signal can be difficult to detect above the noise level but that the compositing of several experiments with different anomalies suggested a tendency for pressure falls on the equatorial side of the anomaly. The induced wind field appears to favour maintenance of the anomalies but there is a large negative feedback from increased turbulent fluxes.

Namias and Cayan (1981) consider that the sensitivity studies of the effects of tropical SST anomaly patterns on the atmosphere have been more convincing. Tropical effects may be easier to detect in models because of the strong, built-in coupling of sea-to-air heat transfer, which modulates cumulus convection with resulting release of latent heat of condensation aloft, this then alters the large-scale air circulation elsewhere as well as in the variable SST area. Rowntree (1979) states that changes in ocean temperatures in the tropics are likely to be of greater importance, especially to the wind field, both locally and remotely, than are anomalies in higher latitudes. He summarizes the reasons for this as follows:

(a) The larger value of the change in equivalent potential temperature for a change in surface temperature with warm water;

(b) the greater depth of convective mixing in the tropics, leading to changes in temperature through a greater depth and to larger surface and pressure changes;

(c) for winds the smaller value of f and consequently larger ratio of wind speed to pressure gradient in low latitudes ($f = 0.25 \times 10^{-4}$ at 10°N, 1.03×10^{-4} at 45°N).

From numerical model experiments, Rowntree (1979) concluded that in the tropics warm water produces low surface pressure with upper tropospheric ridging in the overlying atmosphere. Planetary waves are generated in the upper troposphere with, in the same longitude as the tropical ridge, a subtropical or mid-latitude trough and in some cases (depending on the north–south wave number) a ridge in higher latitudes. Observations in the tropical central Pacific show that warming is associated with considerable increases in cloudiness and rainfall. This was illustrated in the discussion in Chapter 1 on the El Niño effect and the southern oscillation.

7.4 The southern oscillation and El Niño

In his search for precursors to the Indian monsoon rainfall, Walker (1923, 1924, 1928a, b, 1937) uncovered a tendency particularly south of the equator, for a variation in pressure between the eastern and western Pacific over a period of years, an alteration he called the southern oscillation. Bjerknes (1966a, b, 1969, 1972) showed that this alternation in surface pressure was related to variations in rainfall and sea surface temperature (SST) in the equatorial eastern Pacific and, in particular, to the anomalously warm SST off the coast of Peru known as El Niño. Using 24 years of zonal mean sea-surface temperature (SST), Chiu and Newell (1983) have shown that the annual and semi-annual variations account for more than 90 per cent of the variance outside the tropics. In the tropics they found a high persistence and substantial low-frequency variations. A broad region in the tropical Pacific has less than 80 per cent of the total variance explained by the seasonal cycle, and between 7.5°N and 2.5°N the seasonal cycle explains only about 40 per cent of the total variance. Chiu and Newell also found a very strong persistence in the SST of the tropical and subtropical Pacific from 22.5°N to 7.5°S. Wright (1979) and Fleer (1981) have also noted a strong persistence in rainfall in the central equatorial Pacific.

The strength of the Walker circulation varies with the surface water temperature of the Pacific ocean and in turn forms part of a wider tropical fluctuation known as the 'southern oscillation'. According to Berlage (1966), 'the southern oscillation is a fluctuation of the intensity of the intertropical general atmospheric and hydrospheric circulation. This fluctuation, dominated by an exchange of air between the southeast Pacific subtropical high and the Indonesian equatorial low, is generated spontaneously. The period varies between roughly 1 and 5 years and amounts to 30 months on the average.' Associated with the southern oscillation is the 'El Niño' phenomenon when the eastern tropical Pacific Ocean occasionally becomes much warmer than average for periods of a year or more. Historically, 'El Niño' is a phenomenon that manifested itself off Peru at Christmas. Thus the fishermen of the port of Paita gave it the name of 'Corriente del Niño', in English 'Current of the (Christ) Child'. This designation was used by many past authors, although today it is a tacit convention to use the term 'phenomenon' instead of 'current', since it involves a transitory irregularity in the ocean–atmosphere system.

Parker (1983) has recently attempted to satisfy the need for a tabulation of a southern oscillation index that is simple enough to be updated readily by any user. He considers that the mean-sea-level pressure-field index given by Wright (1977), although reliable, is not easily updated because it is based on principal component analysis. Similarly, areally-meaned sea surface temperatures (Rasmusson and Carpenter, 1982) and winds (Wyrtki, 1980) are not simple to update and may also be analysis-dependent in data-sparse areas. Parker considers that the requirement for a simple but reliable index appears to be met best by the index recommended by Chen (1982) for diagnostic studies: Tahiti minus Darwin mean-sea-level-pressure. Parker (1983) has listed this index for 1935 to 1983 (March).

One of the most interesting changes in Parker's listings occurs during 1982 when the southern oscillation entered the 'negative' phase with below normal mean-sea-level pressure at Tahiti, and above normal at Darwin. This corresponds to warmth in the usually cold areas of the equatorial Pacific. The lowest seasonal values for the series for northern summer, autumn and winter occurred in 1982–83 as did the lowest monthly values for August, September, November, and January to March, and the equal lowest monthly values for June and October.

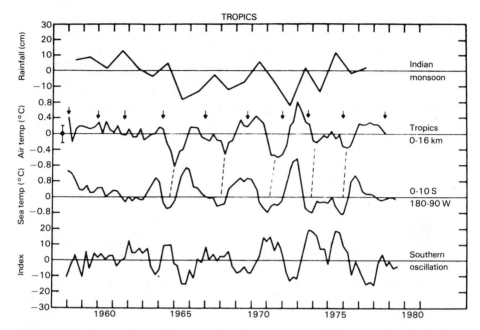

Figure 2.13 Comparison of time variations in southern oscillation index (normalized pressure difference between Tahiti and Darwin), sea-surface temperature, zonally averaged tropospheric temperature in the tropics, and Indian summer monsoon rainfall. Vertical arrows denote time of quasi-biennial west wind maximum at a height of 20 km in the tropics. There is about a 5 per cent chance that the true mean value of seasonal air temperature lies outside the extent of the vertical bar at left (after Angell, 1981).

Figure 2.13 shows the variation since 1958 of SST in the region 0–10°S, 180–90°W, zonally averaged tropospheric temperature in the tropics between 30°N and 30°S, average summer monsoon (June–September) rainfall for India, and a normalized southern oscillation index, where normalization was achieved by dividing the departure from average of the mean monthly pressure difference between Tahiti and Darwin by the standard deviation of the values for that month. The data are plotted by season, with the annual variation eliminated through the use of deviations from long-term seasonal means. Figure 2.13 shows that between 1958 and 1979 there has been an almost exact out-of-phase relation between SST and southern oscillation index, i.e. warm SST has been associated with relatively high pressure at Darwin and/or low pressure at Tahiti. Angell (1981) found that the inverse relation between SST and SOI is significant at the 99.9 per cent level. Figure 2.13 also shows a close relationship between SST and zonally average tropospheric (0–16 km) temperature in the tropics. The curve of Indian summer-monsoon rainfall appears related to the other curves of Figure 2.13. The relation can be expressed as either relatively small rainfall amounts slightly preceding warm SST, or as relatively large rainfall amounts slightly following warm tropospheric temperature in the tropics.

Numerous hypotheses have been proposed to explain interannual changes in Pacific equatorial water temperatures. Bjerknes (1961, 1966a) has examined the data from the eastern tropical Pacific in some detail. He suggested that locally reduced winds will lead to a reduction of sea-to-air heat flux and thus a warming of the near-surface waters. More important, he concluded that an oceanwide weakening of the zonal component of the northeast trades for one or two consecutive years would lead to an increased transport by the north equatorial counter current.

At least three theories have been advanced to explain SST temperature (El Niño) variations along the coast of South America south of the equator. One of the simplest candidate mechanisms is that during El Niños, local reduction of the longshore wind stress can lead to a cessation of upwelling which, in turn, allows local heating to increase SST. These ideas appear in the writings of Wooster (1960), Bjerknes (1961, 1966a) and Wooster and Guillan (1974). Recently, Hickey (1975) and Wyrtki (1975) have suggested that this hypothesis may not be true. Some authors (e.g. Wooster, 1960; Bjerknes, 1961, 1966a; Stevenson *et al.*, 1969; Wyrtki, 1975; Hickey, 1975) have invoked the 'overflow' mechanism to explain major SST changes off Peru. It is hypothesized that the oceanic front between the Galapagos and South America is ruptured, allowing a relatively shallow layer of warm, low salinity water to push south across the equator. The abrupt arrival of this warm water at the coast signals the beginning of El Niño. Bjerknes (1961, 1966a) has suggested that a flux of warm water from the western equatorial Pacific plays an important role in the warming that accompanies El Niño events. Wyrtki (1975) expands this idea by proposing essentially two hypotheses. Firstly, anomalous changes in the zonal component of the southeast trades change the circulation of the subtropical gyre of the South Pacific, particularly the south equatorial current. Higher than usual values of the wind stress result in a buildup of the east–west slope of sea-level and an accumulation of water in the western Pacific. Secondly, as soon as the wind stress of the southeast trades relaxes from its previously high values, the water accumulated in the western Pacific will tend to move back to the east, perhaps as an equatorially trapped Kelvin wave (Lighthill, 1969; Godfrey, 1975; Hurlburt *et al.*, 1976). Warm water accumulates in the eastern Pacific and an El Niño situation results.

Barnett (1977) has attempted to test the above hypotheses. He suggests that the observed interannual fluctuations of SST in the eastern tropical Pacific are consistent with a north equatorial counter current mechanism as hypothesized by Bjerknes (1961, 1966a) and later Wyrtki (1973). Barnett considers that observations of SST fluctuations in the Peruvian coastal zone are not consistent with the idea that locally driven upwelling, by itself, plays a significant role in the large-scale interannual variations in heat balance. Thus variations in SST associated with El Niños are evidently associated with other causes. Also the flow of warm water south across the equatorial front has a significant, but limited, effect on the heat budget of the Peruvian coastal stations. Thus this process cannot account for the major changes observed in SST off Peru. Barnett (1977) found that interannual changes in the zonal components of the trade winds near the dateline and just north of the equator are closely associated with changes in the basinwide sea field and, hence, the zonal oceanic pressure gradient. The wind fluctuations precede those in the sea-level field by 0–2 months. Barnett (1977) comments that the hypotheses of Bjerknes (1966a) and more recently Wyrtki (1975), are compatible with these latter results, but so are the somewhat different mechanistic hypotheses of McCreary (1976) and Hurlburt *et al.* (1976).

The above discussion raises the question as to whether the ocean forces the atmosphere or vice versa. Chiu and Newell (1983) have shown that on a global scale, there is evidence that the major pattern of sea-level pressure change, the southern oscillation, precedes changes in the major mode of SST variations. The zonal mean

SST anomalies then in turn affect atmospheric temperature.

Walker (1923) first pointed out that the southern oscillation phenomenon is the dominant feature in global weather. Eigenvector analysis of global sea-level pressure substantiated Walker's claim (Kidson, 1975). A number of studies have shown strong relationships between tropical weather patterns in particular, and global patterns in general, and El Niño events and the southern oscillation.

The term teleconnections refers to the statistically or empirically determined coupling of large-scale abnormalities of the atmospheric circulation in time and space (Fleer, 1981). Such links were first discovered by Walker (1924) who found linear correlations between sea-level pressure in different parts of the world. Fleer (1981) has undertaken a global survey of rainfall teleconnections. He computed more than 1,000 cross-spectra using several tropical reference stations, and found several interesting relationships which are discussed in Chapter 8. Angell (1981) has found correlations of marginal significance between SST in the equatorial eastern Pacific Ocean and the northern hemisphere, temperate-latitude and United States temperatures, north circumpolar vortex area at 10 km and vortex displacement, and latitude of the subpolar low and subtropical high. He found an equatorward displacement of the subpolar low and subtropical high associated with warm eastern equatorial Pacific SST temperatures. Angell considers that the relationship of greatest interest is probably that between Indian summer-monsoon rainfall and equatorial eastern Pacific SST. That is, small values of Indian summer monsoon rainfall 1–2 seasons earlier. There have been 20 years in which SST was at least 0.8°C above average during the El Niño season (DJF), and during each of these years the summer monsoon rainfall in India two seasons earlier was below average. In 17 of the 20 years between 1958 and 1977, above-average monsoon rainfall has followed (by about two seasons) warm tropospheric temperatures in the tropics, and below-average rainfall has followed cool temperatures. Thus the intensity of the Indian monsoon may be a key to subsequent events, and in particular may lead to better forecasting of the El Niño phenomenon.

References

ANGELL, J. K. 1981: Comparisons of variations in atmospheric quantities with sea surface temperature variations in the equatorial eastern Pacific. *Monthly Weather Review*, 109, 230–43.

ARKIN, P. A. 1982: The relationship between interannual variability in the 200 mb tropical wind field and the southern oscillation. *Monthly Weather Review*, 110, 1393–404.

BARNETT, T. P. 1977: An attempt to verify some theories of El Niño. *Journal of Physical Oceanography* 7, 633–47

—— 1981: Statistical relationships between ocean/atmosphere fluctuations in the tropical Pacific. *Journal of Physical Oceanography* 11, 1043–58.

BERLAGE, H. P. 1966: The southern oscillation and world weather. *Mededeelingen en Verhandelingen* 88. The Hague: Korunklijk Nederlands Meteorologisch Instituut.

BJERKNES, J. 1961: El Niño study based on analysis of ocean surface temperatures, 1935–57. *Bulletin Inter-American Tropical Tuna Commission* 5, 217–303.

—— 1966a: Survey of El Niño, 1957–1958 in its relation to tropical Pacific meteorology. *Bulletin Inter-American Tropical Tuna Commisson* 12, 3–62.

—— 1966b: A possible response of the atmospheric Hadley circulation to equatorial anomalies of ocean temperature. *Tellus* 18, 820–9.

—— 1969: Atmospheric teleconnections from the equatorial Pacific. *Monthly Weather Review* 97, 163–72.

—— 1972: Large-scale atmospheric response to the 1964–65 Pacific equatorial warming. *Journal of Physical Oceanography* 2, 212–17.

BJERKNES, V. 1919: On the structure of moving cyclones. *Geofisiske Publikasjoner*, Oslo 1.

BJERKNES, V. and SOLBERG, H. 1921: Meteorological conditions for the formation of rain. *Geofisiske Publikasjoner*, Oslo 2.

—— 1922: Life cycle of cyclones and the polar front theory of atmospheric circulation. *Geofisiske Publikasjoner*, Oslo 3.

BUDYKO, M. I. 1978: The heat balance of the earth. In Gribbin, J. (ed.) *Climatic Change*. Cambridge: Cambridge University Press, 85–113.

CANE, M. A. 1983: Oceanographic events during El Niño. *Science* 222, 1189–95.

CHANG, C. P. 1970: Westward propagating cloud patterns in the tropical Pacific as seen from time-composite satellite photographs. *Journal of Atmospheric Sciences* 27, 133–8.

CHARNEY, J. G. 1967: The intertropical convergence zone and the Hadley circulation of the atmosphere. In *Proceedings of the WMO/IUGG Symposium on numerical weather prediction in Tokyo. Japan Meteorological Agency Technical Report 67*.

CHARNEY, J. G. and ELIASSEN, A. 1964: On the growth of the hurricane depression. *Journal of Atmospheric Sciences* 21, 68–75.

CHEN, W. Y. 1982: Assessment of southern oscillation sea-level pressure indices. *Monthly Weather Review* 110, 800–7.

CHIU, L. S. and NEWELL, R. E. 1983: Variations of zonal mean sea-surface temperature and large-scale air–sea interaction *Quarterly Journal of the Royal Meteorological Society* 109, 153–68.

CLARK, D. L. 1982: Origin, nature and world climate effect of Arctic Ocean ice-cover. *Nature* 300, 321–5.

CONRAD, V. 1946: Usual formulas of continentality and their limits of validity. *Transactions of the American Geophysical Union* 27, 663–4.

CROMWELL, T. 1953: Circulation in a meridional plane in the central equatorial Pacific. *Journal of Marine Research* 12, 196–213.

DIETRICH, G. and KALLE, K. 1957: *Allgemeine Meereskunde*. Berlin: Gebrüder Borntraeger.

D'OOGE, C. L. 1955: Continentality in the western United States. *Bulletin of the American Meteorological Society*, 36, 175–7.

EGGER, J., MEYERS G. and WRIGHT, P. B. 1981: Pressure, wind and cloudiness in the tropical Pacific related to the southern oscillation. *Monthly Weather Review* 109, 1139–49.

EKMAN, V. W. 1905: On the influences of the earth's rotation on ocean currents. *Ark. Math. Astr. och Fys 2 No. 11*, Stockholm.

FITZ-ROY, R. 1863: *Weather book. A manual of practical meteorology*. London: Longman.

FLEER, H. E. 1981: Teleconnections of rainfall anomalies in tropics and subtropics. In J. Lighthill and R. P. Pearce (eds.) *Monsoon Dynamics* 5–18. Cambridge: Cambridge University Press.

FOBES, C. B. 1954: Continentality in New England. *Bulletin of the American Meteorological Society*, 35, 197–207.

FUJITA, T. T., WATANABE, K. and IZAWA, T. 1969: Formation and structure of equatorial anti-cyclones caused by large-scale cross-equatorial flows determined by ATS-1 photographs. *Journal of Applied Meteorology* 8, 649–67.

GATES, W. L. 1981: The climate system and its portrayal by climate models: a review of basic principles. In Berger, A. (ed.) *Climatic Variations: Facts and Theories*. Dordrecht: Reidel, 3–19.

GEISLER, J. E. 1981: A linear model of the Walker Circulation. *Journal of the Atmospheric Sciences* 38, 1390–400.

GODFREY, J. S. 1975: An ocean spin-down, I: A linear experiment. *Journal of Physical Oceanography* 5, 399–409.

GRAY, W. M. 1968: Global view of the origin of tropical disturbances and storms. *Monthly Weather Review*, 10, 669–700.

HASTENRATH, S. 1977: Hemispheric asymmetry of ocean heat budget in the equatorial Atlantic and eastern Pacific. *Tellus* 29, 523–9.

HASTENRATH, S. and LAMB, P. 1978: On the dynamics and climatology of surface flow over the equatorial oceans. *Tellus* 30, 436–48.

HASTENRATH, S. and KACZMARCZYK, E. B. 1981: On spectra and coherence of tropical climate anomalies. *Tellus* 33, 453–62.

HEYWOOD, G. S. P. 1950: Hong Kong typhoons. *Hong Kong Royal Observatory Technical Memorandum* 3.

HICKEY, B. 1975: Relationship between fluctuation in sea-level wind stress and sea-surface temperature in the equatorial Pacific. *Journal of Physical Oceanography* 5, 460–75.

HOLTON, J. R., WALLACE, J. M. and YOUNG, J. A. 1971: On boundary layer dynamics and the ITCZ. *Journal of Atmospheric Sciences* 28, 275–80.

HOREL, J. D. and WALLACE, J. M. 1981: Planetary-scale atmospheric phenomena associated with the southern oscillation. *Monthly Weather Review* 109, 813–29.

HUANG, J. C. K. 1979: Numerical case studies for oceanic thermal anomalies with a dynamic model. *Journal of Geophysical Research* 84, 5717–26.

HUBERT, L. F., KRUEGER, A. F. and WINSTON, J. S. 1969: The double intertropical convergence zone – fact or fiction. *Journal of Atmospheric Sciences* 26, 771–3.

HURLBURT, H. E., KINDLE, J. C. and O'BRIEN, J. J. 1976: A numerical simulation of the onset of El Niño. *Journal of Physical Oceanography* 6, 621–31.

JULIAN, P. R. and CHERVIN, R. M. 1978: A study of the southern oscillation and Walker Circulation phenomenon. *Monthly Weather Review* 106, 1433–51.

KIDSON, J. W. 1975: Tropical eigenvector analysis and the southern oscillation. *Monthly Weather Review* 103, 187–96.

KUKLA, G. J. 1978: Recent changes in snow and ice. In Gribbin, J. (ed.) *Climatic Change*. Cambridge: Cambridge University Press, 114–29.

LIGHTHILL, M. J. 1969: Dynamic response of the Indian Ocean to the onset of the southwest monsoon. *Philosophical Transactions of the Royal Society* Series A, 256, 45–92.

MALKUS, J. C. 1962: Large-scale interactions. In Hill, M. N. (ed.) *The Sea*, Vol. 1. New York: John Wiley, 88–294.

MCCREARY, J. 1976: Eastern tropical ocean response to changing wind systems with application to El Niño. *Journal of Physical Oceanography* 6, 634–45.

NAMIAS, J. and CAYAN, D. R. 1981: Large-scale air-sea interactions and short-period climatic fluctuations. *Science*, 214, 869–76.

PALMER, C. E. 1951: Tropical meteorology. In Malone, T. F. (ed.) *Compendium of Meteorology*. Boston: American Meteorological Society, 859–80.

PARKER, D. E. 1983: Documentation of a southern oscillation index. *Meteorological Magazine* 112, 184–8.

PERRY, A. H. and WALKER, J. M. 1977: *The Ocean-Atmosphere System*. London: Longman.

PETTERSSEN, S. 1956: *Weather Analysis and Forecasting*. New York: McGraw-Hill.

PHILANDER, S. G. H. 1979: Variability of the tropical oceans. *Dynamics of Atmospheres and Oceans* 3, 191–208.

—— 1983: El Niño southern oscillation phenomena. *Nature* 302, 295–301.

PIKE, A. C. 1971: Intertropical convergence zone studied with an interacting atmosphere and ocean model. *Monthly Weather Review*, 99, 469.

PRESCOTT, J. A. and COLLINS, J. A. 1951: The lag of temperature behind solar radiation. *Quarterly Journal of the Royal Meteorological Society* 77, 121–6.

RAMAGE, C. S. 1974: Structure of an oceanic near-equatorial trough deduced from research aircraft traverses. *Monthly Weather Review* 102, 754–9.

RASCHKE, K., VONDER HAAR, T. H., BANDEEN, W. R. and PASTERNAK, M. 1973: The annual radiation balance of the earth–atmosphere system during 1969–70 from Nimbus 3 measurements. *Journal of Atmospheric Sciences* 30, 341–64.

RASMUSSON, E. M. and CARPENTER, T. H. 1982: Variations in tropical sea-surface temperature and surface wind fields associated with the southern oscillation/El Niño. *Monthly Weather Review* 110, 354–84.

RASMUSSON, E. M. and WALLACE, J. M. 1983: Meteorological aspects of the El Niño/southern oscillation. *Science* 222, 1195–202.

REED, R. J. 1970: Structure and characteristics of easterly waves in the equatorial western

Pacific during July–August, 1967. *Proceedings of Symposium Tropical Meteorology.* Honolulu: American Meteorology Society EII-1 to EII-8.

REITER, E. R. 1983: Surges of tropical Pacific rainfall and teleconnections with extratropical circulation patterns. In Street-Perrot, A., Beran, M. and Ratcliffe, R. (eds.). *Variations in the Global Water Budget.* Dordrecht: Reidel 285–99.

REX, D. F. 1950: Blocking action in the middle troposphere and its effect upon regional climate. *Tellus* 2, 196–211, 275–301.

RIEHL, H. 1954: *Tropical Meteorology.* New York: McGrew-Hill.

ROWNTREE, P. R. 1979: The effects of changes of ocean temperature on the atmosphere. *Dynamics of Atmospheres and Ocean* 3, 373–90.

SASAMORI, T. 1982: Stability of the Walker Circulation. *Journal of the Atmospheric Sciences* 39, 518–27.

SHAW, D. B. 1978: *Meteorology over the Tropical Oceans.* London: Royal Meteorological Society.

SIKDAR, D. N. and SUOMI, V. E. 1971: Time variation of tropical energetics as viewed from a geostationary attitude. *Journal of Atmospheric Sciences* 28, 170–80.

STEVENSON, M., GUILLEN, O. and SANTORO, J. 1969: *Marine Atlas of Pacific Coastal Water of South America.* University of California Press.

SUTCLIFFE, R. C. 1947: A contribution to the problem of development. *Quarterly Journal of the Royal Meteorological Society* 73, 370–83.

—— 1951: Mean upper contour patterns of the northern hemisphere – the thermal synoptic view-point. *Quarterly Journal of the Royal Meteorological Society* 77, 435–40.

SUTCLIFFE, R. C. and FORSDYKE, A. G. 1950: The theory and use of upper air thickness patterns in forecasting. *Quarterly Journal of the Royal Meteorological Society* 76, 189–217.

THE POLAR GROUP, 1980: Polar atmosphere–ice–ocean processes: a review of polar problems in climate research. *Reviews of Geophysics and Space Physics* 18, 525–43.

TUCKER, G. B. 1961: Precipitation over the North Atlantic Ocean. *Quarterly Journal of the Royal Meteorological Society* 87, 147–58.

WALKER, G. T. 1923: Correlation in seasonal variations of weather VIII: A preliminary study of world weather (world weather I). *Memoirs of the India Meteorological Department* 24, 75–131.

—— 1924: Correlation in seasonal variations of weather, IX: A further study of world weather (world weather II). *Memoirs of the India Meteorological Society* 54, 79–87.

—— 1928a: World weather. *Quarterly Journal of the Royal Meteorological Society* 54, 79–87.

—— 1928b: World weather III. *Memoirs of the Royal Meteorological Society* 2, 97–106.

—— 1937: World weather VI. *Memoirs of the Royal Meteorological Society* 4, 119–39.

WATTS, I. E. M. 1959: The effect of meteorological conditions on tide height at Hong Kong. *Hong Kong Royal Observatory Technical Memorandum 8.*

—— 1969: Climates of China and Korea. In Arakawa, H. (ed.) *Climates of Northern and Eastern Asia, World Survey of climatology* Vol. 8. Amsterdam: Elsevier, 1–117.

WEARE, B. C. 1983: Moisture variations associated with El Niño events. In Street-Perrott, A., Beran, M. and Ratcliffe, R. (eds.) *Variations in the Global Water Budget.* Dordrecht: Reidel 273–84.

WELLS, N. 1982: The ocean and climate – an introduction *Weather* 37, 116–21.

WILLIAMS, K. 1970: Characteristics of wind, thermal and moisture fields surrounding the satellite-observed mesoscale trade wind cloud clusters of the western North Pacific. *Proceedings of Symposium Tropical Meteorology.* Honolulu: American Meteorological Society DN-1 to DN-8.

WOOSTER, W. S. 1960: El Niño. *California Cooperative Oceanic Fisheries Investigations Reports* 7, 43–5.

WOOSTER, W. S. and GUILLAN, O. 1974: Characteristics of the El Niño in 1972. *Journal of Marine Research* 32, 387–404.

WRIGHT, P. B. 1977: *The Southern Oscillation – Patterns and Mechanisms of the Teleconnections and the Persistence.* University of Hawaii, Hawaii Institute of Geophysics OCE-76-23173.

—— 1979: Persistence of rainfall anomalies in the central Pacific. *Nature* 277, 371–4.

WYRTKI, K. 1973: Teleconnections in the equatorial Pacific Ocean. *Science* 180, 66–8.

—— 1975: El Niño, the dynamic response of the equatorial Pacific Ocean to atmospheric forcing. *Journal of Physical Oceanography* 5, 572–84.

—— 1980: Comparison of four equatorial wind indices in the Pacific and El Niño outlook for 1981. *Proceedings of the Fifth Annual Climate Diagnostics Workshop*. Washington, DC: US Department of Commerce.

YOSHIDA, K. and MAO, H. L. 1957: A theory of upwelling of large horizontal extent. *Journal of Marine Research* 16, 40-54.

ZEBIAK, S. E. 1982: A simple atmospheric model of relevance to El Niño. *Journal of the Atmospheric Sciences* 39, 2017–27.

3 Glacial Subsystems

Terrestrial snow and ice can be divided into five distinct categories which are seasonal snow, permafrost, mountain glaciers, sea-ice, and continental ice-sheets. According to Untersteiner (1975) the total amount of water in all earthly forms is estimated to be $1,384 \times 10^6$ km^3, and of this 97.4 per cent is sea water; 0.0009 per cent is atmospheric vapour, 0.5 per cent is ground water, most at great depths; 0.1 per cent is contained in rivers and lakes, and 2 per cent is frozen. This last figure means that nearly 80 per cent of the fresh water on earth exists in the form of ice and snow. Today, perennial ice covers 11 per cent of the earth's land surface and an average of 7 per cent of the world ocean. The residence time of solid precipitation in ice masses ranges from 1–10 years for the frozen sea water in the Southern and Arctic Oceans to 10,000–100,000 years in the Antarctic ice-sheet.

The global distribution of snow is shown in Figure 3.1, where it is seen that at the maximum extent in January it covers an area considerably greater than that of all sea-ice and continental ice-sheets combined.

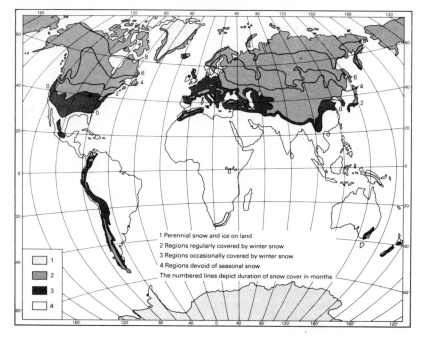

Figure 3.1 Global distribution of snow on land (after Untersteiner, 1975).

At its maximum seasonal extent sea-ice covers nearly 10 per cent of the world ocean, with an average thickness of 2 m in the Arctic and 1 m in the Southern Ocean. In the central Arctic Ocean, a sharp pycnocline exists at a depth of about 25–50 m, forming the lower boundary of the Arctic surface water, with salinities between 32⁰/₀₀ and 34⁰/₀₀ and temperatures close to freezing. The primary factor maintaining this low salinity is the 3,300 km³ continental run-off which adds a layer of approximately 30 cm of fresh water to the Arctic Ocean each year. In contrast in the Southern Ocean, with its less persistent ice cover, higher wind velocities, more intense mixing, and absence of continental run-off, the pycnocline is only about one-tenth of the magnitude of that in the Arctic Ocean and is found at variable depths down to 200 m.

The distribution of glaciers is a complex function of the distribution and amount of accumulating snow and of the nature and amount of incoming energy and its utilization at the glacier surface during the summer ablation period. Thus a particular ice mass exists where, on the average, the amount of snow accumulation equals or exceeds the amount of summer ablation. Temporal changes in the amount and character of snow accumulation or energy input affect this balance and result in glacier growth or shrinkage. Thus the warming trend experienced in the northern hemisphere during the early part of this century led to glacier retreat in most areas; conversely, the downward trend of temperature since about 1940 has caused glaciers to slow this retreat or re-advance. Glaciers are found to exist in a wide range of environments, ranging from the maritime temperate regimes of southern Norway or the North American Cascades on the one hand, to the polar deserts of Canada, northern Greenland and Antarctica on the other.

In general three different major types of glacial climate may be distinguished:
   (i)   A widely distributed high-polar ice-cap climate having summers below 0°C, e.g. north Greenland and Antarctica.
   (ii) A continental type of tundra climate with cool summers (warmest month below 10°C), cold winters (coldest month below – 8°C), wide temperature fluctuations, and small precipitation.
   (iii) A maritime type with small oscillations of temperature and cool summers, e.g. Iceland.

True polar climates are today found in two contrasting areas, one of which is largely an elevated plateau and the other an ocean. The Antarctic ice-sheet contains about 90 per cent of the present glacial ice and is more than seven times greater in area than its counterpart in Greenland. The Antarctic continent has an area of about 14,000,000 km² of which less than 3 per cent is estimated to be free from a permanent ice-sheet. Elevation makes the environment even less inviting because 55 per cent of the surface is above 2,000 m and about 25 per cent is above 3,000 m. In contrast the Arctic Ocean is almost completely land-locked except for one main access point to the warmer waters of the Atlantic between Greenland and Norway. The central Arctic basin is covered by a thin (few metres thick) but permanent ice-pack which extends to the continents during the winter. This contrasts with the Antarctic ice-sheet which is up to 2,400 m thick in places.

Precipitation over the Antarctic ice-sheet falls entirely in solid form and ranges from 40 g cm⁻² yr⁻¹ in a narrow band along the coast to 5 g cm⁻² yr⁻¹ and less over nearly half of the continent. Because of observational difficulties in measuring accumulation and ice export in Antarctica, an accurate assessment of mass balance is virtually impossible.

The present consensus is that the Antarctic ice-sheet today has a zero or slightly positive mass balance. Evidence for past variations of the ice-sheet is sparse, but marks of inland ice-levels 60 m above the present surface exist. Indeed, when compared with the glacial events in the northern hemisphere during the past 15,000 years, the Antarctic ice-sheet appears to have remained virtually unchanged.

1 Polar climates

1.1 Radiation

At high latitudes the total daily solar radiation depends largely on day-length which in turn varies widely with the season of the year. Values of day-length are summarized in Table 3.1 which shows that they vary from 24 hours at 70°N and above in June to zero above the same latitude in December. The radiation climate produced with continuous darkness in winter and continuous daylight in summer is distinctly different from that found at lower latitudes.

Table 3.1 Daily duration of possible sunshine (in hrs and min) on the fifth day of each month at various latitudes (*after Gavrilova, 1963*)

	60	65	70	75	80	85	90
Dec	6 h 43	5 h 02	—	—	—	—	—
Mar	11 h 44	11 h 40	11 h 33	11 h 23	10 h 50	9 h 50	—
Jun	18 h 49	21 h 53	24 h 00	24 h 00	24 h 00	24 h 00	24 h 00
Sept	12 h 55	13 h 07	13 h 26	13 h 57	15 h 10	18 h 15	24 h 00

The duration of the polar night, which is defined as the period when the solar altitude is less that 0° 50′, depends on the latitude, and increases towards the pole. By astronomical calculations which disregard refraction of the sun's rays and twilight, the polar night should last from 179 days at 90° to 24 hours at the Arctic Circle. Solar refraction in practice reduces the duration of the polar night to 175 days at the pole and zero on the Arctic Circle. In actual conditions the polar night is shortened still further by the phenomenon of twilight, which is defined as the period when the solar altitude is between − 6° and 0° 50′. At the equator twilight lasts for only a few minutes, whereas near the pole it may continue for several days in succession, but it is only important for illumination, since the influx of short-wave radiation is negligible. Dates of the beginning and end of the polar day and night are given for various latitudes in Table 3.2.

Table 3.2 Dates of the beginning and end of the polar day and polar night (taking refraction into account) (*after Gavrilova, 1963*)

Latitude	Polar day Beginning	End	Polar night Beginning	End
70 N	17 V	27 VII	25 XI	17 I
80 N	14 IV	30 VIII	22 X	21 II
90 N	19 III	25 IX	25 IX	19 III

In fact the distribution of global radiation is controlled not only by astronomical considerations, but also by the distribution of cloud, which is closely related to the character of the atmospheric circulation processes. According to Gavrilova (1963) the most cloudy regions of the Arctic are the North Atlantic and European sectors where, because of the intense cyclonic activity, the frequencies of overcast skies are very high. The least cloudy regions in the Arctic are the northeastern parts of the Canadian Archipelago and North Greenland, which are subject to the prolonged influence of the polar and Greenland anticyclones. A second minimum is observed in eastern Siberia, where it is low along the coast and decreases inland. Throughout much of the Arctic the greatest number of clear days occurs at the end of the winter and the beginning of spring, while the maximum number of overcast days is observed in the autumn, before the beginning of the polar night. In the North Atlantic, Norwegian, and Greenland Seas there is almost no variation in cloudiness during the year.

A high frequency of low-level clouds is observed during almost the whole year in the Arctic, there being a small increase from winter to summer. These are mainly stratus clouds, for cumulus clouds are rarely encountered, except during the warm period of the year over the continents. The main features of Arctic clouds are their low water content and small depth. Dergach, Zabrodskiĭ, and Morachevskiĭ (1960) discovered that the water content of the clouds most often encountered in the Arctic is only measured in hundredths or tenths of a $g\,m^{-3}$, whereas the water content of clouds at lower latitudes reaches several $g\,m^{-3}$. Similarly, it has been established by Zavarina and Romasheva (1957) that the mean thickness of Arctic low-level clouds averages 350–550 m, while the corresponding value of medium level clouds is 400–450 m. The clouds are particularly thin in the cold season of the year (100–150 m), and thickest in the warm season (up to 1,000 m); however, in 76 per cent of the cases the cloud thickness in the Arctic is less than 700 m. The corresponding mean values for clouds at lower latitudes are as follows: stratus and stratocumulus 700–800 m; altostratus and altocumulus 500–800 m.

Information on cloudiness in the Antarctic is less complete, partly because of the complications introduced by the elevation of the plateau. Schwerdtfeger (1970) states that the circumpolar band of maximum cloud amount is located north of the belt of lowest pressure at sea-level and south of the belt of strongest westerly winds. The situation over the continent is not clear, except that cumulus clouds are occasionally reported along convergence lines.

Details of the distribution of global radiation in the Arctic are contained in Figure 3.2. Minima of global radiation are found during all months over the Norwegian Sea, and Vowinckel and Orvig (1970) suggest that this is a result of the high cloud amount, high cloud density, and the high atmospheric water vapour content. Maxima, observed mostly over eastern Siberia between 140° and 160°E, probably result from the extreme remoteness from open oceans. Secondary maxima, observed over Canada and the Canadian Archipelago, may be caused by the northward extension of the clear skies of the prairies. The thin clouds make the diffuse sky radiation a more important component of the global radiation than is usual in lower latitudes.

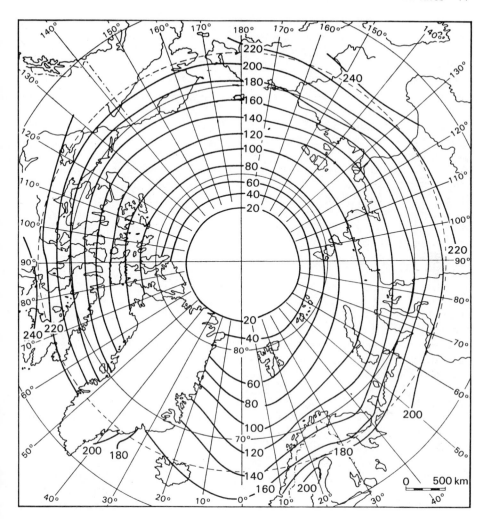

Figure 3.2 The distribution of global radiation in the Arctic (cal cm^{-2} day^{-1}). A: March.

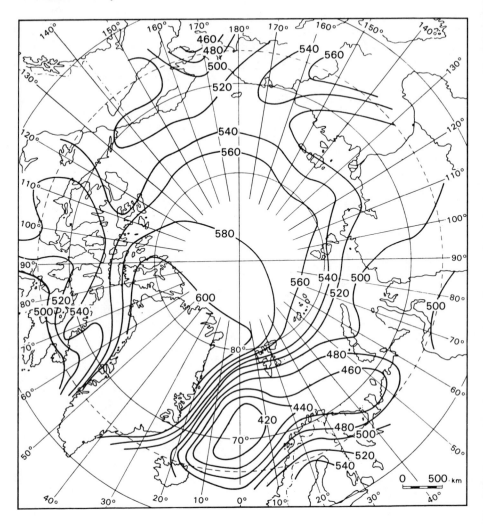

B: June (after Vowinckel and Orvig, 1970).

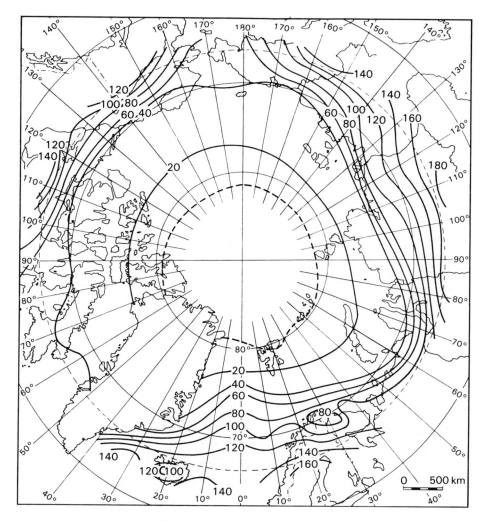

Figure 3.3 Absorbed global radiation in the Arctic (cal cm^{-2} day^{-1}). A: March.

The short-wave radiation actually absorbed by the surface (Figure 3.3) is obtained by multiplying the global radiation by (1 – albedo). Over most parts of the world the albedo values are relatively small and uniform over large areas, but as the albedo of snow and ice is very high, in regions of seasonal snow cover there will be wide variations in albedo. Vowinckel and Orvig (1970) suggest that the smallest Arctic albedos are found in July, August, and September along 65°N, where the absorbed radiation is only 5–6 per cent smaller than the global radiation; but to the north and in winter the albedo becomes a factor of equal or greater importance than all other depletion factors. The charts of absorbed radiation indicate some extremely steep gradients, which are apparent in all months, but especially in spring and early summer. Over the Atlantic Ocean the steep gradients follow the ice limit, but conditions are more complex over the continents because coniferous forests with low albedos extend far northwards beyond the snow-line.

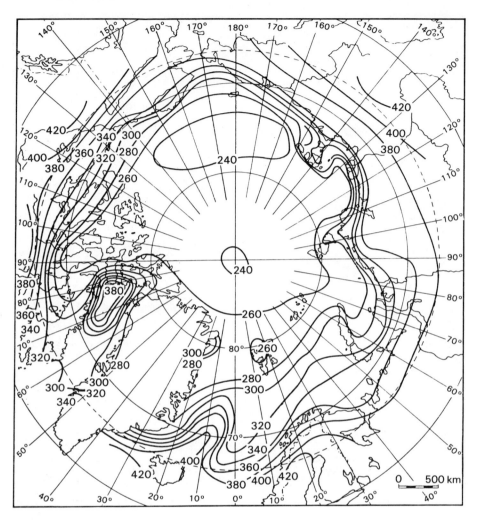

B: July.

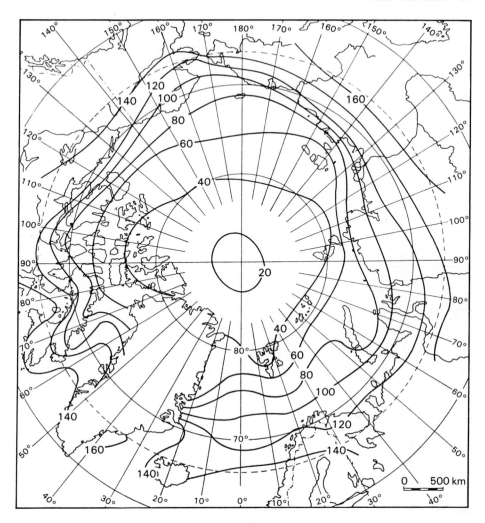

C: September (after Vowinckel and Orvig, 1970).

Long-wave atmospheric counter radiation is particularly important at high latitudes, where the short-wave radiation becomes negligible during the winter. Atmospheric long-wave radiation is the only radiative heat source for the surface during the polar night and even at midsummer the long-wave contribution is higher than that of the solar radiation (Table 3.3).

Table 3.3 Per cent contribution by insolation and atmospheric counter-radiation to total surface radiation income in June (*after Vowinckel and Orvig, 1970*)

	65	70	75	80	85	90
Long wave	60	64	68	69	69	69
Short wave	40	36	32	31	31	31

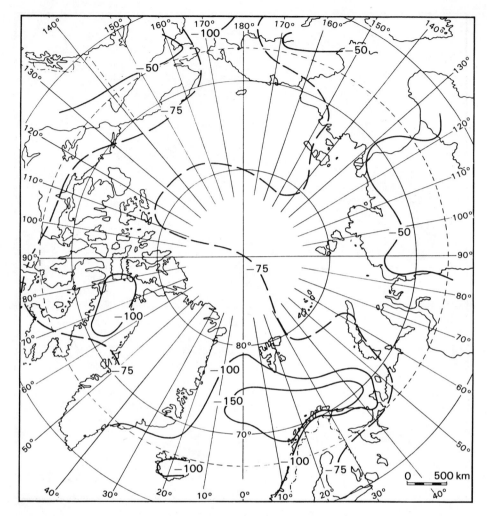

Figure 3.4   Net radiation balance of the Arctic surface (cal cm^{-2} day^{-1}). A: January.

The net radiation balances of the Arctic surface during January and July are shown in Figure 3.4. During winter, the maximum radiation loss is not in the polar basin proper but just to the south, where the ice-free ocean begins. In particular, a major area of loss appears over the Norwegian Sea and the adjoining oceans, forming one of the major radiational heat sinks of the earth's surface. Towards spring, steep gradients arise because of variations in albedo, but in summer the net radiation over the Arctic becomes positive and relatively uniform.

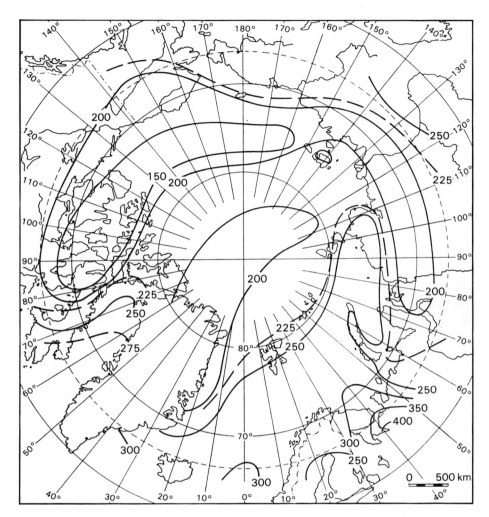

B: July (after Vowinckel and Orvig, 1970).

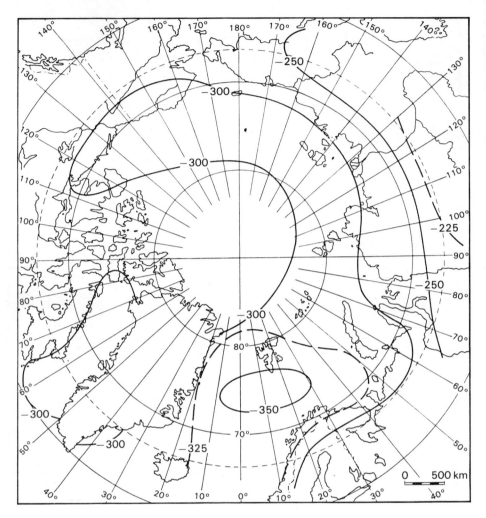

Figure 3.5   Net radiation balance of the earth-atmosphere system in the Arctic (cal cm^{-2} day^{-1}) A: February.

As shown in Figure 3.5, total radiation balances for the earth–atmosphere system during February and August indicate that at all seasons of the year the balance is negative, signifying a continual heat loss to space which must be made up by energy advection from low latitudes. In February, the non-radiative processes must represent 100 per cent of the energy expenditure (that is, disregarding any heat stored in the ground), and 100 per cent of the radiative energy turnover, but the importance of non-radiative processes becomes less towards summer. Vowinckel and Orvig (1970) have calculated the percentage contribution of non-radiative processes to the energy turnover in the Arctic, and some of their results are contained in Table 3.4, which shows that the winter heat advection is extremely powerful.

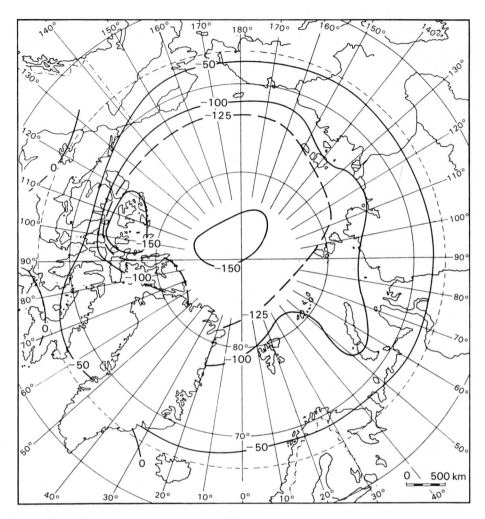

B: August (after Vowinckel and Orvig, 1970).

Table 3.4 The percentage contribution of non-radiative processes to the turnover of energy at the North Pole (*after Vowinckel and Orvig, 1970*)

Jan	Feb	Mar	Apr	May	Jun	Jul	Aug	Sept	Oct	Nov	Dec
100	100	100	48	10	9	22	25	79	100	100	100

Radiation studies in the Antarctic are not as advanced as those in the Arctic, but when allowance is made for the ice-covered high plateau surface, the same general principles hold there as in the Arctic. Very large amounts of solar energy reach the surface of the plateau in summer, and Rusin (1961) has made the interesting comment that 'at the latitudes of the 80s where the coldest point on earth is situated, the

point of maximum monthly amounts of solar energy is also situated'. The high surface albedo and the thin clouds enhance the importance of diffuse sky radiation, so
that the global radiation on cloudy days differs little from that on clear days.

The seasonal variation of surface radiation balance over the continent and the
adjacent ocean has been described by Zillman (1967). The entire region experiences
a surface net radiation gain in January, although values approaching zero are
found over the greater part of the plateau. By April, all the continent and the ocean
southward of 55°S experience a net radiation loss, which is particularly large over
the water oceans surrounding Antarctica. In July, the region of net radiative loss
from the surface has expanded northward to 45°S, with the largest values in the
60–65°S latitude band. The return of the sun in October brings a net radiative gain
to parts of the Ross Sea and the eastern escarpment of the plateau extending as far
south as 85°, but over the remainder of the region the net radiation is still negative.

1.2 Inversions
There are two main types of semi-permanent inversions in the world: the first type
is found in the subtropics and the second in the polar regions. While the inversions
of the subtropics are normally dynamic in origin, the polar inversions are complex
in origin and are maintained not only by the subsidence of warm air but also by
intense surface cooling. Vowinckel and Orvig (1967) consider that temperature
inversions dominate the Polar Ocean practically throughout the year, since no
month averages less than 59 per cent inversion conditions, and in late winter the
frequency reaches 100 per cent. In winter over 80 per cent of the inversions start at
the surface and are not destroyed by the occasional cyclones, for the advection of
warm air seems to further stabilize the surface conditions. A similar phenomenon
is found in the Antarctic where, with the exception of two summer months, an
inversion is an ever-present feature over the high plateau, and a frequent one over
the rest of the continent including the coastal areas.

1.3 Surface temperatures
During most of the year the Polar Ocean is covered by a relatively thin layer of cold
air and the air temperature near the surface is primarily dependent on the temperature of the ice surface. In summer, the prevailing melting of snow and ice holds the
surface temperature close to 0°C, but positive temperatures are usually observed
near the pole in the second half of July. The winter temperatures over the Arctic
pack-ice remain nearly constant for a considerable time and this represents a
balance between heat loss from the ice and snow by radiation, heat conducted
through the ice from the underlying water, and also the heat transported into the
Arctic by intense warm air advection in cyclones. Winter minimum air temperatures occur when the net radiative loss is only balanced by the transfer of heat
through the ice from the underlying water, and according to Vowinckel and Orvig
(1970), this gives temperatures of about $-40°C$, or even down to $-50°C$ over
thick ice. Winter maximum temperatures are nearly a linear function of wind
speed, because of the transport of heat downwards from above.

Antarctic surface temperatures are complicated by the relief of the continent,
large areas of which are elevated above sea level. Taljaard *et al.* (1969) have found
that in January average temperatures vary from $-30°C$ near the centre of the continent to $-2°C$ along the coasts, while in July the coldest areas of the plateau average $-70°C$, and the coasts reach about $-28°C$. The surface temperature in
Antarctica and particularly over the Antarctic plateau decreases very little during
the winter night, from April to August, presenting the so-called coreless winter

phenomenon. Wexler (1959) considers that when the sun sets, the temperature drops rapidly over the continent, but less so over the surrounding oceans thus creating a strong meridional temperature gradient. This initiates the formation of numerous intense cyclones which move vast quantities of warm marine air southward and in this way ventilate large portions of Antarctica above the surface inversion, so slowing the decline of surface temperature.

Mean temperature lapse rates in the troposphere are around $0.65°C$ $100\,m^{-1}$. Because of the presence of strong inversions, temperature lapse rates with altitude along the edges of major ice-sheets can be larger than those observed in the free atmosphere. Putnins (1970) reports lapse rates along the edge of the Greenland ice sheet varying from very low values up to about $1.2°C$ $100\,m^{-1}$. He considers that there is evidence that lapse rates are consistently high along the slope of the ice-cap. Diamond (1958) used a mean lapse rate of $0.7°C$ $100\,m^{-1}$ for the Greenland ice-sheet computed from the annual temperature at five coastal stations and the annual mean in temperatures obtained from snow profile studies at five inland stations at approximately the same latitude. He constructed an annual mean air temperature map of the Greenland ice-cap on this basis. A recent map of annual mean air temperature on the Greenland ice-cap by Robin (1983) probably implies a slightly higher temperature lapse rate.

1.4 Snow cover, precipitation and water balance

During the greater part of the year the Arctic surface is snow-covered since only in July and the first half of August is the snow in a state of rapid melting. The snow cover is less stable on the margins of the Arctic Ocean and vanishes during the summer. According to Gavrilova (1963), on the northern islands of the Soviet Arctic seas the snow cover is established in the first half of September, but at the end of September or the beginning of October in the northern part of the Asian continent, on the coasts of the Laptev, East Siberian and Beaufort Seas, on the southern islands of the Canadian Archipelago, and on the coasts of the Karsk and Chukotsk Seas. In northern Europe the snow cover only appears at the middle or the end of October.

The destruction of the Arctic snow cover begins in the middle of May in the continental areas and on the coast of the Barents Sea, in the first half of June on the coast of the remaining part of the Arctic, and the second half of June on the islands and most northern parts of the continents.

Annual precipitation over the Polar Ocean is meagre, and falls mainly in the form of snow during autumn and late spring. This precipitation is largely frontal in nature and therefore decreases poleward from $250\,mm\,yr^{-1}$ on the margins to about $135\,mm$ in the centre (Vowinckel and Orvig, 1970). The distribution of precipitation over the Antarctic continent is not well known, but the annual accumulation of snow appears to vary from about $60\,g\,cm^{-2}$ along the coast to about $5\,g\,cm^{-2}$ in the centre of the plateau (Giovinetto, 1964, 1968).

The form of ice-sheets partly controls the distribution of precipitation. The steep vertical profile of their margins intensifies the vertical component of the wind velocity, causing precipitation to take place mainly in the marginal areas. The further spreading of ice-sheets replaces the positive feedback between their nourishment and growth (increase in snowfall due to ice-sheet orography) by the negative one (decrease in snowfall due to deviation of cyclone tracks) (Kotliakov and Krenke, 1982).

2 Physics of large ice-sheets

Oerlemans (1981a) considers that large continental ice-sheets may be described by two characteristic quantities. First, some idea exists on the height-to-width ratio; for a large sheet this ratio has an order of magnitude of 10^{-3}. Second, it has been observed that ice-sheets have rather steep edges, which is partly due to the non-linear character of the flow law for ice. These two properties are probably so universal that continental ice-sheets in the past also showed them. Indeed according to Weertman (1961, 1976), the shape of a perfectly plastic ice-sheet is fully determined by its size L according to:

$$H(x) = \sigma(\tfrac{1}{2}L - |x - \tfrac{1}{2}L|)^{\frac{1}{2}} \tag{3.1}$$

where $H(x)$ is the height of the sheet, L the width, and σ a parameter specifying the height-to-width ratio which depends on the surface properties of ice. The parameter σ is approximately given by:

$$\sigma = \frac{4\tau}{3\,\rho g} \tag{3.2}$$

where τ is the yield stress which is equal to the basal shear stress ($\sim (0.5 - 1) \times 10^5$ Pa (1Pa $= 10^{-5}$ bar), ρ is the density of ice, and g is the gravitational acceleration. The value of the constant σ is about 7–14 m.

Equation (3.1) follows directly from the requirement that the shear stress at the bottom of the ice-sheet equals the yield stress everywhere. In the equilibrium case the flux of ice mass through the centre of the sheet is zero, implying that the ice-sheet size can be found by making up the mass balance over the southern half. If the ice-sheet is relatively warm the yield stress will be low and it will have a relatively flatter profile than a very cold ice-sheet.

So far it has been assumed that the bedrock does not react to the load of ice mass. In reality there is a tendency to restore isostatic equilibrium. According to Oerleman (1981b), in the simplest form this may be formulated by

$$\frac{\partial h}{\partial t} = - w(H^* + 2h) \tag{3.3}$$

where h is the height of the bedrock with respect to its equilibrium value (the case of no ice-sheet) and H^* is the elevation of the ice surface above sea-level. The time-scale for adjustment is $\tfrac{1}{2}w$. In the steady state $h = \tfrac{1}{2}H^*$, which corresponds to an isostatic balance if the rock density is three times that of ice.

The parameterization of the mass budget G of ice-sheets is very difficult. Conditions over the Quaternary ice-sheets are badly known and probably are very poorly reflected by present-day conditions in high mountains, or those over the Antarctic and Greenland ice-sheets. Both Andrews and Mahaffy (1976) and Oerlemans (1981a and b) have described the height dependence in the general form of

$$G = 0.73 \times 10^{-3} \times (h - E) - 0.27 \times 10^{-6} \times (h - E)^2 \tag{3.4}$$

where elevation h and E are in metres. E is the height of the equilibrium line (defined by $G = 0$). The maximum value of G is 0.5 m yr^{-1} for $h - E = 1500$ m for larger values of $h - E$, G is kept at this value. A few considerations led to employing a parabolic height dependence of G as formulated by equation (3.4). Observations show that an upper limit to precipitation amounts exists as elevation increases. The reason for this is simply the decreasing water vapour content of the air. For this reason precipitation amounts in the polar regions are small. Character-

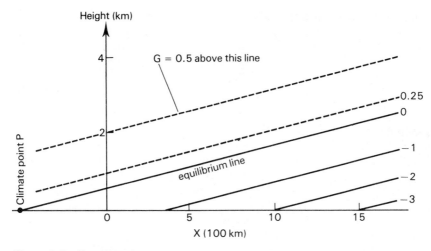

Figure 3.6 Parameterization of the mass-balance conditions for a northern hemisphere ice sheet (see text) (after Oerlemans, 1981b).

istic values of the present yearly precipitation are 0.3 m yr^{-1} in the Antarctic and 0.5 m yr^{-1} in the Arctic.

The parameterization of the mass balance G according to equation (3.4) is shown in Figure 3.6. The point where the equilibrium line intersects the sea-level is often called the 'climatic point', because it serves as an indicator for the prevailing climatic conditions. The slope of the equilibrium line is in the 0.5×10^{-3} to 10^{-3} range. It is seen from Figure 3.6 that accumulation is mainly on the northern and upper parts of the ice-sheet, while ablation is mainly along the lower southern margin.

3 Snow and ice on mountains

General theoretical equations for snowmelt usually express snowmelt rate as a function of wind speed, humidity, rainfall and solar radiation, as well as air temperature. For a monthly parameterization it is possible to neglect all terms except those concerning temperature and solar radiation. For a study of snowmelt during past periods it is particularly important to include solar radiation because it has varied during the ice ages due to changes in the earth's orbit (the Milankovitch mechanism). A suitable parameterization has been obtained by Pollard (1980) and tested on a variety of present-day data. According to Pollard (1980) monthly mean snowmelt may be estimated as follows:

$$\text{Monthly mean ablation} = \max [0; aQ + bT + c] \qquad (3.5)$$

where a, b and c are constants based on crude surface energy balance estimates and actual glacial measurements; Q is the monthly mean solar radiation incident at the top of the atmosphere (W m^{-2})' and T is the monthly mean air temperature (°C). Pollard (1978) arrived at the following values for a, b and c; $a = 0.32$ (g cm^{-2} month^{-1}), $b = 10.0$ (g cm^{-2} month^{-1}) (°C)$^{-1}$ and $c = 47.0$ (g m^{-2} month^{-1}). Therefore monthly mean ablation may be expressed in g cm^{-2} or to a good approximation cm depth rainfall equivalent. Ablation is defined as the reduction in mass of a vertical column of an ice/snow body due to the removal of H_2O out of the column by surface processes. It therefore includes not only melting and run-off but also evaporation. There will probably be significant discrepancies between equation (3.5) and particular examples of ablation, but such discrepancies should not be serious for the ice age problem as long as the summer months during which most ablation occurs are predicted reasonably well. Discrepancies will probably arise mainly because of seasonal variations of surface albedo.

3.1 Present day snow-balance in Britain

Equation (3.5) may be applied to the present-day distribution of snow ablation in Britain (Lockwood, 1982). Values of Q were obtained from the Smithsonian Meteorological Tables (List, 1963) and monthly mean temperatures are approximate values for central England. A temperature lapse rate of $6.5°C\,km^{-1}$ was assumed, since this is about average for the global atmosphere. The resulting mean monthly maximum ablation values at various levels are shown in Table 3.5. The values may be termed 'potential' values in the sense that they represent the maximum snow ablation assuming a level snow/ice cover of great depth. Actual ablation values may be less because of a lack of snow cover.

Bleasdale and Chan (1972) found from an analysis of 6,500 British rainfall stations for the period 1916–50, that the average change of average annual precipitation (Rmm) in Britain with elevation (Hm) is given by:

$$R = 2.42H + 714 \tag{3.6}$$

It is thus possible to calculate the average precipitation at any level. Similarly, if the distribution of monthly mean precipitation at higher elevations is assumed to follow that at the surface, then it is possible to estimate the average monthly precipitation at any given level. Mean monthly snowfall may be estimated by assuming that all precipitation falls as snow when the mean monthly temperature is below $0°C$. For lowland Britain, there is an equal probability of precipitation occurring as rain or snow when the screen temperature is $1.5°C$, implying a freezing level about 250 m above the surface (Murray, 1952; Lamb, 1955). Monthly snowfalls and ablation values allow snow balances to be estimated, and these are shown for levels above 500 m in Figure 3.7. Snow cover is assumed to start at the beginning of the month in which conditions first become suitable (i.e. accumulation exceeds ablation), but in practice it will be delayed until the first heavy snowfall, so the snow cover durations shown in Figure 3.7 are the maximum possible.

From the snow storages shown in Figure 3.7, it is possible to estimate the average number of days with snow lying at each level. The starting dates are a little uncertain since it is assumed that snow cover is established at the earliest possible date, and this may not in fact be so because of lack of snowfall. The relationship between days with snow cover and elevation (Hm) is shown in Figure 3.8, and is given by:

$$\text{Days with snow cover} = 0.155H + 11 \tag{3.7}$$

The slope of the line in Figure 3.8 gives an increase of days with snow lying of one day per 6.46 m rise. Manley (1969) quotes an average value for the central highlands of Scotland of one day per 6.5 m rise, so the approach of the theoretical model to actual conditions is good.

Similarly, Manley (1969) states that the summit plateau of Ben Nevis (1,343 m) is covered by snow from early November until the end of May, and this fits the snow-balance curve for 1,250 m presented in Figure 3.7 very well. Extrapolation of the

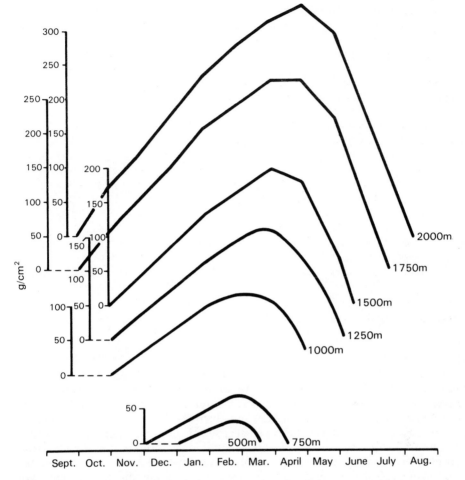

Figure 3.7 Monthly mean snow balances at various levels for present-day Britain (after Lockwood, 1982).

straight line in Figure 3.8 suggests a sea-level value of snow cover of about 10 to 11 days per year, while Table 3.5 suggests a very small number of days per year. The Climatological Atlas of the British Isles. (Meteorological Office, 1952) suggests a value between 0 and 10 days with snow lying would cover much of lowland Britain.

Jackson (1978) had re-evaluated British data using median rather than mean duration since the latter is strongly biased by abnormal values. Given the sea-level median duration of snow cover D_o, the duration at any latitude (D_H) is given by:

$$D_H = D_o \exp (H/300) \text{ for } H \leq 400 \text{ m}$$
$$\text{and } D_H = 3.75 \, D_o \, [1 + (H - 400)/310] \text{ for } H > 400 \text{ m}. \qquad (3.8)$$

Hence, the relationship is linear above 400 m. Jackson notes that data for Vancouver, British Columbia, fit a similar profile. Jackson considers that gradients may be slightly steeper in eastern mountains, and that the sea-level duration of snow cover increases eastward.

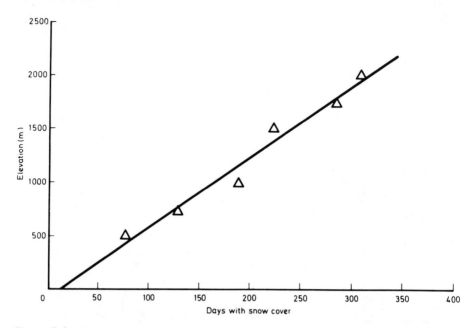

Figure 3.8   Relationship between days with snow cover and elevation in present-day Britain. Triangles mark estimates at various levels (after Lockwood, 1982).

Equation 3.7 suggests that the level at which the snow cover lasts for the whole year is about 2,285 m. This is for a level uniform plain with no shading of solar radiation or drifting of snow. For a variety of reasons which will be explained later this is not the level of the climatic snowline which Manley (1975) considers occurs at about the level where representative open ground is snow-covered for 270–280 days per year. Barry (1981) comments that windblown snow is a strikingly visible element of the climatic environment of many mountain areas, yet the literature specifically addressing the problem is sparse. In the mountain environment, snow distribution as well as drift profiles are strongly affected by meso- and micro-scale topography, including the vegetation structure. Manley (1970) comments that small snow-drift accumulations in steep, narrow and much shaded north-facing gullies may last throughout the year, both on Ben Nevis and on Braeriach in the Cairngorms, where they persist at an altitude of 1,150 m. Up to 1933, these had not been known to completely disappear. In the September of that year, however, they did, and since then there have been at least seven occasions when the last of them melted completely in the late summer, sometimes in September. These snow patches exist because of the shading of the sun's radiation, and also because of the drifting of large amounts of snow from hill slopes and summits into gullies.

If there were no reduction in solar radiation due to slope and shading, then drifting would have to increase snow accumulation by about 4.3 times to allow a permanent snow drift to form at 1250 m in the Ben Nevis region. Bleasdale and Chan (1972) state that actual precipitation values in the western highlands of Scotland exceed the estimates in equation (3.6) by about 600 mm yr^{-1}. Part of the drifting factor of 4.3 may therefore be a correction for winter snowfalls above the country-wide average at 1,250 m.

Table 3.5 Potential mean monthly ablation values for the British Isles at the present time. (Units are g cm^{-2} month^{-1}, monthly mean temperatures to the nearest 0.1°C are shown in brackets) *(after Lockwood, 1982)*

Month	0.32Q−47	Elevation (m)										
		0	250	500	750	1000	1250	1500	1750	2000	2250	2500
J	−25.78	(3.0) 4.2										
F	−0.53	(3.0) 29.5	(1.4) 13.2									
M	30.4	(6.0) 90.4	(4.4) 74.7	(2.8) 57.9	(1.1) 41.7	(−0.5) 25.4	(−2.1) 9.2					
A	61.4	(8.0) 141.4	(6.4) 125.2	(4.8) 108.9	(3.1) 92.7	(1.5) 76.4	(−0.1) 60.2	(−1.8) 43.9				
M	89.29	(11.0) 199.3	(9.4) 183.0	(7.8) 166.8	(6.1) 150.5	(4.5) 134.3	(2.9) 118.0	(1.3) 101.8	(−0.4) 85.5	(−2.0) 69.3		
J	104.7	(15.0) 254.7	(13.4) 238.5	(11.75) 222.3	(10.1) 206.0	(8.5) 189.8	(6.9) 173.5	(5.3) 157.3	(3.6) 141.0	(2.0) 124.8	(0.4) 108.5	(−1.3) 92.3
J	98.59	(16.0) 258.6	(14.4) 242.3	(12.8) 226.1	(11.1) 209.8	(9.5) 193.6	(7.9) 177.3	(6.3) 161.1	(4.6) 144.8	(3.0) 128.6	(1.4) 112.3	(−0.3) 96.1
A	73.80	(16.0) 233.8	(14.4) 217.6	(12.8) 201.3	(11.1) 185.1	(9.5) 168.8	(7.9) 152.6	(6.3) 136.3	(4.6) 120.1	(3.0) 103.8	(1.4) 87.6	(−0.3) 71.3
S	38.2	(14.0) 178.2	(12.4) 161.9	(10.8) 145.7	(9.1) 129.4	(7.5) 113.2	(5.9) 96.9	(4.3) 80.7	(2.6) 64.4	(1.0) 48.2	(−0.6) 31.9	(−2.3) 15.7
O	5.7	(11.0) 115.7	(9.4) 99.4	(7.8) 83.2	(6.1) 66.9	(4.5) 50.7	(2.9) 34.4	(1.3) 18.2	(−0.4) 1.9			
N	−19.1	(6.0) 40.9	(4.4) 24.6	(2.8) 8.4								
D	−29.7	(4.0) 10.3										

The Scottish semi-permanent snow patches probably also owe their existence to the reduction in summer ablation resulting from a decrease in solar radiation. Solar radiation may be decreased by shading from nearby mountains or by the land sloping northwards and thus increasing the angle of incidence. At the earth's surface there are two components to solar radiation: the direct component and the diffuse component. Only the direct radiation component is reduced by steep north-facing slopes, since it can be assumed that the diffuse-beam input from cloudless and cloudy skies is equal for all positions in the sky and therefore does not contribute to spatial variability of solar receipt at the surface. According to both Collingbourne (1976) and Taylor (1976) the ratio of mean daily diffuse solar radiation to mean daily global (total) solar radiation at Eskdalemuir (southeast Scotland, 237 m) varies from about 0.6 in June to between 0.7 and 0.8 in December. This suggests that radiation will not vary greatly on slopes of moderate angle. Daily totals of global solar radiation on a north-facing vertical surface are available for Bracknell (Berkshire) from Collingbourne (1976). These values indicate the radiation totals which might be found in steep north-facing gullies in the Scottish highlands, and probably represent the extreme limits of the problem. Mean daily solar radiation falling on a north-facing surface is about 25 per cent of that falling on a horizontal surface at the same site. Under these conditions the value of Q in equation (3.5) will be small and the ablation rate will be largely determined by the mean temperature. Indeed the net radiation is normally negative. On a north-facing surface at 1,250 m the mean annual ablation is about 240 g cm^{-2}, which is probably approximately equal to the winter snowfall near the summit of Ben Nevis. Thus semi-permanent snow patches near the summits of Scottish mountains can be explained both in terms of drifting and extreme shading on north-facing slopes.

The glaciation level is defined by Østrem (1974) as the critical summit elevation which is necessary to produce a glacier on a mountain. It may be difficult to determine in detail the critical mountain elevation that is necessary to produce a glacier; on north-facing slopes it may be lower than on south-facing slopes. According to Østrem (1974) it has been agreed that the height of the glaciation level shall be determined from the summit elevation of mountains where glaciers are formed. The glaciers themselves will be situated at lower elevations. To determine the height of the glaciation level it is necessary to examine good topographic maps, find the highest mountain in a given area without glaciers (but with favourable topography) and locate the lowest mountain that carries a glacier in the same area. It is then assumed that the critical height lies somewhere between the two mountain summit elevations. According to Andrews *et al.* (1975) the glaciation level can be determined unambiguously and is akin to the 'climatic snowline'.

From the maps of glaciation level published by Østrem (1974) and Andrews *et al.* (1975), climatic parameters (Table 3.6) were estimated at the glaciation level in a number of countries, using suitable nearby climatological stations and published climatological charts. Manley (1975) comments that from a variety of scattered observations in Norway and the western Alps it can be demonstrated that the snowline occurs at about the level where representative open ground is snow-covered for 270–280 days per year. Therefore the number of days with snow lying on open ground was estimated at the glaciation level together with the drift factor, that is, the number of times the estimated ablation exceeds the estimated snowfall. The drift factor includes an unknown contribution due to shading. The drift factor varies from 1.9 to 7.6 with an average of 5.5, and the number of days with snow

Table 3.6 Climatological parameters at glaciation level

Area	Glaciation level (m)	Estimated number of days with snow lying at glaciation level	Factor by which snowfall has to be multiplied to equal estimated ablation
N. Norway			
Mountains near Tromso	1000	247	6.5
Canada and Alaska			
Mountains near Frobisher Bay, NWT	1000	285	7.6
Mountains to east of Prince George, BC	2300	270	5.7
Mountains near Juneau, Alaska	1200	240	5.7
Greenland			
Mountains near Upernavik	750	311	1.9

Climatological data after Bryson and Hare (1974), Orvig (1970) and Wallén (1970) (*after Lockwood, 1982*).

lying from 247 to 311 days with an average of 271 days. Details of the calculations are shown in Table 3.6. Given the large uncertainties in the estimates shown in Table 3.6, they do confirm Manley's comments about the relationship between the snowline and the number of days of snow cover. On this basis the glaciation level should be found roughly between 1,670 and 1,735 m over northern Britain. Charlesworth (1957) estimates the height of the snow-line over Britain as varying from 1,750 m in the south to 1,250 m off the north coast of Scotland with a mean elevation (found over southern Scotland) of about 1,500 m. The slope of the line is around 0.5×10^{-3}. The calculations presented here give only mean values for the central/northern parts of Britain.

4 The history of ice ages through geological time

4.1 Precambrian glaciations

Frakes (1979) comments that a great deal of information exists that much of the earth was glaciated during the late Precambrian, particularly in the timespan from about 950 to about 615 m yr ago. The 950 m yr figure represents the approximate age of the oldest among a series of tillites which signal the end of the long period of carbonate sedimentation which began about 2,000 m yr ago. Tarling (1978) states that at least four major ice ages have been recognized within Precambrian times, although rocks older that 2,800 m yr tend to be metamorphosed to the extent that identification of an original glacial origin is now unlikely. The most striking of these is the youngest, the Varangian, which occurred some 660 to 680 m yr ago. Glacial deposits of this age have been recognized in North America, Greenland, Spitsbergen, Scandinavia, the British Isles, France, USSR, China, India, Australia, Africa and South America (Harland, 1972, Harland and Herod, 1975).

Frakes (1979) considers that the important facts about the Precambrian glaciations include the following:

(1) multiple glaciations took place on several continents;
(2) glaciations apparently occurred at different times on different continents;
(3) the full timespan for episodic glacial activity is at least 350 m yr and possibly as much as 400 m yr;
(4) almost all glacial deposits for which paleomagnetic data are available accumulated at low to low-mid paleolatitudes;
(5) as yet, no definite polar glaciations have been recognized in paleomagnetic studies.

Computer simulations of the evolution of the earth's atmosphere composition and surface temperature have been carried out by Hart (1978) (see Chapter 1). The programme took into account changes in the solar luminosity, variations in the earth's albedo, the greenhouse affect, variation in the biomass, and a variety of geochemical processes. Results indicate that prior to 2 billion years ago the earth had a partially reduced atmosphere, which included N_2, CO_2, reduced carbon compounds, some NH_3, but no free H_2. Surface temperatures were higher than now, due to a large greenhouse effect. After all the reduced gases were oxidized away the situation changed drastically, with the intensity of the greenhouse decreasing and the mean surface temperature falling. By the time free oxygen appeared in the atmosphere about 2 billion years ago, the mean surface temperature had fallen to 280°K and ice-caps of considerable size would be able to form. Since about 1 billion years ago the mean surface temperature has steadily increased, making glaciation less likely.

More relevant to the problem of low-latitude Precambrian glaciations is some work by Hunt (1979a and b) on the effects of past variations in the earth's rotation rate. The rotational history of the earth can be traced back to the late Precambrian era, about 1.5×10^9 years ago. Based on growth patterns from shells and corals, the earth then had at least 800–900 solar days per year implying a rotation rate 2–2.5 times greater than the present value (Mohr, 1975).

Hunt (1979a and b) used a general circulation model in which all radiative terms (including cloud cover, surface albedo, and CO_2 amount) are set at current values. These and other model parameterizations are based on present atmospheric conditions which do not necessarily apply to the extreme conditions of the Precambrian. The variation of the intensity and latitude of the tropospheric westerly jet stream are shown in Figure 3.9. This indicates that for a rotational period of 9 h (late Precambrian era) the jet stream intensity was reduced by about 20 per cent compared to its current value, also the jet was located much closer to the equator at 15° latitude. Hunt comments that the confinement of the jet to low latitudes was associated with a reduction in the overall intensity of the zonal winds in the troposphere as a whole, to a much greater extent than that implied by the 20 per cent reduction in jet intensity. This would have led to the surface wind stress, which is normally proportional to the square of the wind velocity, being reduced by possibly as much as a factor of 2. Hunt considers that perhaps the single most important effect of the reduced surface stress would have been its impact on the oceans. There would have been less wind-induced vertical mixing in the ocean, resulting in a shallow, warmer mixed layer, with presumably a smaller overall heat capacity. A smaller heat capacity would make the oceans less able to reduce the amplitude of seasonal and inter-annual fluctuations of ocean surface temperature. In addition the scales of motion in the ocean would have been reduced by the higher rotation rate of the earth, which, together with the lower stress, should have weakened the strength of the oceanic gyres. This would also have decreased the amount of heat transported

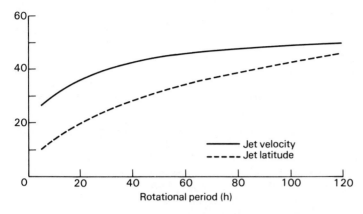

Figure 3.9 Variation with earth's rotation period of the intensity of the mean zonal wind (m s⁻¹) of the tropospheric jet, and of the location of the jet (deg. latitude) (after Hunt, 1979a). Reprinted by permission from *Nature* Vol. 281 p. 190. Copyright © 1979 Macmillan Journals Limited.

polewards by the oceans, the latter being a significant component of the total pole-ward energy flux. According to Hunt the net effect of all these changes should be to produce a general warming of the tropical oceans and atmosphere and a colder polar region, possibly associated with greater transient departures from mean conditions.

Hunt (1979a and b) suggests that the net consequence of the reduced polewards heat fluxes suggested by his model for Precambrian times is a colder polar region and an enhanced equator–to–pole temperature difference. A 20 K colder pole is indicated for a rotational period of 9 h, unless the polar cloud cover was different. This may well have been the case given the reduced water-vapour content of the colder atmosphere and the smaller-scale systems. Although the model implies that the precipitation rate was very small at high latitudes during the Precambrian era, Hunt considers that conditions should have been conducive to the accumulation of ice cover, depending on the dispositions of the continents. The model suggests that equatorial rainfall conditions were similar to those today but that the adjacent zone of high aridity was narrower and closer to the equator. Because of the smaller-scale and less organized nature of meteorological systems suggested for the Pre-cambrian, precipitation generally would have been less intense and more localized.

Hunt (1979a and b) considers that it can be hypothesized that glacial conditions prevailed throughout the Precambrian, solely on rotational considerations. The extent of ice cover would have declined continuously, all else being equal, as the earth's rotation rate decreased and the improved circulation systems warmed high latitudes. Hunt also considers that this explanation can also reasonably account for the termination of the Precambrian ice age 6 × 10⁸ yr ago, when the rotational period of the earth would have been 20 h and thus dynamical conditions less con-ducive to the maintenance of glaciation.

4.2 Tertiary glaciations

According to Frakes (1979), the Cretaceous must be recognized as a time of great warmth over the globe. This conclusion derives from a variety of geological and geophysical evidence which Frakes carefully summarizes. Frakes considers that during the warmest period there was a wide zone of temperatures comparable to modern tropical to subtropical conditions extending to at least 45°N and, possible because of unique current patterns, to 70°S latitude; beyond this zone climates were warm to cool-temperate. Climates analogous to those of modern polar regions are totally undocumented.

Oxygen isotope ratios indicate low surface thermal gradients in the Cretaceous and early Tertiary (Emiliani, 1961; Bowen, 1966; Douglas and Savin, 1975). Some authors consider that in the Cretaceous the deep-water circulation was still very sluggish, often not able to overcome increases in density gradients that developed either due to periodic spreading of low salinity surface waters (Thiede and Van Andel, 1977; Ryan and Cita, 1977) or due to the outflow of highly saline and there-fore dense South Atlantic water (Thierstein and Berger, 1978). Dark organic-rich sediments at intermediate depths on rises and flanks of seamounts are evidence of a pronounced oxygen minimum zone in Cretaceous Pacific waters. Similar 'black shales' on the Atlantic ocean floor indicate that the deep waters of the more restricted North and South Atlantic basins become completely anoxic (Thiede and Van Andel, 1977). Anoxic conditions occur in deep water when the consumption of oxygen by organisms exceeds the rate of oxygen supply. While anoxic conditions prevail organic carbon is accumulated in the underlying sediments at a higher rate. Brass *et al.* (1982) consider that anoxic conditions can also be explained as resulting from a decrease in the ventilation of the deep ocean because of reduced oxygen solubility in the source regions without requiring any change in the circulation rate. This is an immediate consequence that follows from the fact that the solubilities of many gases strongly depend on water temperature and salinity. In the Cretaceous surface water temperature in the polar regions was about 15°C as compared with -2°C at the present time, thus greatly reducing the concentrations of dissolved gases.

Barron and Washington (1982) have carried out atmospheric simulations using a general atmospheric circulation model and realistic Cretaceous geography. They found that paleogeography is an important factor governing the nature of the circulation. The simulated Cretaceous atmospheric circulations are markedly different both from the present-day atmospheric circulation and from the classical hypothesis of a weaker circulation and poleward displacement of circulation features (e.g. the subtropical highs). Much of the paleoclimatic literature simply assumes that a reduced equator-to-pole surface temperature gradient will result in a weaker atmospheric circulation and a weaker wind-driven ocean circulation. Barron and Washington establish that data on the nature of the surface-temperature gradient are insufficient to make a conclusion concerning the intensity of the circulation. Indeed they consider that in the Cretaceous the vertically inte-grated meridional temperature gradient actually increased at low and mid-latitudes, while decreasing at high latitudes.

By the early Tertiary, continued sea-floor spreading had enlarged the Atlantic and Indian Oceans and substantially reduced the size of the Pacific Ocean. This fragmentation of the world ocean led to increasingly inefficient latitudinal energy exchange. Oxygen isotope temperature estimates show a general temperature decline from the mid-Cretaceous into the Paleocene (Douglas and Savin, 1975; Savin *et al*, 1975). After the early Paleocene, the previously rather uniform oceanic microfauna and flora differentiated into several latitudinally arranged assemblages.

Evidence for glaciation during the Eocene in the Antarctic continent is not con-clusive and remains to be confirmed. High surface water temperatures indicated by the oxygen isotope evidence do not suggest the existence of ice rafting of sediments at that time. According to Kennett (1977), if glaciation did occur at sea level at any time within the Eocene, it was probably restricted to the west Antarctic sector.

Kennett (1977) has summarized the Antarctic–Southern Ocean paleoceanographic and glacial characteristics during the Eocene (55 to 38 m yr ago).

1 Cool temperate climate and vegetation found in Antarctica.
2 Glaciation restricted in Antarctica, but possible local glaciation in west Antarctica.
3 Warm surface water temperatures; decreasing sub-Antarctic temperatures (19°C early Eocene to 11°C in late Eocene).
4 Bottom waters relatively warm.
5 Shallow barrier to circumpolar circulation at South Tasman Rise.
6 Australia commenced drifting northward from Antarctica, forming an ocean between the two continents (53 m yr ago).

The early Cenozoic (Paleogene) paleotemperature record of the high southern latitudes (Antarctica and sub-Antarctic) is marked by rather rapid temperature decreases superimposed on a steady climatic cooling commencing in the early Eocene (Shackleton and Kennett, 1975). The sub-Antarctic was marked by warm surface water temperatures which decreased from 19°C in the early Eocene to 11°C by the late Eocene.

Shackleton and Boersma (1981) have presented oxygen isotope paleotemperature estimates for a number of deep-sea sediment samples of Eocene age, covering a wide geographical area. Their results show several interesting features. The overall range of surface temperature values is very low compared with that prevailing today, this agrees with previous studies based on the examination of deep-sea sediment (Savin, 1977). In the North Atlantic, and more markedly in the South Atlantic their data imply the presence of strong western boundary currents. Their data suggest that sea surface temperatures in the tropics were lower than today by up to 5°C, and considerably higher than today in the polar regions. The latitudinal gradient of the Eocene ocean was less than half its present value. This probably implies a vigorous transport of heat polewards by ocean currents.

Oceanic bottom water is formed in small regions because convective buoyancy transfer processes are much more efficient than conductive processes. Moreover, virtually all deep water seems to be formed over continental shelves and in low- and high-latitude marginal seas. Brass *et al.* (1982) suggest that water trapped in marginal seas is made sufficiently dense, by interaction with the atmosphere, for the outflow to drive a turbulent plume. A characteristic of turbulent plumes is the entrainment of environmental fluid into the plume, an idea which is supported by the observation that, although the Mediterranean water at its source is denser that any water in the Atlantic Ocean, it only penetrates to about 1,200 m. Turbulent plume calculations by Brass *et al.* (1982) indicate that the depth of termination and horizontal spreading of the plumes is related directly to their buoyancy flux, which is the product of the density difference between the plume fluid and the surrounding environment and the volume flux of the plume. The plume with the greatest buoyancy flux will form the bottom water and plumes of lesser buoyancy flux will terminate and spread out at intermediate depths. With two plumes, the termination depth of the weaker plume is relatively insensitive to the difference in buoyancy flux between the two until they become nearly equal in strength. According to Brass *et al.* it seems that if the plumes switch (that is, the bottom water plume loses buoyancy flux or the intermediate water plume gains buoyancy flux) a relatively rapid and perhaps catastrophic change occurs when the old minor plume becomes the source for the new bottom water. Brass *et al.* believe that in the Cretaceous, warm water sources were dominant whereas today cold bottom water is produced at high latitudes. This change necessitated at least one transition from warm saline bottom water (WSBW) to cold polar bottom water. However, because of the sensitivity of plume termination depths to small differences in source strengths, several transitions might be expected as a result of variations in buoyancy fluxes from competing deep-water sources during a period in which the source strengths were nearly the same.

Brass *et al.* state that epicontinental seas and marginal seas producing WSBW must be located in the zone of net evaporation (10–40°N and S) associated with the descending branches of the atmospheric Hadley cell circulation. The distribution of Mesozoic and Cenozoic evaporite deposits strongly suggests that this zone has remained almost stationary during the past 120 m yr. Because area is an important factor determining the buoyancy flux from net-evaporation basins, the decrease in the area of net-evaporation marginal seas over the past 100 m yr strongly suggests, according to Brass *et al*, that the late Cretaceous was much more favourable for the production of WSBW.

The first major Antarctic climatic-glacial threshold was crossed about 38 m yr ago near the Eocene–Oligocene boundary. Oxygen isotope changes in deep-sea benthoric foraminifera from the sub-Antarctic (Shackleton and Kennett, 1975; Kennett and Shackleton, 1976) and the tropical Pacific region (Savin *et al.* 1975) indicate that bottom temperatures decreased rapidly by approximately 5°C to approximate present-day levels at the respective water depths of each drilled sequence. This temperature reduction was calculated by Kennett and Shackleton (1976) to have occurred within 100,000 years, which is remarkably abrupt for pre-glacial Tertiary times, and is considered to represent the time when large-scale freezing conditions developed at sea-level around Antarctica, forming the first significant sea-ice.

The psychrosphere, named by Bruun (1957), constitutes the deep and very cold water of the world ocean. Benson (1975) first detected the beginning of psychrosphere near the Eocene–Oligocene boundary, manifested by a major reorganization of deep-water ostracode faunas. Schnitker (1980) suggests that at the end of the Eocene the energy budgets for the Pacific sector and the Atlantic/Indian Ocean sectors of the Antarctic were markedly unequal. Freezing surface temperatures were obtained early in the Atlantic/Indian Ocean sectors and, much like the present, the bulk of the world ocean's bottom water originated in this region. Hiatuses and pulses of sediment redeposition, particularly along the western slopes of the Atlantic and Indian Ocean basins (Van Andel *et al.*, 1977; More *et al.*, 1978) attest to this late Eocene sharp increase in Antarctic bottom water flow.

Continued general climatic deterioration, but in particular the consequences of the separation of Australia from Antarctica, brought about freezing surface conditions in the Pacific sector of Antarctica by the beginning of Oligocene time (Schnitker, 1980). Faunal and sedimentological evidence indicate that a shallow marine connection across the South Tasman Rise came into existence by late Eocene. Cool Antarctic water started to flow from the southern Indian Ocean westward into the South Pacific. By the earliest Oligocene this intrusion had reached sufficient volume to deflect the warm, subtropical water from the Antarctic coast, creating a thermal isolation that allowed cooling to progress rapidly to the freezing point. Schnitker (1980) comments that the ensuing formation of cold bottom water must have been voluminous, judging by the widespread occurrence of hiatuses in the south Pacific Ocean. The circum-Antarctic region was established as the principal source of deep water to the world oceans.

According to Kennett (1977), widespread glaciation probably occurred throughout Antarctica in Oligocene times (38 to 22 m yr BP), but no ice-cap existed. He considers that the east Antarctic ice-cap formed during the middle Miocene (14 to 11 m yr BP) causing world sea-levels to fall by up to 59 m. Since the middle Miocene the east Antarctic ice-cap has remained a semi-permanent feature exhibiting some changes in volume. The most important of these occurred during the latest Miocene (about 5 m yr BP), when ice volumes increased beyond those of the present day.

Kvasov and Verbitsky (1981) have investigated the causes of Antarctic glaciation by means of numerical experiments based on a three-dimensional thermodynamic model for a large ice-sheet. They found that an ice-sheet might have occupied Antarctica at temperatures not exceeding the present ones by more than 15°C. They consider that the annual average temperature near the shoreline in this case could not have been higher than 3°C. All the available evidence indicates that temperatures were higher than this during the Eocene, hence, there was no glaciation of the surface of the southern continent. This does not rule out the possibility of glaciers having existed in mountain massifs. At the boundary between the Eocene and Oligocene (38 m yr BP), a deep strait came into existence south of Tasmania. This immediately affected the oceanic circulation in the southern hemisphere. After the deep passage opened between Australia and Antarctica, the waters of the Southern Ocean had to make an almost round-the-world trip, thereby affording enough time for them to cool considerably. According to Kvasov and Verbitsky (1981) this accounted for a remarkable development of Antarctic glaciation. Cooling brought about the growth of mountain glaciers. Joining, they formed an ice-sheet, 1,200 km in diameter, within the boundary of the Gamburtzer Mountains. The appearance of an ice-sheet in the Gamburtzer Mountains was responsible for further cooling of the climate, which led to glacier growth in the elevated areas. Soon the ice-sheet spread over the eastern plain and the Schmidt plain, thus covering the whole of east Antarctica. Kvasov and Verbitsky suggest that the formation of the east Antarctic ice-sheet in the earliest Oligocene was like a chain reaction, and took about 100,000 years, i.e. it was almost instantaneous in the geological sense. According to Kennett (1977) it was within such a short period of time as 75,000–100,000 yr that such an extreme fall of temperature occurred in the earliest Oligocene in the Southern Ocean. Unlike today, the main ice discharge was accomplished not through iceberg calving, but through melting. So continental glaciation could have led to almost no iceberg deposits in the surrounding seas. So according to Kvasov and Verbitsky (1981) the formation of the Antarctic ice-sheet dates from the earliest Oligocene.

A broad survey of geological, geophysical and glaciological evidence of Quaternary glaciation in Greenland is given by Weidick (1976). The time of the formation of the Greenland ice-sheet is not well known from terrestrial evidence, but the presence of glacial marine sediments in the North Atlantic marine cores first appeared around 3 m yr BP according to Berggren (1972). The oldest moraines in Iceland dated by K Ar methods are approximately 2.6 m yr old (Robin, 1983). Robin considers that since the Greenland ice-sheet is likely to have formed either during or even before the onset of the earliest major Pleistocene glaciation around the North Atlantic, it was probably in existence around 3 m yr BP.

The fundamental climatic event in the Tertiary appears to date from about the Eocene–Oligocene boundary. Before this time there is little evidence of deep cold water in the world ocean and the poles were essentially warm and ice free. After this time extensive cold water existed in the world ocean and an extensive ice-sheet probably existed in the Antarctic. Thus the true global warm climates only existed in the Tertiary prior to the end of the Eocene. From the early Oligocene onwards it may be considered that world climates were in a cold or semi-glacial state. Thus the start of the present glacial climate phase could well be placed at the Eocene–Oligocene boundary. Before the Eocene–Oligocene boundary atmospheric and oceanic circulation conditions were probably different from those observed today. After the boundary they are probably rather similar to present-day conditions. Some aspects of the warm climates appear to have lingered on until the middle Miocene. For example Savin *et al.* (1975) consider that meridional temperature gradients remained low until the middle Miocene when the deep oceans became particularly cold.

Some geochemists are questioning whether paleoceanographs have convincingly verified the fidelity of their climate record over the last 65 million years. Killingley (1983) has recently questioned how much of these isotopically determined temperature trends is due to climate change and how much could have been caused by chemical alteration of the sediment during its burial beneath the sea floors. Even before carbonate sediments turn into limestone under the pressure and heat of burial, forams can gradually dissolve and recrystallize, exchanging oxygen isotopes with the pore water in the sediment as they do. Killingley (1983) has simulated this recrystallization in a mathematical model and duplicated the direction of the observed isotopic trends. Thus at this particular point in time climatic histories of the Tertiary must be treated with some caution.

4.3 The influence of the Antarctic ice-sheet on global climate
The impact of snow cover on climate is mainly caused by its high albedo, small heat conductivity, heat expenditures on snow melting and relatively small roughness at its surface (Kotliakov and Krenke, 1982). Together with the high radiating emissivity of snow this induces low surface temperatures and high temperature inversions above it. Kotliakov and Krenke (1982) have carried out quantitative estimates of glacier effects upon climate. They consider that the Greenland ice-sheet cools the atmosphere at 1,500 m above it by 1°C, the effect being most marked in summer. The heat transfer to the Greenland ice-sheet in a year averages about 2 × 10^{15} kJ/day, which is commensurate with the energy of cyclones crossing the ice-sheet in 12 or 24 hours. According to Kotliakov and Krenke the effects of ice domes much less in size than Greenland's on the global climate are negligibly small.

The present-day influence of the Antarctic ice-sheet on global climate has been discussed by Flohn (1978). He comments that one of the basic features of the global atmospheric circulation is its asymmetry with respect to the equator, a fact which he considers is hardly covered in a satisfactory manner in most textbooks. Flohn comments that the annual average temperature of the upper and middle troposphere above the Arctic is about 11°C warmer than that above the Antarctic. This is the direct result of the differences in the heat budgets of a massive ice-dome and a thin layer of floating sea-ice with many polynyas.

Flohn (1978) illustrates the effects of the differing hemispheric temperature gradients ΔT by reference to the thermal Rossby number R:

$$R \propto \Delta T / \Omega^2 \qquad (3.9)$$

where Ω is the angular speed of the earth's rotation. This dimensionless number depends on the temperature difference between equator and pole, in units of the rotation angular speed of the earth's equator. Temperature gradients in the two hemispheres are similar during the northern winter, but the difference between them reaches nearly 27°C during the northern summer. Then the thermal Rossby number reaches, in the southern hemisphere, more than 250 per cent of its value in the northern hemisphere, while the annual average is more than 140 per cent. According to Lamb (1959) the kinetic energy of the southern westerlies is about 60 per cent larger than that of the northern westerlies. This stronger circulation of the southern hemisphere pushes the meteorological equator towards the northern hemisphere, especially during the northern summer, while during the northern winter the circulation intensity of both hemispheres is about equal. The zonally-averaged annual position of the metereological equator in the Atlantic is about 6°N (Flohn, 1967), varying between 1°N (February–March) and 11°N (August). According to Newell *et al.* (1972) the average position of the northern limit of the southern Hadley cell reaches about 11°S in December–February, but in June–August it is found at about 15°N.

A good approximation (Smagorinsky, 1963) to the subtropical anticyclone latitude at the surface as well as the latitude of the extra-tropical Ferrel/Hadley transition aloft is given by ϕ in the equation:

$$\tan \phi = -\frac{H}{R}\left(\frac{\partial\theta/\partial z}{\partial\theta/\partial y}\right) \tag{3.10}$$

where H is a scale height, roughly the height of the 500 mbar level of the atmosphere; R is the radius of the earth; $\partial\theta/\partial y$ is the north–south gradient of potential temperature; and $\partial\theta/\partial z$ is the vertical lapse rate of potential temperatures. Korff and Flohn (1969) have plotted the latitudes of the subtropical anticyclones against the temperature differences ($\triangle T$) in the layer 300/700 mbar between equator and pole. The points fit a line given by:

$$\text{ctg}\,\phi = 0.0245\,\triangle T + 0.7246 \tag{3.11}$$

Flohn (1978) states that the average annual temperature gradient in the 700/300 mbar layer between equator and pole is 27.3°C in the northern hemisphere and 39.1°C in the southern. Flohn gives the annual average latitudes of the two subtropical anticyclone belts as 37°N and 31°S, fitting the observed temperature gradients well.

4.4 Quaternary glaciations

Figure 3.10 shows that for at least the last 1 m yr the earth's climate has been characterized by an alternation of glacial and interglacial episodes, marked in the northern hemisphere by the waxing and waning of continental ice-sheets and in both hemispheres by periods of rising and falling temperatures. Continental ice-sheets probably first appeared in the northern hemisphere about 3 m yr BP, occupying lands adjacent to the North Atlantic Ocean. Figure 3.10 shows that these fluctuations are found in a number of proxy data records. These include the chemical composition of Pacific sediments, fossil plankton in the Caribbean, and the soil types in central Europe. These 'cycles' identified as A to E by Kukla (1970), may be grouped into a climatic 'regime' covering the last 450,000 years designated α in Figure 3.10. The earlier records (regime β) show higher-frequency fluctuations with less coherence among the various proxy climatic records. Hays *et al.* (1976) have found climatic cycles of 23,000, 42,000 and approximately 100,000 years in south hemisphere ocean-floor sediments during the last 450,000 years, suggesting a connection with the Milankovitch mechanism. The dominant cycle is one of about

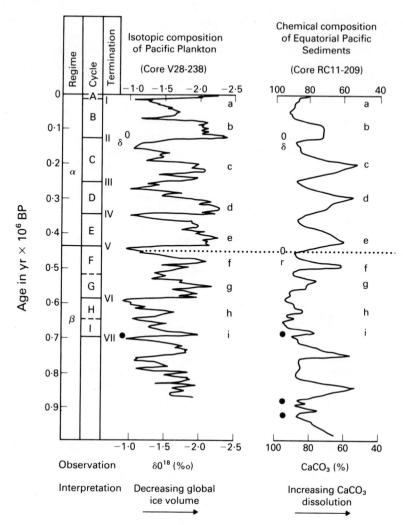

Figure 3.10 Climatic records of the last million years
(a) Oxygen-isotope curve from Pacific deep-sea core interpreted as reflecting global ice volume (Shackleton and Opdyke, 1973);
(b) Calcium carbonate percentage in equatorial Pacific core. Low values are taken to indicate periods of rapid dissolution by bottom water (Hays *et al.*, 1969);
(c) Faunal index reflecting changing composition of Caribbean foraminiferal plankton, calibrated as an estimate of sea surface salinity in parts per thousand (Imbrie *et al.*, 1973). Glacial periods are marked by the influx of plankton preferring higher-salinity waters;
(d) Sequence of soil types accumulating at Brno, Czechoslovakia (Kukla, 1970) (After US National Academy of Sciences).

100,000 years and is seen in the growth and decay of the continental ice-sheets. Indeed, continental ice-sheets appear to develop over a period of 90,000–100,000 years and to terminate rather quickly in about 10,000 years. The interglacial

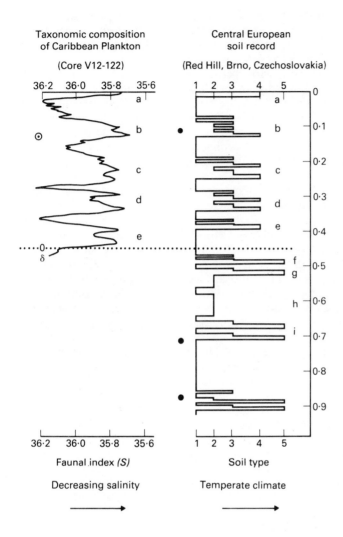

Taxonomic composition
of Caribbean Plankton

(Core V12-122)

Central European
soil record

(Red Hill, Brno, Czechoslovakia)

Faunal index *(S)*

Decreasing salinity

Soil type

Temperate climate

periods between major glaciations lasted about 10,000 years, and were probably never warmer than at the present.

The climatic history of the last 120,000 years is well illustrated by the North Greenland ice-core shown in Figure 3.11. In this diagram temperature is expressed in terms of the amount of [18]Oxygen found in the ice. The snow falling over Greenland was formed of water that evaporated from the Atlantic Ocean. Molecules of sea water containing the commoner form of oxygen, [16]Oxygen, are lighter and evaporate more readily, and remain in vapour form longer than do the small minority that contain heavy oxygen. The proportion of heavy oxygen in the ice therefore gives a general indication of the climate prevailing when the snow fell on Greenland – the less heavy oxygen there is, the colder the climate.

Figure 3.11 shows a warm interval known as the Eemian interglacial followed by a severe glacial period known as the Würm in Europe and the Wisconsin in the

United States. The onset of the Würm/Wisconsin glaciation may be dated about 60,000 to 70,000 years ago, when there was a rapid fall in world mean temperatures to near the lowest levels observed in the Würm/Wisconsin glaciation. The cold period following the initial fall in temperature lasted for only a few thousand years and was followed by a relatively warm period lasting till 30,000 years ago. A further cold period followed, lasting about 10,000 years, in which the lowest temperature and the greatest extent of ice-sheets were attained. This particular cold phase reached its peak between 15,000 and 20,000 years ago. The end of the Würm/Wisconsin glaciation occurred between 13,000 and 10,000 BP, and was marked by a rapid rise of temperature, probably mostly within 2,000 to 3,000 years. Since about 5,000 BP there has been a slight lowering of world mean temperatures.

A study of the changes in ^{18}Oxygen contents of the Greenland ice-core shown in Figure 3.11 suggests that major climatic changes can take place in relatively short time-periods, thus the change from full-glacial to the present interglacial climate took place in a period of a few thousand years. Widespread deglaciation began abruptly about 14,000 years ago, and the waning phases of the continental ice-sheets were characterized by substantial marginal fluctuations. The Cordilleran ice-sheet in North America, which had just attained its maximum extent, melted rapidly and was gone by 10,000 BP. The Scandinavian ice-sheet lasted only slightly longer and retreated at the rate of about 1 km per year between about 10,000 and 9,000 years ago. By 8,500 years ago the ice conditions in Europe had reached essentially their present state, and in North America the ice-sheets had shrunk to about their present extent by about 7,000 years ago. The melting of the ice-sheets was not a continuous process, but marked by sudden warmings and coolings. The normal sequence is Bølling interstadial (warm) – older Dryas (cold) – Allerød interstadial (warm) – younger Dryas (cold), covering a period of less than 2,000 years, with variations in annual temperature of up to 6°C. The cooling prior to the younger Dryas probably lasted less than 300 years.

Flohn (1974) has listed a series of sudden coolings in the climatic record. These events show coolings of the order of up to 5°C per century in contrast to not more than 1°C per century in recent fluctuations. The most dramatic short-lived cooling event was observed in the Greenland ice-sheet record at about 89,000 BP. Here the climate changed from warmer than today into full glacial severity within about 100 yr (Dansgaard *et al.*, 1975). This event has also been found in a stalagmite in a French cave at 97,000 BP with a cooling of the cave by 3°C in a few centuries and an extremely rapid cooling has been described in many cores from the Gulf of Mexico at 90,000 BP. All the dates are slightly uncertain, hence the differences.

The Würm ice age reached its maximum about 18,000 years BP and represents the most dramatic change in the earth's surface for which widespread evidence is still available. Evidence of this paleoclimatic event is found in ocean and lake sediments, in records of soil structure and vegetative cover. In 1971 a consortium of scientists from a number of institutions was formed to study the history of global climate over the past million years, particularly the elements of that history recorded in deep-sea sediments. This study, known as the CLIMAP (Climate, long-range Investigation, Mapping and Prediction) project, is part of the United States National Science Foundation's International Decade of Ocean Exploration Program. One of CLIMAP's goals is to reconstruct the earth's surface at particular times in the past. These reconstructions can then serve as boundary conditions for atmospheric general circulation models.

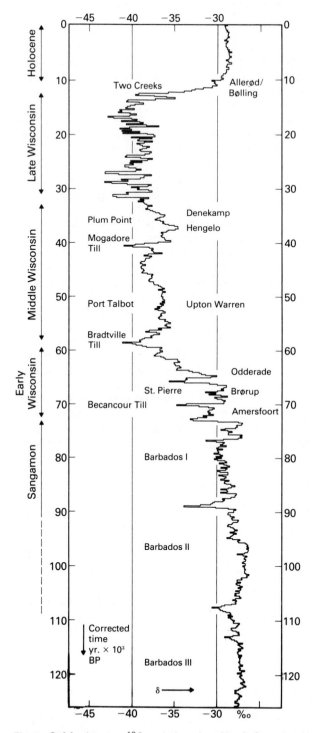

Figure 3.11 Isotope ^{18}O variations in a North Greenland ice-core since 130,000 years BP (after Dansgaard *et al.*, 1971).

4.5 The 18,000 BP continents

Continental geography 18,000 years ago can be approximated by today's configuration with a lowered sea-level, this being the result of the transfer of water from the oceans to the continental ice-caps. CLIMAP project members have made a conservative estimate of sea-level lowering of 85 m, but other authors have estimated lowerings of up to 130 m. The resulting continental outlines are shown in Figure 3.12. Reconstruction of the distributions of vegetation types during the last ice age (Figure 3.12) indicates that desert regions, steppes, grasslands, and outwash plains expanded at the expense of forests, yielding a slight increase in the albedo of land areas not covered by ice. The most striking feature of the world at 18,000 BP shown in Figure 3.12 is the northern hemisphere ice complex, consisting of land-based glaciers, marine-based ice-sheets, and either permanent pack-ice or shelf-ice. This complex stretched across North America, the polar seas, and parts of northern Eurasia, but nevertheless large arctic areas in Alaska and Siberia remained unglaciated. In the southern hemisphere the most striking difference was the winter extent of sea-ice, since changes in land-ice were small. Estimated ice-sheet contours are shown in Figure 3.12 and suggest that some of the land-based ice-sheets reached an approximate thickness of 3 km.

One of the most important conclusions to be drawn from Figure 3.12 is that the ice-age sea surface temperature changes were not in general very large, since the average anomaly over the entire ocean was only $-2.3°C$. The changes at high northern latitudes appear to dominate the pattern, but if the changes are weighted according to the area of ice-free ocean along discrete latitudinal bands, then the maximum effects occur at about 38°N, 6°S and 46°S and are roughly of the same magnitude. This suggests a marked steepening of thermal gradients and a more energetic ocean circulation system, particularly in the North Atlantic and Antarctic. Certainly surface transport was increased and probably a more rapid turnover of surface and intermediate waters was effected. Inferences about the structure of deeper waters are more tenuous.

Figure 3.12 Sea surface temperatures, ice extent, ice elevation, and continental albedo for northern hemisphere summer (August) 18,000 years BP. Contour intervals are 1°C for isotherms and 500 m for ice elevation. Continental outlines represent a sea-level lowering of 85 m. In northern Siberia, dotted lines indicate a recently revised estimate of ice extent (after CLIMAP Project Members, 1976). Albedo values are given by the following key: A: snow and ice; albedo over 40 per cent. Isolines show elevation of the ice-sheet above sea-level in metres. B: sandy deserts, patchy snow, and snow covered dense coniferous forests; albedo between 30 and 39 per cent. C: loess, steppes, and semi-deserts; albedo between 25 and 29 per cent. D: savannas and dry grasslands; albedo between 20 and 24 per cent. E: forested and thickly vegetated land; albedo below 20 per cent (mostly 15 to 18 per cent). F: ice-free ocean and lakes, with isolines of sea surface temperature (°C); albedo below 10 per cent. Copyright 1976 by the AAAS.

A

B

C

D

E

F

5 Milankovitch theory and ice ages

Recently there has been an impressive revival of interest in the astronomical theory of palaeoclimates (Imbrie, 1982; Kominz and Pisias, 1979; Berger, 1978; Hays *et al.*, 1976; Mason, 1976), since geological data appear to give quantitative support to the Milankovitch theory of long-term variations of the earth's orbital parameters and their influence on climate through variations in the insolation available at the 'top of the atmosphere' and its distribution over the earth. The suggestion is that glacial/interglacial transitions are the result of periodic variations in the earth's orbital elements. These orbital variations comprise the longitude of perihelion (22,000 year period), the earth's obliquity (40,000 year period), and variations in orbital eccentricity (100,000 year period).

The present orbit of the earth is slightly elliptical with the sun at one focus of the ellipse, and as a consequence the strength of the solar beam reaching the earth varies about its mean value. At present the earth is nearest to the sun on 2–3 January and farthest from the sun on 5–6 July. This makes the solar beam near the earth about 3.5 per cent stronger than the average solar constant in January, and 3.5 per cent weaker than average in July. Now the gravity of the sun, the moon and the other planets causes the earth to vary, over many thousands of years, its orbit around the sun. Three different cycles are present and, when combined, produce the rather complex changes observed.

First of all there are variations in the orbital eccentricity. The earth's orbit varies from almost a complete circle to a marked ellipse, when it will be nearer to the sun at one particular season. A complete cycle from near-circular through a marked ellipse back to near-circular takes between 90,000 to 100,000 years. When the orbit is at its most elliptical, the intensity of the solar beam reaching the earth must undergo a seasonal range of about 30 per cent.

There are also variations in the longitude of perihelion causing a phenomenon known as precession of the equinoxes; that is to say, within the elliptical orbits the distance between earth and sun varies so that the season of the closest approach to the sun also varies. The complete cycle takes about 21,000 years, so 10,000 years ago the northern hemisphere was in summer when the earth was closest to the sun, rather than in winter as present.

Lastly, there are variations in the earth's obliquity. The tilt of the earth's axis of rotation relative to the plane of its orbit is believed to vary at least between 21.8° and 24.4° over a regular period of about 40,000 years. At present it is almost 23.44° and decreasing by about 0.00013° a year. The greater the tilt of the earth's axis, the more pronounced the difference between winter and summer.

The three mechanisms just described are sometimes known as the Milankovitch mechanism after Milankovitch (1930) who first described them in detail. The Milankovitch mechanism affects only the seasonal and geographical distribution of solar radiation on the earth's surface, yearly totals for the whole earth remaining constant. Surplus in one season is compensated by a deficit during the opposite one, while surplus in one geographical area is compensated by simultaneous deficit in some other zone. Detailed tabulations of the radiation variations due to the Milankovitch mechanism may be found in Vernekar (1972).

5.1 The astronomical theory of climatic change

If the Milankovitch mechanism is to influence climate, a sensitive area and/or season must exist in which heat budgets are modified by changes in incoming radiation to a larger extent than in the rest of the globe and/or during the rest of the year.

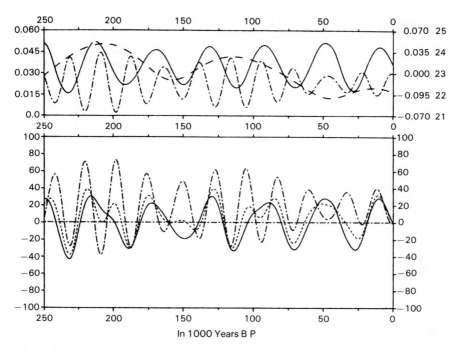

Figure 3.13 Long-term variations of astronomical parameters defining insolation climates and of isolation for various latitudes. The upper art shows eccentricity, *e* (dotted line), the deviations of precessional parameters Δ (*e* sin ω̄) from its 1950 AD value (dash-dot line), and obliquity ε (full line). The left-hand scale is related to *e*, the right-hand scales, respectively, to Δ (*e* sin ω̄) and ε. The lower part shows the deviations of solar radiation (cal cm^{-2} day^{-1}) from their 1950 AD values for the caloric northern hemisphere summer half-year at 80°N (full line), 65°N (dotted line), and 10°N (dash-dot line) (after Berger, 1978).

The location of any core area has to take into account both the climatic zone and latitude in which it lies. Thus, Bernard (1962) and Berger (1978) have pointed out that the astronomical control factor at high latitudes is the obliquity, whereas for low latitudes it is the eccentricity-preccession, *e* sin w̄, that prevails. Here *e* is the eccentricity of the orbit, given by ((semimajor axis)²-(semiminor axis)²)$^{0.5}$, and w̄ is the longitude of the perihelion relative to the moving vernal equinox. This is graphically evident from Figure 3.13, which represents the long-term variation of the deviations of solar radiation from their 1950 AD values for the caloric northern hemisphere summer at 10°, 65°, 80°N (lower part of figure), and the long-term variation of the obliquity, eccentricity and deviations of the precessional term from its 1950 value (upper part). Clearly, insolation at 80°N shows a quasi-periodicity which corresponds to the obliquity, while insolation at lower latitudes (e.g. 10°N) exhibits a 20,000 year period related to the precessional term. There is mounting evidence that at least some of the major glaciations during the Quaternary era began more or less simultaneously in regions where a small anomaly in the area of snow cover might have initiated a positive feedback mechanism, i.e. increased albedo leading to tropospheric cooling and a cold upper vortex leading to increased snowfall. Kukla (1975) claims that the northern high and mid-latitude continents are such sensitive areas, and that the autumn is the sensitive season. Flohn (1974) states that the key to the initiation of a northern glaciation (either complete or incomplete) is the summer climate of Labrador–Ungava, including Keewatin and probably Baffin Island. Williams (1978) has suggested that the Laurentide ice-sheet originated with extensive perennial snow cover, and that the snow cover affected climate so as to aid ice-sheet development. He considers that the combined effects of orbital variations of the earth and snow-cover expansion would make climatic conditions favourable for the establishment of ice-caps in northern Keewatin and northern Baffin Island.

Berger (1978) has also commented that the astronomical elements of the earth's orbit are the only external parameters (i.e. outside the atmosphere continents–oceans–cryosphere system) which can account directly and simultaneously for a decrease of insolation in high latitudes and an increase in tropical regions. This is particularly important because, at least during the initiation step of a glacial stage, the subtropical oceanic sources of water vapour (Peixoto and Oort, 1983; Adam, 1973; 1975) must be warm enough to allow sufficient evaporation which, through the activated general circulation, will be transported to high latitudes in order to feed the growing ice-sheets on the continents. This is a point that has been made strongly in a paper by Ruddiman and McIntyre (1979). They claim that isotopic data from deep-sea cores with high accumulation rates delineate two very large and rapid phases of ice growth during the last glaciation. The first occurred at about 115,000 yr BP, and marks the last glacial inception, that is the increase of global ice above modern values. The second, at roughly 75,000 yr BP, can be regarded as the time of temperature-latitude transition into glacial conditions. During the first transition the sea surface in the subpolar Atlantic maintained temperatures as high as today's or 1° or 2°C warmer until about halfway into the ice-growth interval. Sea-surface temperatures again remained high during most of the second period of rapid glacier ice accumulation. Ruddiman and McIntyre speculate that, along with low summer insolation, a warm subpolar North Atlantic Ocean may be a necessary condition for rapid and extensive northern hemisphere ice growth. They also support the contention that a vigorous meridional atmospheric circulation directed northward along an anomalously strong surface thermal gradient off the east coast of North America is the circulation regime most compatible with the process of rapid glaciation over North America. Figure 3.13 shows that both periods of rapid ice accumulation correspond to marked minima in the summer solar radiation at both 80° and 65°N. Also the periods of ice accumulation were preceded by periods when the solar radiation at 10°N was at a maximum, thus allowing heat to accumulate in the tropical oceans and circulate northwards.

The seasonal formation of ice in both polar oceans may allow the Milankovitch mechanism to influence global climate. According to Berger (1978) the annual mean ice cover over the Arctic today is about $9.4 \times 10^6 \text{ km}^2$, distributed as $7.1 \times 10^6 \text{ km}^2$ in summer and $11.4 \times 10^6 \text{ km}^2$ in winter. In southern polar latitudes the mean ice cover of $12 \times 10^6 \text{ km}^2$ is divided into only $2.5 \times 10^6 \text{ km}^2$ in summer and $22 \times 10^6 \text{ km}^2$ in winter. This clearly means that, in the Arctic Ocean, 70 per cent of the ice is prevented from melting during the summer season and, in the sub-Antarctic oceans, most of the ice is formed during the winter. As a consequence, it is the minimal melting of snow during the northern summers and the enormous amount of ice formed during southern winters which can account for the accumulation rate in the respective polar regions. Cool northern summers and cool southern winters therefore seem to be important control factors as far as the pack-ice is concerned.

Berger and also Flohn (1978) comment that this is in perfect agreement with the Milankovitch requirement of cool summers in high northern latitudes, both cool northern summers and cool southern winters occurring at the same astronomical time.

Recently presented geological evidence appears to give strong support to the astronomical theory of climate change. In particular Hays *et al.* (1976) have found climatic cycles of 23,000, 42,000 and approximately 100,000 years in southern hemisphere ocean-floor sediments during the last 450,000 years, suggesting a connection with the Milankovitch mechanism. The dominant cycle is one of about 100,000 years (50 per cent of the climatic variance) and is seen in the growth and decay of the continental ice-sheets. Indeed, continental ice-sheets appear to develop over a period of 90,000–100,000 years and to terminate rather quickly in about 10,000 years. However, Kominz and Pisias (1979) have commented that the degree to which the climatic record of glacial and interglacial fluctuations is actually a direct result of orbital controls has not yet been determined. In addition, the orbital parameters that are expected to have a significant effect on climate are not always those that are actually reflected in the climatic record. In many cases most of the variance of the palaeoclimatic records obtained from deep-sea cores is centred at frequencies equal to periods of about 100,000 years. This corresponds to the periodicity of variations in the eccentricity of the earth's orbit. According to the astronomical theory of climatic change, however, eccentricity influences the earth's isolation only by modifying the magnitude of the precessional effect. Thus, it should play, at most, a minor role in climatic change.

Hays *et al.* (1976) assumed that all the variance in the peaks in the power spectrum of their climatic record, which correspond to the periodicities of precession, obliquity and eccentricity, could be attributed to forcing. As a result, they concluded that about 80 per cent of the variance in their 450,000 year climatic record was due to orbital forcing. To determine the degree to which the ^{18}Oxygen record is governed by the earth's orbital parameters, Kominz and Pisias (1979) assumed a linear relationship between ice volume record and the tilt and precession, and they then applied a two-input, single-output linear time series model. The resulting calculations indicate that 12 per cent of the variance of the ^{18}Oxygen record is due to forcing by the earth's tilt and 3 per cent is due to forcing by the precession of the equinox. Thus, about 15 per cent of the variation in the global climate can be explained in terms of simple linear forcing.

Kominz and Pisias conclude that more than 75 per cent of the variance in the oxygen isotope record for the last 730,000 years is not linearly related to orbital variation. The spectrum for the residual time series is a red noise spectrum and displays a simple inverse square relationship with respect to frequencies. They comment that such a spectrum is predicted by many possible stochastic models. Kominz and Pisias summarize their conclusions as follows; (i) less than 25 per cent of the variation in global ice volume during the last 730,000 years is related to the tilt and precessional variations of the earth's orbit; (ii) there is no evidence for a linear relationship between eccentricity and global climates; and (iii) the spectrum of global ice volume with the linear effect of the earth's orbital variations removed is a red noise spectrum with a simple form.

Hays *et al.* (1976) comment that unlike the correlations between climate and the higher-frequency orbital variations (which can be explained on the assumption that the climate system responds linearly to orbital forcing), an explanation of their correlation between climate and eccentricity probably requires an assumption of non-linearity. Flohn (1978) states that from a climatological viewpoint, the observed dominant role of the 100,000 year cycle of Hays *et al.* cannot be correlated with a simultaneous variation of eccentricity, but Flohn (1978), Berger (1978) and Wigley (1976) consider that it might be expected from the non-linear interaction between the precessional frequencies. From the 23,100 and 18,800 year periods deduced by Hays *et al.* from the numerical values of the precessional parameter over the past 468,000 years, Wigley (1976) computed a beat of 101,000 years. From the series expansion of *e* sin *w*, Berger (1977a) found beats of 95,000 and 123,000 years. So the 100,000 year cycle of growth and decay of ice-sheets could be related to non-linear interaction between precessional peaks. Kominz and Pisias (1979) did not consider the non-linear models that have been suggested to explain the dominant 100,000 years component in the spectrum of climatic change. They do comment though that any non-linear model relating eccentricity to this component of climatic change must satisfy the observations that the coherence at this frequency is very small and possibly zero.

Variations in the orbital parameters during the past several hundred thousand years can be computed with great precision (see Vernekar, 1972), and the resulting changes in the incident radiation at the top of the atmosphere are easily obtained. A number of authors (Weertman, 1976; Suarez and Held, 1976; Pollard, 1978) have suggested that the Milankovitch theory could be tested using simple climatic models.

The astronomical theory of the ice ages has been investigated by Pollard, Ingersoll and Lockwood (1980) using a simple climate model which includes ice-sheets explicitly. The model has previously been outlined in Pollard (1978); there are two distinct parts corresponding to the two distinct time-scales of the global seasonal weather and the long-term ice-sheet response.

Following North (1975), the weather through one year over a spherical globe is described by a zonally averaged, one-level energy-balance equation for sea-level air temperature T:

$$\frac{\partial}{\partial t}[CT(x, t)] - \frac{\partial}{\partial x}\left[(1 - x^2)\frac{D}{R^2}\frac{\partial}{\partial x}(CT)\right] + [A + BT] =$$
$$Q(x, t)(1 - \alpha) + S \qquad (3.12)$$

Here x is sin (latitude) and t is time. All dependent variables are defined as 1 month running-means, so daily correlations are effectively assumed constant. Boundary conditions are $(1 - x^2)^{1/2}\partial T/\partial x = 0$ at the poles $x = \pm 1$. The variables and their values for the 'standard' model are:

C = 4.6×10^7 J m^{-2}°C^{-1}, a constant seasonal heat capacity of the atmosphere–land–ocean system (equivalent to a layer of liquid water 11 m thick).

D = 0.501×10^6 m^2 s^{-1}, a linear diffusion coefficient acting over the whole thickness of the layer represented by C.

R = 6.36×10^6 m, radius of the earth.

A = 207 W m^{-2}, $B = 1.9$ W m^{-2}°C^{-1}, net infrared radiation coefficients.

Q = zonal mean insolation åt the top of the atmosphere, computed for any given era from orbital elements using a solar constant of 1360 W m^{-2}.

α = $r\alpha_c + (1 - r)\alpha_f$, earth–atmosphere albedo. $\alpha_c = 0.62$ represents areas covered by seasonal snow or ice-sheet, and $\alpha_f = 0.31 + 0.08 [(3x^2 - 1)/2]$ represents areas free of seasonal snow and ice-sheet.

r = 1 north of 75°N to represent perennial Arctic Ocean sea-ice, and $r = 1$ south of 70°S to represent a fixed Antarctic ice-sheet. At all other latitudes $r = 0.6$

when covered by seasonal snow or ice-sheet, and $r = 0$ when free of seasonal snow and ice-sheet.

S = 1.27 [J m^{-2} s^{-1}] per [g cm^{-2} month^{-1}], representing latent heat of fusion released or required at each latitude by the varying amount of seasonal snow cover. The annual mean of S at each latitude is zero.

Most of these parameterizations (discussed for instance in Coakley (1979)) have found general use in many annual mean energy-balance models and are based on annual mean data.

Seasonally varying snow cover on land and ice-sheet surfaces is modelled diagnostically by parameterizing monthly snowmelt and snowfall as functions of the current air temperature T and insolation Q. For snowmelt equation (3.5) was used, and snowfall was assumed if the temperature was below 0°C.

Ice-sheets are incorporated into the weather model following Weertman's (1976) simple treatment. Ice-sheet flow under its own weight is approximated to be perfectly plastic, which constrains the model ice-sheet profiles to always remain parabolic as in equation (3.1).

The model ice-sheet, representing the Laurentide and Scandinavian ice-sheets of past eras, is constrained to extend equatorward with its northern tip fixed at 75°N (corresponding to the Arctic Ocean shoreline). Where the margins of the real northern hemispheric ice-sheets reached continental coastlines, further advance was prevented by rapid ice-shelf and iceberg calving into the ocean, but their equatorial extent and over-all volume were probably limited more by ablation on their southern flanks. Therefore, as in Weertman (1976), the long-term variation in the model ice-sheet size is controlled by the net accumulation (snowfall) minus ablation (mostly snowmelt) on its southern half only. Also, since its profile is constrained by equation (3.1), any change in size is determined simply by the total ice volume added to or removed from the entire southern half.

Model ice-age curves were generated for the last several 100 K years by computing the seasonal climate as above once every 2 K years, with insolation calculated from actual earth orbit perturbations. The change in ice-sheet size for each 2 K year time step depends only on the net annual snow budget integrated over the whole ice-sheet surface. In these model runs, the equatorward tip of the northern hemisphere's ice-sheet oscillates through $\sim 7°$ in latitude, correctly simulating the phases and approximate amplitude of the higher frequency components (43 K year and 22 K year) of the deep-sea core data (Hays *et al.*, 1976). However, the model failed to simulate the dominant glacial-interglacial cycles (~ 100 to 120 K years) of this data.

Oerlemans (1980), 1981b) has carried out experiments with a northern hemisphere ice-sheet model that shows that the 100 K year cycle and its sawtooth shape may be explained by ice-sheet/bedrock dynamics alone. According to Oerlemans this cycle seems to be an internally generated feature and is not forced by variations in the eccentricity of the earth's orbit. There is a tendency to restore isostatic equilibrium under ice-sheets, and in the simplest form this may be formulated by:

$$\frac{\partial h}{\partial t} = -w(H^* + 2h) \tag{3.13}$$

where h is the height of the bedrock with respect to its equilibrium value (the case of no ice-sheet); H^* is the elevation of the ice surface above sea-level, and the time scale for adjustment is $\frac{1}{2}w$.

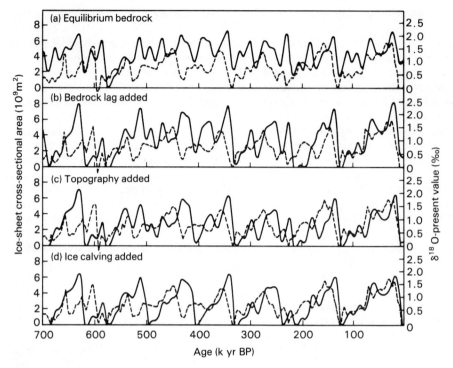

Figure 3.14   Total cross-sectional area of ice versus time for various model versions (solid curves). The corresponding approximate ice volume can be obtained by multiplying by a typical east–west ice sheet (~ 3000 km). The dotted curve in each panel is an oxygen isotope deep-sea core record minus its present value (after Pollard, 1982). Reprinted by permission from *Nature* vol. 296 p. 335. Copyright © 1982 Macmillan Journals Limited.

In the steady state $h = \frac{1}{2}H^*$, which corresponds to an isostatic balance if the rock density is three times that of ice. In the numerical experiments carried out by Oerlemans (9181b), the bedrock sinks or rises to restore the isostatic balance, and this mechanism has a time-scale of $t = \frac{1}{2}w$. Now Oerlemans found that in the case of no bedrock adjustment ($t = \infty$) the ice-sheet grows to a steady state in about 40 K years. If isostatic adjustment with a long time-scale is included ($t = 3$ K years), the picture changes drastically. The ice-sheet grows during about 50 K years, but then the bedrock has sunk so much that a large part of the ice surface comes below the equilibrium line which causes the ice-sheet to decay rapidly. For smaller time-scales for bedrock sinking, this feature occurs earlier.

Birchfield, Weertman and Lunde (1982) have examined the response of a combined global zonal averaged energy balance model and a continental ice-sheet model to insolation anomalies produced by earth orbital perturbations. They found that the presence of a high-latitude plateau significantly increases the sensitivity of the climate to the insolation perturbations. The sensitivity is maximized when the elevation of the summer snowline is near the elevation of the plateau, as appears to be the case with Baffin Island today. Both Birchfield, Weertman and Lunde (1981) and Oerlemans (1980, 1981b) have obtained encouraging agreement with some features of the glacial cycle by using a simple ice-sheet model with a realistic time lag in response of the bedrock to the ice load. Pollard (1982, 1983) has extended their basic model, first by including topography to represent high ground in the north. Improved results (Figure 3.14) can then be obtained but only with unrealistic parameter values and for some aspects of the record. Pollard obtained further improvements by crudely parameterizing possible calving at the equatorward ice-sheet tip during deglaciation by proglacial lakes and/or marine incursions. The resulting ice-volume curves agree fairly well with the observed records and their power spectra over the past 700 K years.

Prell and Start (1982) have shown that two distinct climatic regimes occurred within the Quaternary. In both modes, spectra of indices show significant concentration of variance in frequency bands identified with orbital variation. However, the partitioning of climatic variance among the various orbital frequencies is distinctly different between the two modes. For example, the late Quaternary mode has significant concentrations of variance at periods around 100 K and 23 K years, whereas the early Quaternary mode spectra have more power at 41 K and 19 K years. So there are still major problems in the numerical modelling of ice-sheet variations.

References

ADAM, D. P. 1973: Ice ages and the thermal equilibrium of the earth. *Journal of Research of the United States Geological Survey* 1, 587–96.
— 1975: Ice ages and the thermal equilibrium of the earth. *Quaternary Research* 5, 161–71.
ANDREWS, J. T., BARRY, R. G., DAVIS, P. T., DYKE, A. S., KOHAFFY, M., WILLIAMS, L. D. and WRIGHT, C. 1975: The Laurentide ice-sheet: problems of the mode and speed of inception. In WMO *Proceedings of the WMO/IAMAP Symposium on long-term climate fluctuations*. Geneva: World Meteorological Organization, 87–94.
ANDREWS, J. T. and MAHAFFY, M. A. W. 1976: Growth rate of the Laurentide ice-sheet and sea level lowering (with emphasis on the 115,000 BP sea level low). *Quaternary Research* 6, 167–83.
BARRON, E. J. and WASHINGTON, W. M. 1982: Cretaceous climate a comparison of atmospheric simulations with the geologic record. *Palaeography, Palaeoclimatology, Palaeoecology* 40, 103–33.
BARRY, R. G. 1981: *Mountain Weather and Climate*. London: Methuen.
BENSON, R. H. 1975: The origin of the psychrosphere as recorded in changes of deep-sea ostracode assemblages. *Lethaia* 8, 69–83.
BERGER, A. C. 1977a: Support for the astronomical theory of climatic change. *Nature* 269, 44–5.
— 1977b: Power and limitation of an energy-balance climate model as applied to the astronomical theory of palaeoclimates. *Palaeogeography, Palaeoclimatology, Palaeoecology* 21, 227–35.
— 1978: Long-term variations of caloric insolation resulting from the earth's orbital elements. *Quaternary Research* 9, 136–67.
BERGGREN, W. A. 1972: Late Pliocene-Pleistocene glaciation. In *Initial Reports of the Deep-Sea Drilling Project* 12. Washington: US Government Printing Office, 953–63.
BERNARD, E. 1962: *Théorie astronomique des pluviaux et interpluviaux du Quaternaire Africain*. Brussels: Académie Royale des Sciences d'Outre-Mer, Classe des Sciences Naturelles et Médicales, Nouvelle Serie 12.
BIRCHFIELD, G. E., WEERTMAN, J. and LUNDE, A. T. 1981: A palaeoclimatic model of northern hemisphere ice-sheets. *Quaternary Research* 15, 126–42.
— 1982: A model study of the role of high-latitude topography in the climatic response to orbital insolation anomalies. *Journal of the Atmospheric Sciences* 39, 71–86.

BLEASDALE, A. and CHAN, Y. K. 1972: Orographic influences on the distribution of precipitation. In *Distribution of Precipitation in Mountainous Areas*. Geilo Symposium, Norway 31 July – 5 August, 1972. Vol. 2. Geneva: World Meteorological Organization, 322–33.

BOWEN, R. 1966: *Paleotemperature Analysis*. Amsterdam: Elsevier.

BRASS, G. W., SOUTHAM, J. R. and PETERSON, W. H. 1982: Warm saline bottom water in the ancient ocean. *Nature* 296, 620–1.

BRUUN, A. F. 1957: Deep sea and abyssal depths. *Geological Society of America Memoir* 67, 641–72.

CHARLESWORTH, J. K. 1957: *The Quaternary Era*. London: Edward Arnold.

CLIMAP PROJECT MEMBERS 1976: The surface of the ice-age earth. *Science* 191, 1131–7.

COAKLEY, Jr, J. A. 1979: A study of climate sensitivity using a simple energy balance model. *Journal of the Atmospheric Sciences* 36, 260–9.

COLLINGBOURNE, R. A. 1976: Radiation and sunshine. In Chandler, T. J. and Gregory, S. (eds.) *The Climate of the British Isles*, London: Longman, 74–95.

DANSGAARD, W., JOHNSON, S. J., CLAUSEN, H. B. and LANGWAY, C. C. Jr 1971: Climates record revealed by the Camp Century ice core. In Turekian, K. K. (ed.) *The Late Cenozoic Glacial Ages*. New Haven: Yale University Press, 37–56.

— 1975: Speculations about the next glaciation. *Quaternary Research* 2, 396–8.

DERGACH, A. C., ZABRODSKIĬ, G. M. and MORACHEVSKIĬ, V. G. 1960: Opyt Kompleksnogo issledovaniya oblakov tipa st-sc i tumanov v Arktike (Experience in the complex investigation of St-Sc type clouds and of fogs in the Arctic). *Akademiia Nauk SSSR, Izvestica, Ser. Geofiz*. 1. 107.

DIAMOND, M. 1958: Air temperature and precipitation on the Greenland ice cap. *US Army Corps Engineers, Snow, Ice, Permafrost Research Establishment. Research Report 43*.

DOUGLAS, R. G. and SAVIN, S. M. 1975: Oxygen and carbon isotope analysis of Tertiary and Cretaceous microfossils from Shatsky Rise and other sites in the North Pacific ocean. In *Initial reports of the Deep sea Drilling Project*, 32, Washington, DC: US Government Printing Office, 509–20.

EMILIANI, C. 1961: Cenozoic climate changes as indicated by the stratigraphy and chronology of deep-sea cases of Globigerina ooze facies. *New York Academy of Sciences* 95, 521–36.

FLOHN, H. 1967: Bemerkungen zur Asymmetrie der atmosphärischen Zirkulation. *Annalen der Meteorologie Neue Folge* 3, 76–80.

— 1974: Background of a geophysical model of the initiation of the next glaciation. *Quaternary Research* 4, 385–404.

— 1978: Comparison of Antarctic and Arctic climate and its relevance to climatic evolution. In Van Zinderen Bakker, E. M. (Ed.) *Antarctic Glacial History and World Palaeoenvironments*. Rotterdam: Balkema, 3–14.

FRAKES, L. A. 1979: *Climates throughout Geologic Time*. Amsterdam: Elsevier.

GAVRILOVA, M. K. 1963: *Radiation Climate of the Arctic*. Jerusalem: Israel Program of Scientific Translations, 1966.

GIOVINETTO, M. B. 1964: The drainage systems of Antarctica: accumulation. In Mellor, M. (Ed.) Antarctic snow and ice studies. *American Geophysical Union Antarctic Research Series* 2, 127–55.

— 1968: *Glacier landforms of the Antarctic coast, and the regimen of the inland ice*. Thesis, University of Wisconsin, Madison.

HARLAND, W. B. 1972: The Ordovician ice age. *Geological Magazine* 109, 451–6.

HARLAND, W. B. and HEROD, K. N. 1975: Glaciations through time. In Wright, A. E. and Moseley, F. (eds.) *Ice Ages: Ancient and Modern. Geological Journal special issue* 6, 189–216.

HART, M. H. 1978: The evolution of the atmosphere of the earth. *Icarus* 33, 23–39.

HAYS, J. D., IMBRIE, J. and SHACKLETON, N. J. 1976: Variations in the earth's orbit: pacemaker of the ice ages. *Science* 194, 1121–32.

HAYS, J. D., SAITO, T., OPDYKE, N. D. and BUCKLE, L. H. 1969: Pliocene-Pleistocene sediments of the equatorial Pacific, their paleomagnetic, biostratigraphic and climatic record. *Bulletin Geological Society of America* 80, 1481–514.

HUNT, B. G. 1979a: The effects of past variations of the earth's rotation rates on climate. *Nature* 281, 188–91.

— 1979b: Influence of the earth's rotation rate on the general circulation of the atmosphere. *Journal of the Atmospheric Sciences* 36, 1392–408.

IMBRIE, J. 1982: Astronomical theory of Pleistocene ice ages: a brief historical review. *Icarus* 50, 408–20.

IMBRIE, J. and IMBRIE, K. P. 1979: *Ice Ages: Solving the Mystery*. Short Hills, NJ: Enslow Pubs.

IMBRIE, J., VAN DORK, J., and KIPP, N. G. 1973: Paleoclimatic investigation of a late Pleistocene Caribbean deep-sea core: comparison of isotopic and faunal methods. *Quaternary Research* 3, 10–38.

JACKSON, M. C. 1978: Snow cover in Great Britain. *Weather* 33, 298–309.

KENNETT, J. P. 1977: Cenozoic evolution of Antarctic glaciation, the circum-Antarctic ocean, and their impact on global paleoceanography. *Journal of Geophysical Research* 82, 3843–60.

KENNETT, J. P. and SHACKLETON, N. J. 1976: Oxygen isotope evidence for the development of the psychrosphere 38 m.y. ago. *Nature* 260, 513–15.

KILLÍNGLEY, J. S. 1983: Effects of diagenetic recrystallization on $^{18}O/^{16}O$ values of deep-sea sediments. *Nature* 301, 594–7.

KOMINZ, M. A. and PISIAS, N. G. 1979: Pleistocene climate: deterministic or stochastic? *Science* 204, 171–3.

KORFF, H. C. and FLOHN, H. 1969: Zusammenhang zwischen dem Temperatur-Gefälle Äquator–Pol und den planetarischen Luftdruckgurteln. *Annalen der Meteorologie Neue Folge* 4, 163–4.

KOTLIAKOV, V. M. and KRENKE, A. N. 1982: Data on snow cover and glaciers for the global climatic models. In Eagleson, P. S. (ed.) *Land Surface Processes in Atmospheric General Circulation Models*. Cambridge: Cambridge University Press, 449–61.

KUKLA, G. J. 1970: Correlations between loesses and deep-sea sediments. *Geol. Foren, Stockholm Forh* 92, 148–80.

— 1975: Missing link between Milankovitch and climate. *Nature* 253, 600–3.

— 1976: Revival of Milankovitch. *Nature* 261, 11.

KVASOV, D. D. and VERBITSKY, M. Ya. 1981: Causes of Antarctic glaciation in the Cenozoic. *Quaternary Research* 15, 1–17.

LAMB, H. H. 1955: Two-way relationship between the snow or ice limit and 1,000–500 mb thickness in the overlying atmosphere. *Quarterly Journal of the Royal Meteorological Society* 81, 172–89.

— 1959: The southern westerlies: a preliminary survey, main characteristics and apparent associations. *Quarterly Journal Royal Meteorological Society* 85, 1–23.

LIST, R. J. 1963: *Smithsonian Meteorological Tables*. Washington: The Smithsonian Institute.

LOCKWOOD, J. G. 1982: Snow and ice balance in Britain at the present time, and during the last glacial maximum and late glacial periods. *Journal of Climatology* 2, 209–31.

MANLEY, G. 1969: Snowfall in Britain over the past 300 years. *Weather* 24, 428–37.

— 1970: The climate of the British Isles. In Wallen, C. C. (ed.) *Climates of Northern and Western Europe*. Amsterdam: Elsevier, 81–133.

— 1975: Fluctuations of snowfall and persistence of snow cover in marginal-oceanic climates. In WMO *Proceedings of the WMO/IAMAP Symposium on Long-Term Climatic Fluctuations*. Geneva: World Meteorological Organization, 183–7.

MASON, B. J. 1976: Towards the understanding and prediction of climate variations. *Quarterly Journal of the Royal Meteorological Society* 102, 473–98.

METEOROLOGICAL OFFICE, 1952: Climatological Atlas of the British Isles. London: HMSO.

MILANKOVITCH, M. 1930: Mathematische Klimalehr and astronomische Theorie der Klima-schwankungen. In Köppen, W. and Geiger, R. (ed.) *Handbuch der Klimatologie* 1, Berlin: Teil A.

MOHR, R. E. 1975: Measured periodicities of the Biwabik (Precambrian) stromatolites and their geophysical significance. In Rosenberg, G. D. and Runcorn, S. K. (eds.) *Growth*

Rhythms and the History of the Earth's Rotation. London: Wiley, 43–56.

MOORE, T. C., VAN ANDEL, T. H., SANCETTA, C. and PISIAS, W. 1978: Cenozoic hiatuses in pelagic sediments. *Micro-paleontology* 24, 113–38.

MURRAY, R. 1952: Rain and relation to the 1,000–700 mb and 1,000–500 mb thickness and the freezing level. *Meteorological Magazine* 81, 5–8.

NEWELL, R. E. KIDSON, J. W., VINCENT, D. G. and BOER, G. J. 1972: *The General Circulation of the Tropical Atmosphere* 1. Cambridge, Mass: MII Press.

NORTH, G. R. 1975: Theory of energy-balance climate models. *Journal of Atmospheric Sciences* 32, 2033–43.

OERLEMANS, J. 1980: Model experiments on the 100,000 yr glacial cycle. *Nature* 287, 430–2.

— 1981a: Modeling of Pleistocene European ice-sheets: some experiments with simple mass-balance parameterizations. *Quaternary Research* 15, 77–85.

— 1981b: Some basic experiments with a vertically integrated ice-sheet model. *Tellus* 33, 1–11.

ØSTREM, G. 1974: Present alpine ice cover. In Ives, J. D. and Barry, R. G. (eds.) *Arctic and Alpine Environments.* London: Methuen, 225–50.

PEIXOTO, J. P. and OORT, A. H. 1983: The atmospheric branch of the hydrological cycle and climate. In Street-Perrott, A., Beran, M. and Ratcliffe, R. (eds.) *Variations in the Global Water Budget.* Dordrecht: Reidel, 5–65.

POLLARD, D. 1978: An investigation of the astronomical theory of ice ages using a simple climate ice-sheet model. *Nature* 272, 233–5.

— 1980: A simple parameterization for ice-sheet ablation rate. *Tellus* 32, 384–8.

— 1982: A simple ice-sheet model yields realistic 100 K yr. glacial cycles. *Nature* 296, 334–8.

— 1983: Ice-age simulations with a calving ice-sheet model. *Quaternary Research* 20, 30–48.

POLLARD, D., INGERSOLL, A. P. and LOCKWOOD, J. G. 1980: Response of a zonal climate ice model to the orbital perturbations during the Quaternary ice ages. *Tellus* 32, 301–19.

PRELL, W. C. and START, G. G. 1982: Isotopic and carbonate evidence for two climatic modes in the Quaternary. *EOS* 63, 1297.

PUTNINS, P. 1970: The climate of Greenland. In Orvig, S. (ed.) *Climates of the Polar Regions.* Amsterdam: Elsevier, 3–128.

ROBIN, G. de Q. 1983: The Greenland ice-sheet. In Robin, G. de Q. (ed.) *The Climatic Record in Polar Ice Sheets.* Cambridge: Cambridge University Press, 38–42.

RUDDIMAN, W. F. and MCINTYRE, A. 1979: Warmth of the sub-polar North Atlantic Ocean during northern hemisphere ice-sheet growth. *Science* 204, 173–5.

RUSIN, N. P. 1961: *Meteorological and radiation regime of Antarctica,* Jerusalem: Israel Program for Scientific Translations 1964.

RYAN, W. B. F. and CITA, M. B. 1977: Ignorance concerning episodes of ocean-wide stagnation. *Marine Geology* 23, 117–215.

SAVIN, S. M. 1977: The history of the earth's surface temperature during the past 100 million years. *Annual Review of Earth and Planetary Sciences* 5, 319–55.

SAVIN, S. M., DOUGLAS, R. G. and STEHLI, F. G. 1975: Tertiary marine paleotemperatures. *Geological Society of America Bulletin* 86, 1499–510.

SCHNITKER, D. 1980: Global paleoceanography and its deep water linkage to the Antarctic glaciation. *Earth-science reviews* 16, 1–20.

SCHWERDTFEGER, W. 1970: The climate of the Antarctic. In Orvig, S. (ed.) *Climates of the Polar Regions.* Amsterdam: Elsevier, 253–355.

SHACKLETON, N. and BOERSMA, A. 1981: The climate of the Eocene ocean. *Journal of the Geological Society* 138, 153–7.

SHACKLETON, N. J. and KENNETT, J. P. 1975: Late Cenozoic oxygen and carbon isotope changes at DSP Site 284, implications for glacial history of the northern hemisphere and Antarctica. In *Initial reports of the Deep Sea Drilling Project* 29, Washington, DC: US Government Printing Office.

SHACKLETON, N. J. and OPDYKE, N. D. 1973: Oxygen isotope and paleomagnetic stratigraphy of equatorial Pacific core V28–238: oxygen isotope temperatures and ice volumes on a 10^5 and 10^6 year scale. *Quaternary Research* 3, 39–55.

SMAGORINSKY, J. 1963: General circulation experiments with the primitive equations

(Appendix B). *Monthly Weather Review* 91, 159–62.

SUAREZ, M. J. and HELD, I. M. 1976: Modeling climatic response to orbital parameter variations. *Nature* 263, 46–7.

TARLING, D. H. 1978: The geological-geophysical framework of ice ages. In Gribbin, J. (ed.) *Climatic Change*. Cambridge: Cambridge University Press, 3–24.

TALJAARD, J. J., VAN LOON, H., CRUTCHER, H. L. and JEUNE, R. L. 1969: *Climate of the Upper Air*. Part 1. *Southern Hemisphere* 1. *Sea-Level Pressures and Selected Heights, Temperatures and Dew-points*. NAVAIR 50-16-15. Washington DC: US Government Printing Office.

TAYLOR, J. A. 1976: Upland climates. In Chandler, T. J. and Gregory, S. (Eds.). *The Climate of the British Isles*. London: Longman, 264–87.

THIEDE, J. and VAN ANDEL, T. H. 1977: Paleoenvironment of anaerobic sediments in late Mesozoic south-Atlantic ocean. *Earth and Planetary Science Letters* 33, 301–9.

THIERSTEIN, H. R. and BERGER, W. H. 1978: Injection events in ocean history. *Nature* 276, 461–6.

US NATIONAL ACADEMY OF SCIENCES 1975: Understanding climatic change, a program for action. Washington, DC.

UNTERSTEINER, N. 1975: Sea-ice and ice-sheets and their role in climatic variations. In GARP, *The Physical Basis of Climate and Climatic Modelling*, 206–24. Geneva: WMO.

VAN ANDEL, T. H., THIEDE, J., SCLATER, H. G. and MAY, W. W. 1977: Depositional history of South Atlantic ocean during last 125 million years. *Journal of Geology* 85, 651–98.

VERNEKAR, A. D. 1972: *Long-period Global Variations of Incoming Solar Radiation*. *Meteorological Monographs* 12. Boston: American Meteorological Society.

VOWINCKEL, E. and ORVIG, S. 1967: The inversion over the polar ocean. In Orvig, S. (ed.), WMO—SCAR—ICPM. *Proceedings of Symposium on Polar Meteorology, Geneva 1966*. WMO Technical Note 87, 39–59.

— 1970: The climate of the north polar basin. In Orvig, S. (ed.), *Climates of the Polar Regions*. Amsterdam: Elsevier, 129–252.

WEERTMAN, J. 1961: Equilibrium profile of ice caps. *Journal of Glaciology* 3, 953–64.

— 1976: Milankovitch solar radiation variations and ice age ice-sheet sizes. *Nature* 261, 17–20.

WEIDICK, A. 1976: Glaciation and the Quaternary of Greenland. In Escher, A. and Stuart Watt, W. (eds.) *Geology of Greenland*. Copenhagen: Geological Survey of Greenland. 431–58.

WEXLER, H. 1959: Seasonal and other temperature changes in the Antarctic atmosphere. *Quarterly Journal of the Royal Meteorological Society* 85, 196–208.

WIGLEY, T. M. L. 1976: Spectral analysis and the astronomical theory of climatic change. *Nature* 264: 629–31.

WILLIAMS, C. D. 1978: Ice-sheet initiation and climatic influences of expanded snow cover in Arctic Canada. *Quaternary Research* 10, 141–49.

ZAVARINA, M. V. and ROMASHEVA, M. K. 1957: Moshchnost' oblakov nad arkticheskimi moryami i Tsentral' noi Arktikoi. (Thickness of clouds over Arctic seas and over the central Arctic). *Problemy Arktiki* 2.

ZILLMAN, J. W. 1967: The surface radiation balance in high southern latitudes. In Orvig, S. (ed.), WMO-SCAR-ICPM *Proceedings of Symposium on Polar Meteorology*, Geneva 1966. WMO Technical Note 87, 142–74.

4 Arid subsystems

1 Characteristics of desert climates

1.1 Arid climates

The important arid zones of the world are situated around latitudes 30°N and °S, where large areas are dominated by dynamic anticyclones, but it must not be assumed that arid areas cannot occur elsewhere, because aridity is normal and it is atmospheric precipitation which needs special explanation. In Figure 4.1, which indicates the generalized pattern of precipitation over the earth, the arid zones in the subtropics stand out very clearly, and it is interesting that these zones extend over the subtropical oceans as well as over the subtropical landmasses. This indicates that a lack of surface water is not the prime cause of lack of rainfall, for indeed the actual cause of general aridity in the subtropics is the widespread subsidence of the atmosphere in the descending limbs of the Hadley cells. The subsidence effectively suppresses extensive low-level convergence, and it is therefore almost impossible for extensive cloud systems to form and rain to fall.

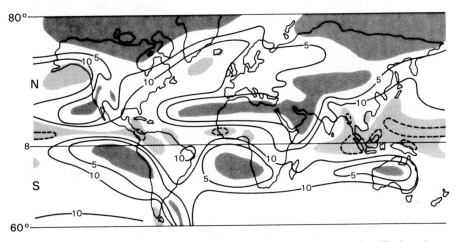

Figure 4.1 World distribution of annual precipitation (decimetres), simplified to show only major regimes; oceanic rainfall estimated. Areas with precipitation below 2.5 and above 20 decimetres shade; dashed lines: 30 decimetres (after Riehl, 1965).

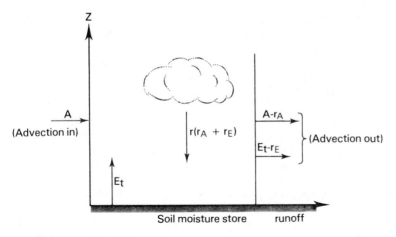

Figure 4.2 The hydrological cycle in the atmosphere over a small land area.

In Figure 4.2 which explores the water balance over a small land area, rainfall (r) within the unit area is partly formed of moisture advected (r_A) into the area by the winds and partly of moisture which has evaporated (r_E) within the area:

$$r = r_A + r_E \tag{4.1}$$

The local evapotranspiration (E_t) partly supplies moisture for local rainfall and is partly advected (C) out of the unit area by the winds:

$$E_t = r_E + C \tag{4.2}$$

Finally, the rainfall must either flow out of the unit area as rivers or ocean currents, or be used in local evapotranspiration.

A study of atmospheric moisture content shows that the air over the great deserts is normally very dry. This dryness of the air is partly a result of the lack of local evapotranspiration and partly due to the lack of horizontal moisture advection. Subsidence within anticyclones does not extend right to the surface, since normally the warm, dry subsiding air is insulated from the surface by a shallow layer of relatively cool air. The properties of this surface layer, which may be several hundred metres thick, are often completely different from those of the subsiding air, and it is normally maintained by horizontal advection from a source region outside the main anticyclone. If the air forming the surface layer originates over the sea, it may be moist and even contain layer cloud which can give rise to light rain or drizzle. Within the tropics, the surface moist layer has already been identified with the layer below the trade wind inversion.

Most of the moisture advection within the tropical atmosphere takes place in the shallow layer below the trade wind inversion. The high-level winds over the subtropical anticyclones are extremely strong, but they transport little water vapour because of their extreme cold, and therefore do not help in the formation of cloud. At low levels, winds are lighter and hence inefficient in transporting water vapour; the low-level winds in the anticyclonic cores are also generally divergent. The dryness of the air above the major deserts results partly, therefore, from a lack of advection of water vapour.

The atmospheric moisture balance model (Figure 4.2) suggests that local evapotranspiration has to be considered as well as general moisture advection, and this means that the nature of the local surface has to be taken into account. If the surface consists of the open ocean, then evaporation will be high because of the high net radiation; the surface layer will become moist and this moisture will be available for use by occasional synoptic disturbances. In contrast, if the underlying surface is completely arid, there will be no local evapotranspiration and no moisture gain by the surface layers of the atmosphere. Thus it is possible to distinguish two general types of arid climate; a maritime arid climate and a continental arid climate. The maritime anticyclones are the source regions of the trade winds and the evaporated water is carried towards the equator as described in previous chapters. The moisture in the layer below the anticyclonic inversion can be made use of by any synoptic disturbances which may occur. On the other hand, over large arid continents all the net radiation is available for heating the air and soil, and as a result surface temperatures are high and humidity low. If surface convergence does occur over an arid landmass, there is not normally enough water vapour in the air to form cloud and rainfall; moisture must be imported from outside the desert areas if rain is to be formed. Conditions are therefore more extreme in large deserts than over the oceans.

1.2 Deserts

No desert area is completely rainless, though in extreme cases several years may elapse between individual storms. The subtropical anticyclones undergo only minor seasonal variations, so the desert cores are liable to remain almost rainless, but rainfall is apt to increase and become more seasonal towards the poleward and equatorial limits of the deserts. Rainfall within desert regions normally results from disturbances arising outside the true desert areas. For instance, upper cold pools and troughs from the middle latitudes bring rainfall to the poleward margins of the deserts; similarly, disturbances forming near the equator can bring rainfall to the equatorward margins. As middle-latitude troughs are most intense and nearest to the equator in winter, the poleward margins of the deserts usually experience a winter rainfall maximum. The equatorial trough tends to move north–south with the sun; in contrast to the poleward margins, the equatorial limits of the desert areas normally experience summer rainfall maxima. It is therefore possible to recognize two distinct rainfall regimes in desert areas; these are discussed in more detail later.

On the equatorward side of the subtropical highs there are generally easterly winds, and where these originate over landmasses they are extremely dry. This dryness is carried over the oceans to leeward of the landmasses. The trade winds pick up moisture by local evaporation from the ocean surface as they travel westward across the oceans and because of this, deep surface moist layers are most likely to be found in the western parts of the tropical oceans. As tropical disturbances depend on the release of latent heat for their maintenance, they tend to become more frequent towards the western ends of the subtropical oceans where deep moist layers exist. The western edges of the oceans receive large amounts of rainfall from disturbances and therefore form a break in the normal subtropical arid belt.

It is clear from the atmospheric moisture balance model (Figure 4.2) that where topographical factors are added to those caused by the general circulation, the aridity is greatly increased and in these areas the most severe desert conditions in the world can be found. It has already been noted that in the tropics the moisture available for rain formation is trapped in a shallow layer below an inversion. The depth of this moist layer varies, but it is usually between 1 and 2 km. If a mountain barrier projects through the moist layer, it will interrupt the surface flow and the surface moist layer will not penetrate behind the mountain range. Even if the mountains do not completely block the moist layer, the reduction in moisture advection to the lee of the range can still be substantial. For example, under tropical conditions an airstream with a surface moist layer 2 km deep, and which has its lowest kilometre blocked by mountains, will lose between 60 and 70 per cent of its precipitable water content. Dryness can be increased by subsidence of air from near the inversion down the lee slopes of the range, and for this reason mountain-enclosed inland basins are often extremely arid; Death Valley, USA is a good example.

2 Radiation and temperature

The energy balance of a desert differs from that of the surrounding vegetated areas in that the surface is dry and therefore there is little or no evaporation. Northern summer radiation budgets of the surface–atmosphere system have been investigated using Nimbus III data. The balance is positive almost everywhere to the north of 10°S with a maximum in the cloudless zones of the northern areas of the subtropical oceans. Negative radiation balances are observed over the desert regions of North Africa and the Middle East. There are several reasons for these negative balances during summer over the northern deserts. Because the surface is dry there is no evaporation and surface temperatures become high with an associated large infrared radiation loss. The lack of clouds, and of water vapour in the air, allows unusually large amounts of radiation to escape from the surface to space. Lastly, the albedo of desert tends to be high, probably between 30 and 40 per cent, as compared with vegetated surfaces. Because of this negative radiation imbalance, energy must be imported into the desert region to fill the deficit and maintain the temperature of the surface. This extra energy is supplied by the atmospheric circulation. The air that descends into the deserts originally rose near the equator, and supplies the extra heat from the equatorial regions. It is therefore necessary to discuss energy transport in the atmosphere.

Equation (1.4) for the total energy content of a unit mass of air can be used to study the energy balance of the tropical atmosphere. Outside southern Asia, the mean north–south circulation of the tropical atmosphere can be considered as taking the form of two simple cells, with rising air near the equator and sinking air over the subtropical deserts. The low-level circulation of these cells forms the northeast and southeast trade winds. Sinking air in the subtropics increases its temperature according to the dry adiabatic lapse rate, thus resulting in clear skies and low relative humidities. Subtropical deserts are largely a result of atmospheric subsidence leading to cloudless and rainless conditions. Large areas of the subtropics consist of ocean, and the clear skies result in a plentiful supply of solar radiation reaching the surface where it is mostly used to evaporate sea water. Water vapour evaporated over the subtropical oceans is mixed through the lower layers of the atmosphere by turbulence and convection and carried towards the equator by the trade winds. Near the equator, the trade winds enter the equatorial trough and here ascent takes place in localized weather systems and in particular in thunderstorms. In the thunderstorms, the latent heat released by the condensing water vapour is converted into sensible heat which in turn is transformed into potential energy by

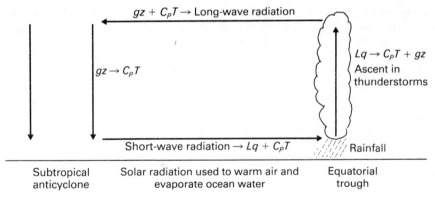

Figure 4.3 Energy conversions in the Hadley cell.

the rising air mass (Figure 4.3). In this way, the total potential energy (sensible heat + potential energy) of the rising air in the thundercloud is increased by the release of latent heat and is then exported at high levels in the atmosphere into the subtropics and also into middle latitudes. In this manner the radiation deficit over the deserts is made up.

In the desert region there is a continual conversion of net radiation into sensible heat ($C_p T$), this contrasts with a vegetated region where there is a conversion of net radiation into both sensible heat and latent heat (Lq). Thus the air may be cooler in a vegetated area, but its total energy content ($C_p T + Lq + gz$) may be similar to or greater than that in a desert area. Now the important factor controlling the amount of absorbed short-wave radiation and therefore the net energy available for sensible heat transfer and evaporation is the albedo. The total energy ($C_p T + Lq$) imparted to the atmosphere is increased in the lower layers when the albedo is decreased. The albedo of deserts is considerably greater than the albedo of vegetated surfaces, so the actual net energy imported to the atmosphere over deserts is less than that imported over vegetated areas. Charney *et al*. (1976) have used this fact to suggest that convective rainfall will be decreased by a vegetated area being turned into a desert by drought or overgrazing. The important quantity determining convective precipitation is the negative gradient of total energy content with altitude and this negative gradient is increased when the albedo is decreased. Over low albedo, vegetated surfaces, evaporated water vapour is soon converted into sensible heat by rainfall, thus warming the atmosphere. Therefore at the desert boundaries the moist air over the vegetated surfaces will be warmer aloft than drier air over the high-albedo desert surfaces. The desert air will therefore tend to sink at high levels relative to the air over the vegetated surfaces even though temperatures are very high at the desert surface. According to Charney *et al*. (1976) an extension towards the equator of desert conditions will cause an extension of the associated sinking air which will in turn tend to intensify the change towards desert conditions.

The highest values of the diurnal temperature range are experienced in deserts, where the soil is generally dry and the atmosphere contains little water vapour. The hotter parts of the Sahara experience mean maxima of 45°C in the hottest month, and values nearly as high are probable in the Great Sandy Desert in western Australia, while Death Valley, California has a mean maximum temperature of

47°C in July. As explained earlier, these very high temperatures occur because nearly all the net radiation is available for heating the air and soil. The low water vapour content in the air allows relatively large long-wave radiation losses from the surface, leading to low temperatures at night. Deacon (1969) has illustrated this for the Sahara Desert, where in April the average cloud amount over considerable areas is less than one okta and the total precipitable water content only about 1.2 cm, while the global average for 20°N is nearly 4 cm. As a result the mean daily range of temperature reaches 20°C over wide regions and as much as 22.5°C over the sandier areas. In July, a month with an equally small cloud amount, greater amounts of dust and moisture in the atmosphere cause the diurnal range to be nearly 5°C smaller than in April.

Priestley (1966) has considered the limitation of temperature by evaporation in hot climates, and concluded that with extremely few exceptions 33°C is the highest mean maximum temperature (from monthly statistics) attained by the air over any extensive freely evaporating surface. A study by Deacon (1969) of temperature and rainfall data at Alice Springs, which is in the desert near the centre of the Australian landmass, provides a good illustration of this effect. At Alice Springs the rainfall occurs mainly as brief heavy showers, so its main effect in reducing temperature will be via evaporation, since the influence of cloud amount on radiation can be neglected. Deacon showed that the January mean maximum temperature at Alice Springs is not only influenced by the January rainfall but also by that of the previous month. Working with rainfalls on a logarithmic scale, he obtained partial correlation coefficients from 50 years of data as follows:

January mean maximum temperature and January rainfall; December rainfall constant 0.66.

January mean maximum temperature and December rainfall; January rainfall constant 0.40.

Net radiation values in the poleward regions of the subtropical deserts undergo large seasonal variations, and lead to large seasonal variations in temperature. The high summer temperatures at Aswan have been mentioned, but in January and February the night temperature can fall as low as 5°C. Indeed, large areas of the Sahara have recorded absolute minimum temperatures of 0°C or below, indicating that frost is not unknown. At Alice Springs night temperatures regularly fall below 0°C in June, July and August, and the absolute minimum temperature is about -7°C.

3 Rainfall

The nature of desert rainfall, mentioned in the introductory section, can be explored further by a study of North Africa. The Mediterranean and North Africa lie on the southern flank of the middle latitude westerlies and are strongly influenced by them. Kirk (1964) states that, particularly during the colder months, the weather is essentially determined by the large-scale fluctuations of the main westerly flow pattern, and that the long waves and their modes of behaviour are of fundamental importance. There are two main jet streams in the westerlies of the northern hemisphere: the northern jet stream is related to the polar front, while the southern subtropical westerly jet is not associated with surface fronts on most occasions. The northern jet stream undergoes irregular quasi-cyclic variations connected with index cycles, and the associated weather patterns are discussed in Chapters 1 and 2. With a high zonal index, there is a strong zone of flow in middle and high latitudes and bad weather tends to be confined to this fairly narrow band; but at low zonal index, wave amplitude in the westerlies is large and blocking activity is at a maximum, extending bad weather into the subtropics. The middle latitude and subtropical jet streams combine or approach each other in periods of low zonal index, thus complicating the upper wind and vorticity fields.

In the Mediterranean, Middle East and adjacent areas, cyclones appear that are completely different from the warm-sector depressions formed along the polar front. These Mediterranean cyclones are associated with domes of cold air which have been cut off from the main cold air to the north and are surrounded by warm air. Cutoff cold pools, as these domes are known, form mostly at times of low index flow in middle latitudes and are most frequent in winter. Once a cutoff cold pool is formed in the Mediterranean, a corresponding surface cyclone soon appears, and convergence into the low leads to shower activity. The cold pools have a life period of from 2–12 days and usually remain stationary or drift slowly eastward.

The summer rains in the transition zone along the southern margins of the Sahara may be of two kinds, the one possibly associated with local random convection, and the other with organized shower activity occurring in connection with extensive atmospheric disturbances. The nature of the atmospheric disturbances is not clear, but they appear to be related to perturbations in the equatorial westerlies.

It is characteristic of rainfall in North Africa that much of it comes in the form of showers. Although these showers are normally localized, on some occasions they are accompanied by such heavy precipitation that they cause destructive floods in otherwise arid areas. Sometimes the mean monthly or mean annual rainfall will be precipitated by one storm, and numerous examples of this are contained in the meteorological literature. No desert area is completely immune from destructive rainfall; a good example of this in Tunisia and Algeria has been described by Winstanley (1970). On 23 September 1969, a cyclonic storm developed over the central Mediterranean and during the next few days it slowly retrogressed westwards, moving across northwestern Libya and southern Tunisia, and finally dissipated over northeastern Algeria. It was accompanied by widespread and heavy rains, which caused such devastation in Tunisia and northeastern Algeria that nearly 600 people were killed and a further quarter of a million were made homeless. The rainfall during the 10 day period, 20–29 September, was several times greater than the mean for the whole month over large areas of central and southern Tunisia and northeastern Algeria. At Biskra in Algeria, the rainfall on 27 and 28 September alone was over twice the mean annual fall of 148 mm; the annual fall has only once exceeded the monthly total for September 1969, namely 329.8 mm in 1934. Winstanley comments that at many places in North Africa, September, rather than in the middle of the wet season, is the month with the greatest mean rainfall intensity. This is probably because in the early autumn the vapour content of the air over the Mediterranean is at a maximum and conditions very suitable for heavy rainfall given the right synoptic conditions. This particular cyclonic storm appears to have resulted from the amalgamation of mid-latitude and subtropical circulation systems.

Similar intense rainfalls are recorded in desert regions elsewhere in the world. In central Australia, for example, Alice Springs has an average annual rainfall of about 267 mm, the highest daily fall recorded being of the order of 145 mm, while the probable maximum daily rainfall in the Alice Springs region is about 450 mm (Wiesner, 1970). The intensity and irregularity of desert storms tend to make mean annual rainfall statistics for desert areas rather meaningless.

3.1 Rainfall variability

Apart from its bearing on rates of erosion, the extreme variability of rainfall is of importance in dry–land farming. Wallén (1967, 1968) has discovered that the minimum annual rainfall at which dry-land farming is possible in the Middle East varies from region to region, mainly because the variability of annual rainfall (P) is higher in some than in others. For simple investigations of the variability of precipitation, Wallén applied the following parameter:

$$IAV_{rel} = \frac{100 \, \Sigma_{r=2}^{r=n} (P_{r-1} - P_r)}{\overline{P}\,(n-1)} \qquad (4.3)$$

where IAV_{rel} is the interannual variability relative to the mean value. He calculated this parameter for a variety of stations in the Middle East and then plotted the calculated values against mean rainfall. A scatter of points was obtained, but it was possible to fit a curve indicating the general decrease of variability with the increase of annual precipitation. Wallén fitted further curves to his diagram separating stations where dry-land farming is possible from those where it is not possible. The intersection of these two curves and the primary curve gave the approximate values of the precipitation conditions at the limit for dry-land farming in the Middle East, namely 240 mm of annual rainfall with an interannual variability of 37 per cent.

Landsberg (1951) has published a diagram showing the probability of obtaining, at various places, certain amounts of rainfall in an individual year in relation to the mean annual rainfall. By using this diagram as a basis for his calculations and allowing for a crop failure because of insufficient rain in 2 years out of 10, Wallén found that the minimum amount of rainfall in an individual year that would permit dry-land farming was different in the various parts of the Middle East. He calculated that the amount was as low as 180 mm in the area of regular rainfall in Jordan, southwestern Syria, and the 'Fertile Crescent', and as high as 210 mm in the Zagros mountains and 230 mm in the agricultural region of northern Iran. Further, by making use of these minimum values, it is possible to estimate the corresponding average annual rainfall and thus draw a boundary (Figure 4.4) on the

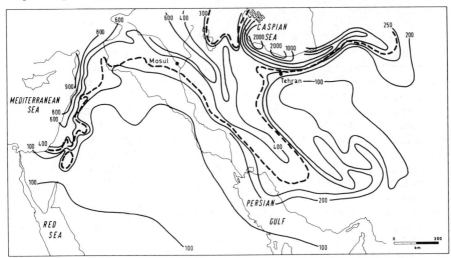

Figure 4.4 Annual rainfall and limits for regular dry-land farming in part of the Middle East. Dashed line indicates theoretical outer limit for dry-land farming; rainfall amounts are in mm (after Brichambaut and Wallén, 1963 and Wallén, 1967, 1968).

map of mean annual rainfall of those areas with no possibilities of dry-land farming.

In trying to define a wet season or a season of reliable rainfall it is necessary to consider not only the amount of rainfall but also its variability. In the Middle East, which has winter rainfall, Wallén considers that the start and end of the reliable rainfall season can be defined to correspond to a monthly rainfall of 25 mm and a variability of 100 per cent.

For the Middle East, Wallén discovered only small variations in the start of the wet season, which is usually in November or early December. The end of the wet season is, however, extremely variable. It comes as early as the beginning of April in the semi-arid parts of Jordan, while in the 'Fertile Crescent' the season lasts until the end of April or the beginning of May, and in Iraq and Iran it may last until the end of May or even the middle of June.

4 Evapotranspiration and water balance

The soil water balance equation is generally written in the form

$$P - O - U - E_a + \Delta W = 0 \tag{4.4}$$

where ΔW is the change in soil water storage (initial minus final) during the period, and $P, O, U,$ and E_a are the precipitation, run-off, deep drainage, and actual evapotranspiration respectively.

This equation can be more specifically written as

$$\int_{t_1}^{t_2} [(P - O) - E_a - V_z]\, dt = \int_{t_1}^{t_2} \int_0^z \frac{\partial \theta}{\partial t} \quad dz\, dt \tag{4.5}$$

where $(t_2 - t_1)$ is the time interval over which the measurements are made; z is the depth to the lowest point of measurement; V_z is the net downward flux of water at depth z (cm s^{-1}), and θ is the volumetric soil water content (cm^3 water/cm^3 soil).

The amount of water in the soil available for plant growth is generally considered to be that retained at soil water potentials lower than 15 atmospheres, and the difference in volumetric water contents at 0.3 atmospheres and 15 atmospheres is commonly referred to as the available water range. The significance of 15 atmospheres as a lower limit of a soil water availability depends on the type of plant under consideration, but it appears reasonable for most crops. Water extraction in arid plant communities does not cease when the available soil water storage is depleted but continues until a fairly consistent minimum value is reached, which represents an additional 'survival' storage capacity, estimated to be approximately 0.025 cm/cm depth. Slatyer (1968) has suggested for Alice Springs that the available water range of the soils is approximately 0.075 cm water/cm soil depth and that the rooting depth of the vegetation approximates 150 cm, giving a water storage capacity of approximately 112.5 mm. If the survival storage capacity is added to this amount, the total storage capacity in the range available for plant growth rises to 150 mm.

Following rainfall and soil water recharge in an arid region, plant growth recommences in the case of a dormant perennial vegetation, or commences in the case of an annual pasture or sown crop. If the soil water supply is adequate, leaf area increases progressively with growth rate and community evapotranspiration may reach values dependent primarily on atmospheric conditions. At this stage the rate of evapotranspiration is commonly referred to as potential evapotranspiration and under many conditions it can be predicted, with a fair degree of accuracy, from meteorological data alone. In communities with widely spaced plants, the limited extent of the transpiring leaf surface may prevent this stage from being reached. Progressive reductions in actual to below potential rates will occur as soil water storage is reduced.

The nature of water balance in a semi-arid area is well illustrated by work carried out at Alice Springs in central Australia. At Alice Springs the long-term average annual rainfall is 267 mm with 191 mm falling in the summer months (October–March inclusive). The rainfall is described by Winkworth (1970) as sporadic, with occasionally up to 90 per cent of the annual total being recorded in 2 months of the year. In the average year 16 groups of 1–3 wet days supply 171 mm of rain and 2 groups exceeding 3 days account for the remaining 96 mm, with long dry periods separating the wet periods. Over 80 per cent of the wet periods have rain-falls of less than 13 mm, and so it can be expected that an appreciable proportion of rain is directly evaporated, since the evaporation from the free water surface of a standard Australian tank varies from 90 to 300 mm per month.

Winkworth found that after significant rainfall, the soil profile at Alice Springs was wetted rapidly down to 30 or 40 cm, reaching field capacity at these depths in 3–4 days. Downward movement to deeper layers continued for a further 7–10 days, so that water contents at 41 cm and below were increasing or constant while the upper layers had begun to lose water. Depletion of stored water normally started 6–12 days after rain in the upper 30 cm of soil and one week later at greater depths. The duration of periods of available soil water ranged from 25–141 days. Moisture extraction usually continues until all available water has been removed from the whole of the wetted profile, and the soil then remains at low moisture content until the process is repeated at the next effective rainfall. As mentioned earlier, during the dry periods the soil water potential became greater than 15 atmospheres and at 10 cm reached 550 atmospheres at times, while at other depths it ranged from 70 to 170 atmospheres.

Fitzpatrick, Slatyer and Krishnan (1967) have estimated, for Alice Springs, the likelihood of commencement of growth or non-growth periods of various dura-tion. At Alice Springs, growth periods of 30 days duration or longer can only be expected three times in 2 years, and of 90 days only once in 4 years. The probability of a year without plant growth is extremely small.

4.1 Southern Sahara

Cochemé and Franquin (1967) have studied the agroclimatology of the semi-arid region south of the Sahara in detail, and their work illustrates another approach to water-balance problems in such regions. The zone of study was bounded to the north by the limits of dry-land farming corresponding more or less to the isohyets of 400 to 500 mm of mean annual rainfall, while the approximate southern limits were where the length of the wet season exceeds that of the dry season. The belt of territory thus investigated in West Africa was 6,000 km long and about 1,000 km wide, bounded to the west by the Atlantic Ocean and to the east by the border between Sudan and Chad. This zone is situated between the tropical anticyclonic belt and the equatorial convergence. It is virtually free from maritime influence from the west, and shows very little geographical relief.

In this particular region, the climate tends to be latitudinally uniform, but to vary along the meridians. As a result of this climatic zonation, the native vegetation is also arranged zonally and consists of sparsely wooded steppe in the north chang-ing southwards to relatively dry woodlands and savannahs. Mean annual rainfall is

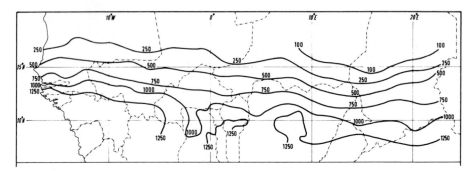

Figure 4.5   Semi-arid area in West Africa south of the Sahara. Mean annual rainfall (mm) (after Cochemé, 1968).

depicted in Figure 4.5 and, as suggested, the pattern is basically zonal with a slight dip southward from the Atlantic coast to Lake Chad, the pattern being reversed further east. The rains come in the summer, are heaviest on average in August, with their duration varying from about 6 months in the south to 2 in the north.

For the period 1953–62, monthly mean potential evapotranspiration was calculated assuming an albedo of 0.25 for 35 stations in or near the area by Cochemé and Franquin using Penman's formula. As noted before, the meteorological variables in Penman's formula are distributed zonally except for the wind, which shows a maximum in north Mali and on the coast of Senegal, and a minimum in south Chad. Therefore, mean annual potential evapotranspiration is also distributed zonally, and is higher in the north where it reaches 2,200 mm, than in the south where it is below 1,500 mm. The demand for water increases from south to north faster than the rainfall decreases, and thus the ratio rainfall/potential evapotranspiration varies from about 0.8 in the south to 0.2 in the north.

The annual march of the water balance at a typical station is shown in Figure 4.6. There are two maxima to the potential evapotranspiration curve, before and after the wet season (the former being higher), and two minima, in August and winter. During the dry season the soil moisture is almost completely exhausted and therefore the actual evapotranspiration will be considerably less than the estimated potential values. Cochemé and Franquin made some assumptions in their study about the relationship of the actual to the potential evapotranspiration. During the humid period (Figure 4.6), when rainfall exceeds potential evapotranspiration, it is assumed that actual and potential evapotranspiration are equal and that there is a water surplus; clearly this is the period of maximum vegetative growth. Because at the beginning of the wet season the soil will be dry and actual evapotranspiration will approach the potential as the soil moisture increases, two periods are defined: the preparatory period when the actual evapotranspiration increases from 0.1 to 0.5 of the potential, and the intermediate period when it varies from 0.5 to the full potential rate. A similar intermediate period occurs at the end of the wet season when the soil starts to dry and actual evapotranspiration falls to 0.5 of the potential. The humid period plus the two intermediate periods are defined as the moist period. At the end of the humid period it is assumed that up to 100 mm of water are stored in the ground and that this water is available for use in evapotranspiration. It is therefore possible to define a third period: the moist period plus reserve. The length of the various periods, for a selection of stations varying from the wettest to the driest, is shown in Figure 4.7. The two intermediate periods are of fairly con-

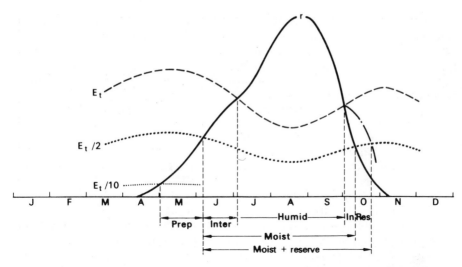

Figure 4.6 Semi-arid area in West Africa south of the Sahara. Graphical comparison of rainfall r, with E_t, $E_{t/2}$ and $E_{t/10}$ to define preparatory, intermediate, moist, moist with reserve, and humid periods (after Cochemé, 1968).

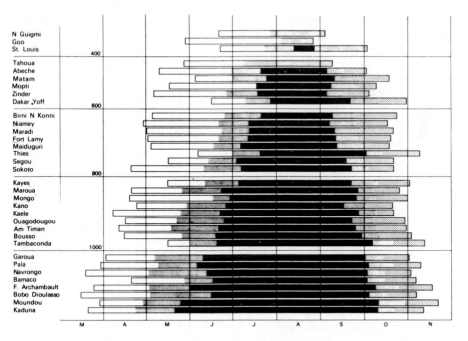

Figure 4.7 Length of moisture periods defined in Figure 4.6 at a selection of stations in West Africa. Stations arranged in order of increasing rainfall and in classes 200 mm apart. The preparatory period is blank, the humid period is darkest, and the intermediate shaded; the shading at extreme right represents reserve addition to moist period (after Cochemé, 1968).

stant duration, 25 days for the prehumid and 15 for the posthumid; but the humid period on the other hand increases from zero on the monthly scale in the north to 140 days in the south. To the moist period may be added up to 20 days for the soil moisture reserve, giving a range of about 55 to 200 days for the moist period, including reserves. Since temperature is always adequate, the length of the moist period with reserve seems to be the most important agroclimatological factor, because the main adaptation of semi-arid crops to their environment involves the matching of the growing cycle and its main biological events to the moist period with reserve and its main climatic events.

Five-year means of annual rainfall in the Sahel region of West Africa expressed as percentages of the 1905–70 average are shown in Figure 4.8. Series of dry years are recorded for 1900–3, 1909–16 (especially 1913–14), 1931, 1940–4, and then 1968–73. These do not correspond with the series of dry years on the subtropical, Mediterranean edge of the Sahara which are for the most part the reverse. They have not been of the same severity in the whole of the Sahel, except for the most recent drought. Rainfall for the period 1931–60 is taken almost universally as the standard for comparison for the Sahel region. This is unfortunate, since Figure 4.8 shows that it was a period of greater than 'average' rainfall in West Africa.

During the period 1968–72 rainfall along the desert fringe was only 40 to 60 per cent of the 1931–60 average, with the largest percentage decreases in monthly rainfall occurring at the beginning and end of the wet season. Winstanley (1975) gives a number of other interesting facts about the latest Sahelian drought. From 1967 to 1972–3 the discharge of the Niger and Senegal Rivers declined by 50 to 70 per cent respectively, and Lake Chad was reduced in area by 65 per cent. The level of the

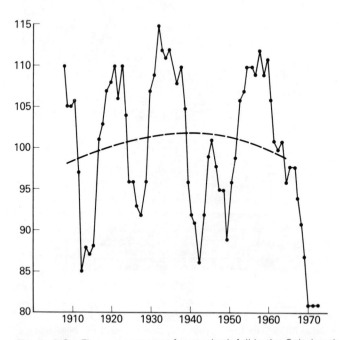

Figure 4.8   Five-year means of annual rainfall in the Sahel region of West Africa expressed as percentages of the 1905–70 average. The dashed curve shows the long-term trend (after Mason, 1976).

water table also fell and the annual floods on the inland deltas of the Niger and Senegal Rivers virtually disappeared. In 1974, rainfall became sufficient again and the pastures revived quickly.

4.2 North Africa and Middle East
Similar studies to those in section 4.1 have been undertaken for North Africa and the Middle East by Wallén (1967, 1968), the major difference from the previous study being that these are winter rainfall areas. Potential evapotranspiration was calculated using Penman's method and values were found to be surprisingly similar over large parts of the Middle East, the main exceptions being in low latitudes where the incoming radiation was higher than further north. Water-balance diagrams were constructed and theoretical dates calculated when there would be no more water available in the soil for active plant growth. The limits for dry-land farming established by the water-balance method agree well with those established from rainfall conditions only (Figure 4.4). Indeed, Wallén concludes that in large areas of similar winter evapotranspiration and soil conditions the degree of aridity is mainly determined by the amount of rainfall and its variability.

The values of the ratio rainfall/potential evapotranspiration for the winter season are considered by Wallén to give a reasonable idea of the degree of aridity throughout the region. For instance, Hassekieh, a station which agricultural experience shows is on the limit of dry-land farming, has a ratio of 0.80, while Kamishlieh, situated in one of the best agricultural districts of the whole area, has a ratio of 1.16. There are similarities between the basic sub-regions and other parts of the world. The climate of the semi-arid zones with dry-land farming in east Jordan and southwest Syria is like that of south and southwest Australia. Likewise, the climate of the semi-arid zones of the Syrian wheat belt and of the semi-arid valleys in southern Zagros of Iran is related to that of the San Joaquin Valley of California. There are several common features between the climate of lower Mesopotamia and that of the Imperial Valley in California. Also, the Azerbaidjan area of northern Iran corresponds with the Anatolian Plateau throughout the year and with the high plateau of Spain and North Africa in summer. Wallén suggests that these climatic similarities could be important when considering the exchange of crop varieties between areas.

5 African rainfall and teleconnections

Arid and semi-arid lands, where interannual rainfall variability is inherently extreme, dominate the African continent, and severe droughts are a common occurrence in nearly every region (Nicholson and Chervin, 1983). The meteorological causes of such droughts are still unclear, but numerous studies of the recent one in the Sahel suggest that major changes in large-scale tropical circulation patterns were involved (Krueger and Winston, 1975; Kanamitsu and Krishnamusti, 1978; Kidson, 1977). Some studies (Hastenrath, 1978; Fleer, 1981; Doberitz, 1969; Bjerknes, 1969) have shown that such large-scale changes in tropical circulation patterns result in synchronous rainfall anomalies throughout large areas of the tropics.

The relationship between rainfall fluctuations in the semi-arid subtropics of Africa has been investigated by Nicholson and Chervin (1983). The more important of their findings are as follows.
   (a) The markedly synchronous occurrence of droughts in the subtropics of both hemispheres.

(b) The extreme spatial coherence of rainfall anomalies.
(c) Synchronous fluctuations along both tropical and temperate margins of the semi-arid/arid zones.
(d) The persistence of anomalies in the sub-Saharan zones.
(e) The apparently 'normal' location of the intertropical convergence zone during most drought years.

Nicholson and Chervin consider that major rainfall fluctuations in Africa are associated primarily with factors modifying the intensity or frequency of disturbances, e.g. changes in the intensity of the Hadley circulation or factors modifying energetics and stability.

6   The Indian monsoon and global deserts

According to the continental reconstructions of Smith and Briden (1977), the Indian subcontinent was in collision with Asia sometime around the late Eocene. This collision led to extensive mountain building which may have wide-spread climate implications. Hahn and Manabe (1975) have considered the role of mountains in the south Asian monsoon circulation. They used numerical models to study the southwest monsoon and conducted experiments with smoothed mountain topography and with all the mountains removed. Comparison of the simulation with mountains and the simulation without mountains reveals that the presence of mountains is instrumental in maintaining the south Asia summer low-pressure system as the summer continental low forms far to the north and east in the simulation without mountain topography. The general conclusion is that mountain effects help to extend a monsoon climate further north into the Asian continent. The most significant difference in the precipitation patterns between the two models is found north of the intertropical convergence zone. In the non-mountain model, a desert-like climate is simulated in south Asia resulting from dry northwesterly continental air flowing southward toward the rainbelt. However, in the mountain-model, moist southwesterly flow at the surface moves northward toward the south Asian low-pressure belt, resulting in substantially more rainfall over south Asian continental regions. Large amounts of rainfall are found particularly along the slope of the Tibetan Plateau with most of India receiving more rainfall than in the non-mountain model in July.

In the surface flow field, the intertropical convergence zone can be located at approximately the latitudes 10–15°N for both models in July. In the non-mountain model, the intertropical covergence zone at the surface is clearly identified by a line of convergence where northwesterly flow from Asia meets the maritime southwesterly flow originating in the southern hemisphere. The intertropical convergence zone extends from the Indian ocean westward along 10°N across the Somali Republic and Ethiopia, suggesting that the areas should receive adequate summer rainfall. In the mountain-model the characteristics of the intertropical convergence zone are somewhat different. It is identified by a line of confluence (rather than convergence) imbedded in the maritime southwesterly flow. The rain area does not extend westward into Africa, where there are only isolated wet areas. Thus it appears that large areas of northeast Africa would have considerably more rainfall if the Tibetan Plateau was absent.

Frakes (1979) has plotted the global distributions of evaporites, lignites, reefs, bauxite, cherts and laterite profiles for the late Eocene. At this time he shows lignites in what is now India and evaporite deposits in what is now North Africa and the Middle East. This parallels closely the present-day distribution of rainfall and suggests strongly that the full Indian Monsoon circulation was established at this time. The date when the Indian monsoon became established is extremely uncertain. During the Paleocene the Indian subcontinent lay south of the equator and there was an extensive ocean between it and Asia. These conditions may not have been suitable for the development of a full-scale monsoon circulation, resulting in the Paleocene in a drier southern Asia and a wetter North Africa. Against this, Manabe and Wetherald (1980) found an increase in monsoonal precipitation in their warm high CO_2 atmosphere, and this may imply that even in the Paleocene, which was warm, the Indian monsoon was established.

There may be grounds for suspecting that in the early Tertiary extensive deserts were limited. According to Smith and Briden (1977), North Africa was further south than at present and this would have restricted the extent of desert conditions. Australia was also too far south to have had a desert climate. Leakey (1981) infers from the remains of ape-like creatures that thick forest carpeted much of the tropics and what is now Eurasia before the early Miocene. According to Zinderen Bakker (1978) it can be assumed that in early Oligocene, the north and west coastal regions of Africa were covered with a humid tropical vegetation which graded further inland into subtropical woodland and savanna. These favourable conditions deteriorated from Oligocene times onward mainly as a consequence of two factors, i.e. the continued drop in world temperature and the changes in the palaeogeography of the region. Similarly, Siesser (1978, 1980) considers that Namibia was mostly wooded grasslands until middle Miocene times, suggesting that the early spasmodic conditions of the Benguela current did not cause significant aridification. He suggests that the major cooling and upwelling of the Benguela current in early late Miocene times initiated the aridification of the Namibian Desert. Thus on the warm earth of the early Tertiary deserts may have been relatively reduced in extent, and become more widespread with palaeogeographical changes and cooling.

Davies *et al.* (1977) have described four distinct oceanic sedimentation episodes in the Cenozoic. The Paleocene–early Eocene has very low rates, with a very high proportion of carbonate sediment. The middle Eocene has higher rates, with a much larger non-carbonate component. The late Eocene–early Miocene is again characterized by very low rates with a high carbonate content. The middle Miocene–Recent has very high sedimentation rates and a relatively low proportion of carbonate. Davies *et al.* consider that these changes imply gross fluctuations of river input to the oceans on a global scale, and also imply that precipitation on most continents for much of the time must have been much less than at present. This is in contrast to the ideas being presented above about a densely forested earth before the Miocene. There is a considerable amount of evidence which has been summarized by Bosch and Hewlett (1982) that forest cover reduces run-off. They consider that pine and eucalypt forest cause on average about 40 mm change in annual water yield per 10 per cent change in forest cover. Deciduous hardwoods are associated with about 25 mm change in yield per 10 per cent change in cover, while 10 per cent changes in brush or grasslands seem to result in about 10 mm change in annual yield. Thus the low oceanic sedimentation rates before the Miocene could well be associated with continents more densely vegetated than at present and thus having low run-off rates. The increase in oceanic sedimentation rates in the Miocene could be associated with a decrease in continental vegetation cover and an associated increase in run-off.

The Würm/Wisconsin glaciation was marked by hyperaridity in the Sahara.

During this period, Saharan sands reached as far south as the Zaire basin and from the southern hemisphere Kalahari sands migrated northward to cross the Zaire River near Kinshasa. These old dunes are today largely buried beneath tropical vegetation and are partly degraded. Fairbridge (1976), considers that the dune sands of the Sahara evolved progressively during the Pliocene and Quaternary, expanding over greater and greater areas with each glacial desiccation, but interrupted to a large extent during each interglacial phase when there were widespread humid conditions that involved valley development and lake growth.

Following the glacial aridity there was a late Pleistocene phase of very heavy precipitation in central and subtropical Africa, which began around 13,000 BP and continued until about 8,000 BP. Since this pluvial period there has been a general trend towards increased aridity. Similar long-term trends are observed in the other semi-arid regions of the world. In particular it has been observed in the Kalahari region, eastern Brazil, western India, Iran, central Asia and Australia.

7 Closed lakes and the climatic history of North Africa

Closed lakes have long been recognized as good indicators of climatic change (Tetzlaff and Adams, 1983). Most approaches coupling lake level with climatic conditions make use of the water balance of the lake and its surface heat balance (Kutzbach, 1980, 1983). The water balance equation of a closed lake may be expressed as:

$$R/A + (P - E) - I = \Delta V \tag{4.6}$$

Here the total runoff R from the catchment area is regarded as a point source, whereas the precipitation P onto the lake and the evaporation E depend on the magnitude of the lake's surface area A. In the case of Lake Chad, Tetzlaff and Adams assumed that the loss by seepage I is proportional to the area of the lake. Hence, changes of the water volume ΔV enable conclusions to be drawn about climatic processes, provided that the transfer function is known. If no changes in water volume occur, then the equilibrium area A_E of the lake is given by

$$A_E = R / (E + I - P) \tag{4.7}$$

The surface energy balance of a lake is given by

$$\Delta S = Q + H + LE + B - \epsilon \sigma T_L^4 \tag{4.8}$$

where ΔS = net heat storage in the water body,
Q = absorbed incoming radiation,
H = flux of sensible heat,
LE = flux of latent heat,
L = heat of vaporization of water,
B = heat flux into the ground,
$\epsilon \sigma T_L^4$ = long-wave outgoing radiation from the lake surface at temperature $T_L K$, with
ϵ = emissivity of the water = 0.97,
σ = Stefan–Boltzmann constant.

Tetzlaff and Adams (1983) have used the above equations in a simple model to simulate water level in Lake Chad, in the Sahel zone of Africa. They successfully simulated lake level for the period of the Sahel drought. The calculations showed that the fall in the level of Lake Chad resulted mainly from a substantial decrease in run-off rather than increased evaporation. Further, a data set for the early Holocene (9,000 yr BP) indicated a 15 per cent decrease in annual evaporation compared with the present-day value of 2,200 mm yr^{-1}. A sensitivity analysis showed an increase of about 1,200 per cent (i.e. 6.4×10^{11} m^3), compared to the present (4.8×10^{10} m^3), in the total annual inputs from run-off and precipitation during the early Holocene.

Kutzbach (1980, 1983) has also used combined hydrological and energy-balance models of Paleolake Chad to estimate past precipitation. Two versions of the model were developed. The first version is one in which precipitation and lake area are linearly related. This version requires specification first of the area, net radiation and Bowen ratio of the lake, and second of the paleovegetation, net radiation and Bowen ratio of the surrounding basin. In the second version of the model the relationship between precipitation and lake area is non-linear because the run-off ratio and Bowen ratio of the basin are made functions of precipitation. As the lake increases in area in response to increased precipitation, this version of the model allows for further increases in run-off (from basin into the lake) as the vegetation changes from steppe to savanna and swamp.

Modern Lake Chad fills about 1 per cent of its basin and the vegetation in the basin is broadly characterized as desert and steppe in the north; savanna and moist woodland predominate in the south. The linear model of Kutzbach yields an estimate of annual precipitation of 254 mm yr^{-1} if desert–steppe values are used for the basin parameters. If a more realistic partitioning of the basin into lake fraction, savanna–swamp fraction and desert–steppe fraction is used, along with the corresponding estimates for net radiation and the Bowen ratio, the linear model yields a precipitation estimate of 378 mm yr^{-1}. The larger of the two estimates is closest to the modern value of basin-average precipitation of approximately 350 mm yr^{-1}. The paleoenvironmental evidence (Street and Grove, 1976) indicates that the entire basin did not have desert–steppe vegetation during the early Holocene. If the portion of the basin not covered by lake was half savanna–swamp and half desert–steppe, the precipitation is estimated to have been 658 mm yr^{-1} in the early Holocene. Thus Kutzbach suggests that the basin-averaged precipitation of Lake Chad was at least 650 mm yr^{-1} for certain times between 10,000 and 5,000 yr BP. The non-linear model yields precipitation estimates for Paleolake Chad ranging from 597 mm yr^{-1} to 648 mm yr^{-1}.

Kutzbach (1980) comments that the increased precipitation for Paleolake Chad, compared to modern Lake Chad, is accompanied by increased evaporation from the basin (480 mm yr^{-1} compared to 282 mm yr^{-1}). However, the basin evaporation does not increase as much as basin precipitation and therefore the runoff from basin to lake is augmented. Kutzbach considers that total run-off (run-off weighted by lake area) is almost a factor of 10 larger for Paleolake Chad than for modern Lake Chad. Because lake evaporation (per unit area) is assumed to be the same for Paleolake Chad and modern Lake Chad, the 14-fold increase in the area of Paleolake Chad implies that the total supply of water to the lake (run-off plus precipitation) was 14 times greater at that time. Grove and Pullan (1963) have calculated that the total water supply to Paleolake Chad was at least 16 times greater than at present.

Lake-level fluctuations are one of the most abundant and widely distributed sources of palaeohydrological and palaeoclimatic information for low-latitude continental areas over the late Quaternary (Street-Perrott and Roberts, 1983). Lake levels and associated data have been used to reconstruct the general circulation of the tropics in late Pleistocene and Holocene times (i.e. Nicholson and Flohn, 1980). Street and Grove (1979) and Street-Perrott and Roberts (1983) have published detailed accounts of African lake levels. According to Street-Perrott and Roberts, the tropical lake–level record since the last global ice-volume maximum, at 18,000 yr BP, can be divided into phases of relative stability separated by rapid transitions that were approximately synchronous over wide areas, although the direction of change varied with latitude and, sometimes, with elevation. Throughout Africa and western Eurasia, five major time divisions (called A–E) can be recognized. These are described by Street-Perrott and Roberts as follows.

Phase E (pre 17,000 yr BP)
At 18,000 yr BP (Figure 4.9), lakes stood at high levels around the southern and eastern margins of the Mediterranean and in the Arabian peninsula. In contrast, water levels in Africa were low or intermediate and falling, apart from a few equatorial and montane lakes. Street-Perrott and Roberts comment that the dearth of data points available for the 18,000 yr BP horizon reflects the intensity and widespread nature of intertropical aridity. Both Gates (1976a,b) and Manabe and Hahn (1977) have used global general circulation models to simulate the climate of this time when the northern ice-sheets were at their maximum extent. Both models estimate significantly lower precipitation rates for July and August over the tropical and subtropical continents than are observed today. Because tropical Africa and southern Asia receive a significant fraction of their precipitation during the summer monsoon this probably indicates drier annual conditions.

Phase D (17,000–12,500 yr BP)
The pattern of lake levels at this junction shows the most striking degree of hemispheric symmetry to be found within the whole period of the record. By 14,000–13,000 yr BP, low levels were recorded throughout almost the whole study area, except Sinai, Tibesti and southwestern Africa. Street-Perrott and Roberts consider that the interval 14,000–12,500 yr BP represents one of the most arid periods in the entire late Quaternary, drier even than at the glacial maximum.

Phase C (12,500–10,000 yr BP)
Phase C was a transitional period in the lake-level record.

Phase B1 (10,000–7,500 yr BP)
There was a major change in the pattern of lake-level status around 10,000 yr BP, almost all the intertropical lakes having reached high levels by 9,000 yr BP. Maps of lake-level trend indicate that water levels respond first near the equator and subsequently rose progressively northwards towards the central Sahara. At its maximum, the belt of lakes with high water-levels extended from 4°S to 33°N.

Phase B2 (7,500–5,000 yr BP)
The levels of the intertropical lakes which fell at the end of phase B1 rose again between 7,300 and 6,800 yr BP and then generally remained high until around 5,000 yr BP.

Phase A (5,000 yr BP to present)
The African belt of high lake-levels disintegrated rapidly after 5,000 yr BP, a modern pattern of lake levels being established by 2,500 yr BP. The present pattern

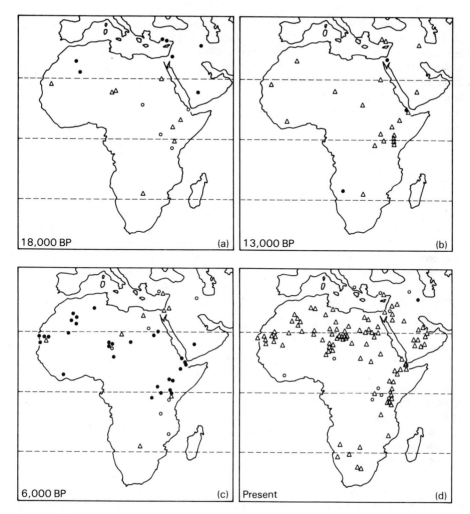

Figure 4.9 Lake-level status at (a) 18,000 yr BP; (b) 13,000 yr BP; (c) 6,000 yr BP; (d) at the present day: Dots, high; circles, intermediate; triangles, low (after Street-Perrott and Roberts, 1983).

(Figure 4.9) includes no tropical lakes with high water-levels and only a few of intermediate status, clustered in a narrow band between 2°S and 13°N which spans the present meteorological equator.

During the interval 12,500–5,000 yr BP, the otherwise orderly northward migration of the African belt of lakes with high levels was twice interrupted by episodes in which the pattern abruptly disintegrated (Street-Perrott and Roberts, 1983). A similar, and even more general, expansion of arid conditions occurred during the period 14,000–12,500 yr BP, when moisture convergence over the entire Afro-Arabian landmass must have been strongly suppressed. According to Flohn (1979), abrupt climatic events of this type, on a time-scale of 10^2–10^3 yr, are likely to result from (a) solar events, (b) clustering of explosive volcanic eruptions or (c) surges of major ice-sheets. To this list Street-Perrott and Roberts add (d) pulses of melt-water released into the oceans during glacial retreat (Berger *et al.*, 1977; Flohn and Nicholson, 1980; Ruddiman and McIntyre, 1981a). Of the four mechanisms cited above, Street-Perrott and Roberts consider that oceanic meltwater 'spikes' and episodic volcanism appear to provide the most plausible explanation for the observed lake-level minima. There is clear evidence of strong pulses of meltwater release into the northern North Atlantic during ice-sheet break-up (Ruddiman and McIntyre, 1981b). The most important of these occurred around 16,000–13,000 yr BP when the entire North Atlantic south to at least 50°N was flooded by meltwater and icebergs in summer and probably covered by sea-ice in winter. A weaker pulse of meltwater release at 11,000–10,000 yr BP has been attributed by Ruddiman and McIntyre (1981b), to the break-up and outflow of large ice shelves in the Arctic. A third peak of meltwater release might be expected to have accompanied the rapid disintegration of the Laurentide ice-sheet after 7,900 yr BP, when the sea invaded Hudson's Bay (Denton and Hughes, 1981). Street-Perrott and Roberts consider, therefore, that both the arid period culminating around 14,000–13,000 yr BP and the temporary collapse of the belt of high lake-levels between 10,800 and 10,200 yr BP coincided with episodes of strong cooling of the North Atlantic by icebergs and meltwater, while the abrupt recession centred on 7,400 yr BP may possibly have been linked to the evacuation of ice from Hudson's Bay.

7.1 African monsoons and variations in the earth's orbital parameters

Changes in the earth's orbital elements (obliquity, precession and eccentricity) during the late Pleistocene and early Holocene may have influenced monsoon cir-culations through their effect on the seasonal cycle of solar radiation (Kutzbach, 1981, 1983; Kutzbach and Otto-Bliesner, 1982; Rossignol-Strick, 1983). At 9,000 yr BP, obliquity was 24.23° (cf. the present-day value of 23.45°), perihelion was July 30 (January 3) and eccentricity was 0.0193 (0.0167); these factors combine to produce an increase in solar radiation in July and a decrease in January. The seasonal radiation changes were about 7 per cent of modern values over a broad band of latitudes.

Kutzbach (1981, 1983) and Kutzbach and Otto-Bliesner (1982) have used a low-resolution model of the general circulation of the global atmosphere to investigate the relationships between earth orbital changes and African monsoons. In contrast to the climate model experiments for glacial maximum conditions (Gates, 1976a,b; Manabe and Hahn, 1977), where surface boundary conditions are specified from studies of the geologic record, the boundary conditions for 9,000 yr BP are set at modern values. The use of modern boundary conditions provides a clear test of the sensitivity of the model climate to the solar radiation changes. Moreover, the modern ocean surface temperatures should be a fair approximation to the condi-tions at 9,000 yr BP.

Preliminary results from the Kutzbach model agree with many palaeoclimatic observations. Over the African–Eurasian land mass both the low-level cyclonic inflow of air and the high-troposphere anticyclonic outflow of air are stronger at 9,000 yr BP than at present. At the surface increased southwesterly winds carry moisture into West Africa and India. In the upper troposphere, the tropical easterly jet stream is more than 10 m s^{-1} stronger during July 9,000 yr BP than in modern July over the western Indian Ocean and equatorial Africa. Along 12°N over Africa, the precipitation is about 0.6 cm day^{-1} for June to August 9,000 yr BP, compared to about 0.55 cm day^{-1} now, an increase of 10 per cent. Between 23° and 35°N over Arabia, India and southeast Asia, the precipitation is about 0.75 cm day^{-1} for June to August 9,000 yr BP and about 0.5 cm day^{-1} now, an increase of 50 per cent.

Kutzbach (1983) considers that the pattern of increased precipitation simulated by his model and also of increased precipitation-minus-evaporation across North Africa and Asia at 9,000 yr BP agrees broadly with the maps of high lake-level reported by Street and Grove (1979) and Street-Perrott and Roberts (1983). Therefore the suggestion of these and other authors that the increased rainfall should be attributed to enhanced monsoonal circulations are broadly confirmed. Rossignol-Strick (1983) has shown that during the past 464,000 yr, African monsoons signalled by the East Mediterranean sapropels were heaviest always and only when a northern summer monsoon index, computed from the orbital variations of insolation, reached maximum values. This occurred during all the interglacials, but also twice during glacial periods. Thus tropical aridity is not the single climatic pattern of glacial periods.

Kutzbach (1983) sounds a final warning on this topic in commenting that while radiation changes associated with long-period orbital variations may contribute to an explanation of the changes in monsoon climates during the early Holocene, they could not have been the sole cause. Thus shorter-period orbital influences, different levels of volcanic or solar activity or internal processes must also have played a role.

8 Desertification

There are three basic requirements for significant land-surface evapotranspiration: moisture in the soil; vegetation, to transfer the moisture from the soil to the interface with the atmosphere; and energy, to convert that moisture to water vapour. During the temperate summer and in warm climates much of the energy for evaporation comes from radiational heating of the surface and therefore depends on surface albedo. In cool seasons and climates much of the energy for evaporation comes from sensible heat advection from the atmosphere. Now in nature the albedo depends on the vegetation, which in turn partly depends on the soil moisture. In fact both the albedo and the soil moisture can independently influence the atmosphere, yet albedo depends partly on soil moisture. The advantage of numerical calculations is that these two factors can be made independent of one another. Thus it is possible to either keep the soil moisture constant and let albedo change or let the soil moisture change but keep the albedo constant.

8.1 Vegetation, albedo and climate
Many recent modelling attempts have been concerned with the effect of changed albedo, and other consequences of the degradation of vegetation cover. The albedo feedback hypothesis was introduced by Otterman (1974, 1975) who reasoned that the destruction of vegetation and exposure of soil would increase albedo, and hence lower surface temperatures. This would in turn lower the sensible and latent heat fluxes to the atmosphere, and suppress convective shower formation. By examing NOAA satellite radiometer measurements of surface temperature over an area in the Middle East he was able to show that in summer day-time conditions the western Negev, with about 35 per cent vegetation cover, was warmer than nearby northern Sinai, with 10 per cent cover. Otterman's interpretation was challenged by Jackson and Idso (1975), who claimed that denuded soils in the Sonoran Desert of Mexico and the United States were always warmer than vegetated soils. They

concluded that the denuding of soil may have thermal and climatic effects just the opposite of those postulated by Otterman. Otterman (1977) also suggested that protected steppe areas have a low albedo due to dark plant debris acumulating on the crusted soil surface, whereas the same type of terrain, when overgrazed, exhibits the higher albedo of trampled, crumbled soil. Charney (1975a, 1975b) noted that the central and northern Sahara, eastern Saudi Arabia and southern Iraq actually have a negative radiation balance at the top of the atmosphere on hot summer days, in spite of the intense input of solar radiation through the cloudless atmosphere. This deficit arises because:

  (i) Outgoing terrestrial radiation is enhanced by high surface temperatures due to lack of evaporation (agreeing with Jackson and Idso rather than Otterman), clear skies and low humidities.

  (ii) The high surface albedo (about 35 per cent over sandy deserts).

Thus during the summer the North African deserts are heat sinks relative to the rest of the hemisphere. Charney's theory (1975a) postulates that overgrazing decreases the radiation balance making it less positive or more negative. In order to maintain thermal equilibrium and in the absence of advection the air must descend, compress and warm adiabatically thus decreasing the relative humidity and precipitation. Even though Charney applied his theory to deserts with negative radiation balance at the top of the atmosphere, the theory is also valid for deserts with a positive radiation balance. The main assumption is that adiabatic warming (descending motion), rather than advection (horizontal motion), is the main compensation for the decrease in the radiation balance. Charney, Stone and Quirk (1975) have used a limited sector radiative-dynamical model of North Africa, which suggested that an increase of albedo would enhance meridional overturning such that the rate of subsidence in tropical latitudes could be more than doubled in the summer season. They simulated the effect of a change of albedo from 15 to 30 per cent over North Africa and found sharp reductions of cloud and rain following the increase. A similar model has also been applied to the United States Dust Bowl area and to the Thar desert of Rajasthan, with very similar results (United Nations Conference on Desertification, 1977).

Sud and Fennessy (1982) have recently used a general circulation model to investigate surface albedo changes. In the first integration, called the control run, the surface albedo was normally prescribed, whereas in the second integration, called the anomaly run, the surface albedo was modified in four regions: the Sahel in Africa, the western Great Plains in the United States, the Thar Desert border in the Indian subcontinent, and northeast Brazil in South America. This experiment is similar to that of Charney *et al.* (1977); however, it was performed with the GLAS model with vastly different boundary forcings and several changes in the physical parameterizations. An analysis of two simulations for July shows that in the Sahel and the Thar Desert border regions the current results suggest reduced precipitation with increased surface albedo in accordance with Charney *et al.* (1977) and Charney (1975). The semi-arid northeast Brazil, which had a winter circulation, also conformed to Charney's (1975) hypothesis. There was a general lack of albedo impact on precipitation in the Great Plains. This experiment provides support for Charney's hypothesis (1975) regarding the influence of surface albedo on mean monthly climatology in the subtropical desert margin regions.

Ellsaesser (1976) used a two-dimensional model (longitudinally averaged) for an earth in which 30 per cent of the 15–25°N latitude belt had an albedo increase of 14 to 35 per cent. In the latitude of the Sahara their model predicted a 22 per cent decrease of rainfall. The same model applied to the case where the entire tropical rainforest was removed (hence raising equatorial albedo from 7 per cent to 25 per cent) predicted a small increase in precipitation in the 5 to 25° latitude belt (Potter *et al.*, 1975). Sagan *et al.* (1979) have considered the influence of anthropogenic albedo changes on the earth's climate. They consider that in the tropics man has caused climatic modifications by both desertification and the clearance of tropical forest. A change from savanna to desert causes an increase in albedo from about 16 per cent to about 35 per cent, and a similar change from tropical forest to fields or savanna causes a change in albedo from about 10 per cent to about 16 per cent. According to Sagan *et al.* (1979), over the past few millennia, desertification has increased global albedo by 4×10^{-3}, and the clearance of tropical forest by 1×10^{-3}. Over 25 years from the present the changes are 6×10^{-4} due to desertification and 3.5×10^{-4} due to tropical deforestation. They further suggest that if the global albedo changes from its value of 30 per cent by 1 per cent (1×10^{-2}), a surface temperature change of about 2°C will result. During the past several thousand years the earth's temperature could therefore have been depressed by about 1°C, due primarily to desertification, which might have significantly augmented natural processes in causing the present climate to be about 1 or 2°C cooler than the climatic optimum. Land use changes during the past 25 years have increased global albedo by about 1×10^{-3} (35 per cent of which is due to tropical forest clearance and 60 per cent due to desertification) and this would have depressed global temperature by about 0.2°C. Observations show that since 1940 the global mean temperature has declined by about 0.2°C, despite an accelerated increase in the carbon dioxide content of the atmosphere.

Using a statistical dynamical climate model, Potter *et al.* (1980) and Potter *et al.* (1981) have computed the combined impact of desertification of the Sahara and deforestation of the tropical rainforest. The model is similar in many ways to general circulation models in that it uses the basic conservation equations to calculate zonal mean fields of pressure, temperature, wind and humidity. The model has nine pressure layers and a 10° latitude grid. The surface area of each latitude zone is divided into land (of differing types and elevations) and ocean (partially covered by ice). For the experiment, two model equilibrium states were achieved: a control case with standard surface conditions corresponding to a pristine environment; a perturbed case corresponding to the suggestion of Sagan *et al.* (1979) concerning albedo changes on tropical deforestation and desertification. The surface bulk transfer coefficients, the thermal and hydrologic characteristics were not altered; the only surface parameter modified for those areas perturbed was the surface albedo which, after modification, was held constant. Figure 4.10 shows the temperature difference (perturbed minus control) between the two cases. While the model computed a surface cooling of 0.6°C for the northern hemisphere, the global mean of about 0.2°C was substantially less than the 1°C suggested by Sagan *et al.* (1979). There seem to be two important factors in the chain of events leading from surface modification to cooling in the northern hemisphere. The first is the direct reduction in surface absorption in the zones of increased surface albedo. The change at 20°N was sufficiently strong to strengthen the temperature gradient between the Sahara and the equator and this caused the northern Hadley cell to strengthen and shift southwards. The second factor cooling the northern hemisphere was the ice-albedo feedback due to increased sea-ice coverage, primarily at 70°N. In the perturbed case, the southern Hadley cell shifted southwards with the ascending branch weakening slightly and the southern descending branch weakening more. The weakened descending branch permitted increased convective

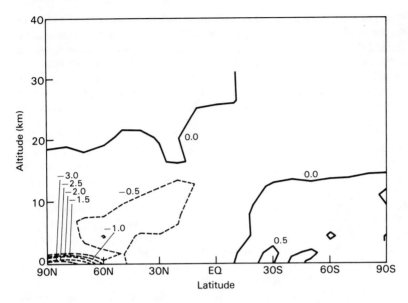

Figure 4.10 Latitude–height temperature difference (perturbed minus control). Contour interval is 0.5 K (after Potter *et al.*, 1980, 1981). Reprinted by permission from *Nature*. Copyright 1981 Macmillan Journals Limited.

activity, leading to clouds with more vertical structure and less horizontal extent. The total cloud coverage thus decreased at 30°S, allowing increased insolation. However, other than the additional solar flux at 30°S, there seems little reason for the slight warming of the southern hemisphere.

8.2 Vegetation, soil moisture and climate
Albedo depends on the vegetation cover and therefore on the wetness of the ground. Moreover Idso *et al.* (1975) have shown that the albedo of bare soil is linearly related to the water content of the uppermost layers of the soil with values ranging from about 14 per cent if the soil is saturated to 30 per cent if it is dry. The mechanism described by Charney (1975a and b) may therefore be thought of as dependent on the soil moisture. However, soil moisture content may, additionally, affect atmospheric conditions directly by influencing the evaporation and hence the proportion of the net radiation which is available as latent heat.

There have been several numerical experiments on the influence of soil moisture on the atmosphere. Manabe (1975) ran a mathematical model of the atmosphere for June–August with all the land wet. He particularly examined the results over Africa along 15–30°E. Precipitation increases over the Sahara were about 2–3 mm day^{-1}, only slightly less than the evaporation increases, while south of this, in 10–15°N, rainfall was increased by 5–6 mm day^{-1}, much more than the evaporation. Just south of the equator over Africa rainfall decreased, apparently because of changes in the meridional circulation. Walker and Rowntree (1977) applied a limited-area eleven-level atmospheric model to West Africa to test the effect of variable soil moisture. As might be expected, the model predicts large variations of surface temperature between dry surface conditions and those for wet. Five to ten days after the beginning of the experiment the dry land surface temperature in

Sahelian latitudes was predicted to be over 18°C more than the wet surface. This tends to confirm the Jackson–Idso position rather than that of Otterman as regards vegetation and surface temperature. Walker and Rowntree also predicted large differences in the behaviour of atmospheric disturbances. With a zonally symmetric north–south variation of surface conditions similar to that found near longitude 0°, the model atmosphere resembled that of West Africa with depressions developing and moving west in the easterlies generated by the thermal contrast between the dry land north of and moist land south of 14°N. Though rainfall was associated with these lows it was mostly confined to the areas with a moist underlying surface so that the initial aridity north of 14°N was maintained. The atmospheric evolution with a moist land surface was much more like that over a tropical ocean, the easterly waves being maintained by the latent heat release associated with the disturbances. The disturbances and rainfall were no longer restricted to a narrow latitudinal belt, so maintaining the initially moist surface. Walker and Rowntree suggest that in the tropical atmosphere, soil moisture can play a major role both in the short-range forecasting period of 1–2 days and over longer periods of 2–3 weeks or more.

Walker and Rowntree have made a good case for saying that Sahelian rainfall feeds on Sahelian rainfall, not so much because of local evaporation, as of the behaviour of baroclinic disturbances over the moist surface. Unfortunately the time period of their experiment is small (20 days) so they may have just observed a transient response to a large shock such as a widespread and rapid flood. Their 20-day integration does not conclusively demonstrate that a statistical steady state is achieved. Shukla and Mintz (1982) have used a global atmospheric general circulation model to investigate this problem further. They placed two different constraints on the land-surface evapotranspiration: in one case the evapotranspiration is always set equal to the potential evapotranspiration calculated by the model (this is the evapotranspiration when the soil is moist and completely covered by vegetation); in the other case no evapotranspiration is allowed to take place. Starting from an observed atmospheric state on 15 June, the calculations for the dry-soil and wet-soil cases were carried forward for 60 days. The results discussed here are the time-averaged fields for July, the month when the northern hemisphere extratropics has the maximum potential evapotranspiration.

In the wet-soil case, the precipitation over Europe and over most of Asia is about 4 mm day^{-1} and does not differ much from the calculated potential evapotranspiration. In contrast in the dry-soil case, Europe and most of Asia have almost no precipitation. Only over southeast Asia and India, in the dry-soil case, is there transport of water vapour from the ocean which produces heavy rain and in this case the precipitation most closely resembles the observed summer rainfall. Over most of North America the precipitation in the wet-soil case is between 3 and 6 mm day^{-1}, but in the dry-soil case the precipitation over most of the continent is reduced to about 1 mm day^{-1} or less. Over South America, the rainfall near the equator in the wet-soil case is about 6 mm day^{-1}, and in the dry-soil case the rainfall is almost as large, all of it being water transported from the ocean. Across Africa, at about 10°N, the precipitation in the wet-soil case is about 4 mm day^{-1} larger than the local evapotranspiration, but north and south of the rain band the precipitation is about 2 to 3 mm day^{-1} smaller than the evapotranspiration. In the dry-soil case, there is a band of rain of 3 to 4 mm day^{-1} at about 14°N, and this precipitation is about the same as the amount by which the precipitation exceeded the evapotranspiration in the wet-soil case. North of about 20°S the land surface in the dry-soil case is about 15° to 25°C warmer than in the wet-soil case. According to Shukla and Mintz there are two reasons for this: (i) there is no evaporative cooling of the land surface and (ii) there is a large increase in the heating of the ground surface by solar radiation. This is because the calculated cloudiness is less when there is no land-surface evapotranspiration.

Shukla and Mintz state that if their calculations are indeed applicable to nature, the implication for forecasting extratropical summer rainfall is clear. In about the month of May the continental rainfall changes from large-scale upglide condensation type to the cumulus convection type. If, after this change takes place, there is a large amount of moisture stored in the soil, the summer months that follow can have a large or small amount of rainfall, depending on the circulation conditions. If the soil is dry, so that there is little or no evapotranspiration to keep the atmosphere planetary boundary layer moist, the remaining summer months will have little rainfall. Shukla and Mintz consider that surface evapotranspiration, which requires moisture in the soil, is a necessary (though not sufficient) condition for extratropical summer precipitation.

References

ALI, F. M. 1953: Prediction of wet periods in Egypt four to six days in advance. *Journal of Meteorology* 10, 478–85.

ÅNGSTRÖM, A. 1936: A coefficient of humidity of general applicability. *Statens Met-Hydrgr. Anstalt Meda.* 11, Stockholm.

BERGER, W. H., JOHNSON, R. F. and KILLINGLEY, J. S. 1977: 'Unmixing' of the deep-sea record and the deglacial meltwater spike. *Nature* 269, 661–3.

BERKOFSKY, L. 1976: The effect of variable surface albedo on the atmospheric circulation in desert regions. *Journal of Applied Meteorology* 15, 1139–44.

BJERKNES, J. 1969: Atmospheric teleconnections from the equatorial Pacific. *Monthly Weather Review* 97, 163–72.

BOSCH, J. M. and HEWLETT, J. D. 1982: A review of catchment experiments to determine the effect of vegetation changes on water yield and evapotranspiration. *Journal of Hydrology* 55, 3–23.

BRICHAMBAUT, G. P. DE and WALLÉN, C. C. 1963: A study of agroclimatology in semi-arid and arid zones of the Near East. *Technical Note 56.* Geneva: World Meteorological Organization.

CHARNEY, J. 1975a: Dynamics of deserts and drought in the Sahel. In *The Physical Basis of Climate and Climate Modelling.* GARP Publication Series 16. Geneva: World Meteorological Organization 171–6.

—— 1975b: Dynamics of deserts and drought in the Sahel. *Quarterly Journal of the Royal Meteorological Society* 101, 193–202.

CHARNEY, J. G., STONE, P. H. and QUIRK, W. J. 1975: Drought in the Sahara: a biogeographical feedback mechanism. *Science* 187, 434–5.

—— 1976: Drought in the Sahara: insufficient biogeographical feedback? *Science* 191, 100–2.

CHARNEY, J. G., QUIRK, W. J., CHOW, S. H. and KORNFIELD, J. 1977: A comparative study of the effects of albedo change on drought in semi-arid regions. *Journal of the Atmospheric Sciences* 34, 1366–85.

COCHEMÉ, J. 1967: Agroclimatology of the Sudano–Sahelian zone. *WMO Bulletin* 16, 201–9.

—— 1968: Agroclimatology survey of a semi-arid area in West Africa south of the Sahara. In *Agroclimatological methods. Proceedings of the Reading Symposium.* Paris: UNESCO 235–48.

COCHEMÉ, J. and FRANQUIN, P. 1967: An agroclimatological survey of a semi-arid area in Africa south of the Sahara. *Technical Note 86.* Geneva: World Meteorological Organization.

DAVIES, T. A., HAY, W. W., SOUTHAM, J. R. and WORSLEY, T. R. 1977: Estimates of Cenozoic oceanic sedimentation rates. *Science* 197, 53–5.

DEACON, E. L. 1969: Physical processes near the surface of the earth. In Flohn, H. (ed.) *General climatology 2.* Amsterdam: Elsevier, 39–102.

DENTON, G. H. and HUGHES, T. J. 1981: *The Last Great Ice-Sheets.* New York: Wiley – Interscience.

DOBERITZ, R. 1969: Cross spectrum and filter analysis of monthly rainfall and wind data in the tropical Atlantic region. *Bonner Meteorologische Abhandlungen* 11, Bonn.

ELLSAESSER, H. W., MACCRACKEN, M. C., POTTER, G. L. and LUTHER, F. M. 1976: An additional model test of positive feedback from high desert albedo. *Quarterly Journal of the Royal Meteorological Society* 102, 655–66.

EMBERGER, C. 1955: *Une classification biogeographique des climats. Recueil des travaux.* Faculté Sciences de l'Université de Montpellier. Fasicule 7.

FAIRBRIDGE, R. W. 1976: Effects of Holocene climatic change on some tropical geomorphic processes. *Quaternary Research* 6, 521–56.

FITZPATRICK, E. A., SLATYER, R. O. and KRISHNAN, A. I. 1967: Incidence and duration of periods of plant growth in central Australia as estimated from climatic data. *Agricultural Meteorology* 4, 389–404.

FLEER, H. 1981: Large-scale tropical rainfall anomalies. *Bonner Meteorologische Abhandlungen* 26, Bonn.

FLOHN, H. 1964: Investigations on the tropical easterly jet. *Bonner Meteorologische Abhandlungen* 4, Bonn.

—— 1979: On time-scales and causes and abrupt palaeoclimatic events. *Quaternary Research* 12, 135–49.

FLOHN, H. and NICHOLSON, S. 1980: Climatic fluctuations in the arid belt of the 'Old World' since the last glacial maximum; possible causes and future implications. *Palaeoecology of Africa* 12, 3–21.

FRAKES, C. A. 1979: *Climates throughout Geologic Time.* Amsterdam: Elsevier.

GATES, W. C. 1976a: Modelling the ice-age climate. *Science 191, 1138–44.*

—— 1976b: The numerical simulation of ice-age climate with a global general circulation model. *Journal of the Atmospheric Sciences* 33, 1844–73.

GAUSSEN, H. 1963: Bioclimatic map of the Mediterranean zone. UNESCO *Arid Zone Research* 21.

GROVE, A. T. and PULLAN, R. A. 1963: Some aspects of the Pleistocene paleogeography of the Chad Basin. In Howell, F. C. and Bourliere, F. (eds.) *African Ecology and Human Evolution* No. 36. Chicago: Adeline. 230–45.

HAHN AND MANABE, 1975: The role of mountains in the south Asian monsoon circulation. *Journal of the Atmospheric Sciences 32, 1515–41.*

HASTENRATH, S. 1978: On modes of tropical circulation and climate anomalies. *Journal of the Atmospheric Sciences* 35, 2222–31.

IDSO, S. B. and DEARDORFF, J. W. 1978: Comments on the effect of variable surface albedo on the atmospheric circulation in desert regions. *Journal of Applied Meteorology* 17, 560.

IDSO, S. B., JACKSON, R. D., REGINATO, R. J., KIMBALL, B. A. and NAKAYAMA, F. S. 1975: The dependence of bare-soil albedo on soil water content. *Journal of Applied Meteorology* 14, 109–13.

JACKSON, R. D. and IDSO, S. B. 1975: Surface albedo and desertification. *Science* 189, 1012–13.

KANAMITSU, M. and KRISHNAMUSTI, T. N. 1978: Northern summer tropical circulations during drought and normal rainfall months. *Monthly Weather Review* 106, 331–47.

KIDSON, J. W. 1977: African rainfall and its relation to the upper air circulation. *Quarterly Journal of the Royal Meteorological Society* 103, 441–56.

KIRK, T. H. 1964: Discontinuities with reference to Mediterranean and North African meteorology. *WMO Technical Note 64*, Geneva, 35- .

KRUEGER, A. and WINSTON, J. 1975: Large-scale circulation anomalies over the tropics during 1971–72. *Monthly Weather Review* 103, 465–73.

KOPPEN, W. 1931: *Die Klimate der Erde, Berlin*: Walter de Gruyter.

KUTZBACH, J. E. 1980: Estimates of past climate at Paleolake Chad, North Africa based on a hydrological and energy-balance model. *Quaternary Research* 14, 210–23.

—— 1981: Monsoon climate of the early Holocene: climate experiment the earth's orbital parameters for 9,000 years ago. *Science* 219, 59–61.

—— 1983: Monsoon rains of the late Pleistocene and early Holocene: patterns, intensity and possible causes of changes. In Street-Perrott, A., Beran, M. and Ratcliffe, R. (eds.) *Variations in the Global Water Budget*. Dordrecht: Reidel, 371–89.

KUTZBACH, J. E. and OTTO-BLIESNER, B. C. 1982: The sensitivity of the African–Asian monsoonal climate to orbital parameter changes for 9,000 yr BP in a low-resolution general circulation model. *Journal of the Atmospheric Sciences* 39, 1177–88.

LANDSBERG, H. 1951: A study of the Hawaiian rainfall. *Meteorological Monographs 3*.

LEAKEY, E. C. 1981: *The Making of Mankind*. London: Michael Joseph.

MANABE, S. 1975: A study of the interaction between the hydrological cycle and climate using a mathematical model of the atmosphere. Summary of presentation at Meeting on Weather–Food Interactions, Massachusetts Institute of Technology, May 9–11, 1975.

MANABE, S. and HAHN, D. G. 1977: Simulation of the tropical climate of an ice-age. *Journal of Geophysical Research* 82, 3889–911.

MANABE, S. and WETHERALD, R. T. 1980: On the distribution of climatic change resulting from an increase in CO_2 content of the atmosphere. *Journal of the Atmospheric Sciences* 37, 99–118.

MARTONNE, E. DE 1926: L'indice d'aridité. *Bulletin Association de Geographes Français* 9.

MASON, B. J. 1976: Towards the understanding and prediction of climatic variations. *Quarterly Journal of the Royal Meteorological Society* 102, 473–98.

NICHOLSON, S. E. 1980: The nature of rainfall fluctuation in subtropical West Africa. *Monthly Weather Review* 108, 473–87.

—— 1981: Rainfall and atmospheric circulation during drought periods and wetter years in West Africa. *Monthly Weather Review* 109, 2191–208.

NICHOLSON, S. E. and CHERVIN, R. M. 1983: Recent rainfall fluctuations in Africa – interhemispheric teleconnections. In Street-Perrott, A., Beren, M. and Ratcliffe, R. (eds.) *Variations in the Global Water Budget*. Dordrecht: Reidel, 221–38.

NICHOLSON, S. E. and FLOHN, H. 1980: African environmental and climatic changes and the general atmospheric circulation in late Pleistocene and Holocene. *Climatic Change* 2, 313–48.

NORTON, C. C., MOSHER, F. R. and HINTON, B. 1979: An investigation of surface albedo variation during recent Sahel drought. *Journal of Applied Meteorology* 18, 1252–62.

O'KEEFE, P. and WISNER, B. 1975: African drought – the state of the game. In Richards, P. (ed.), *African Environment: Problems and Perspectives*. London: Internation African Institute, 31–9.

OTTERMAN, J. 1974: Baring high-albedo soils by overgrazing: a hypothesized desertification mechanism. *Science* 186, 531–3.

—— 1975: Reply to Jackson, R. D. and Idso, S. B. Surface albedo and desertification. *Science* 189, 1013–15.

—— 1977: Anthropogenic impact on the albedo of the earth. *Climatic Change* 1, 137–57.

PEREIRA, H. C. 1972: The influence of man on the hydrological cycle. In IASH, *World Water Balance* 3, 553–69. Gentbrugge.

POTTER, G. L., ELLSAESSER, H. W., MACCRACKEN, M. C. and LUTHER, F. M. 1975: Possible climatic impact of tropical deforestation. *Nature* 258, 697–8.

POTTER, G. L., ELLSAESSER, H. W., MACCRACKEN, M. C., ELLIS, J. S. and LUTHER, F. M. 1980: Climatic change due to anthropogenic surface albedo modification. In Bach, W., Pankrath, J. and Williams, J. (eds.) *Interactions of Energy and Climate*. Dordrecht: Reidel, 317–26.

POTTER, G. L., ELLSAESSER, H. W., MACCRACKEN, M. C. and ELLIS, J. S. 1981: Albedo change by man: test of climatic effects. *Nature* 291, 47–9.

PRIESTLEY, C. H. B. 1966: The limitation of temperature by evaporation in hot climates. *Agricultural Meteorology* 3, 241–6.

RASCHKE, K., VONDER HAAR, T. H., BANDEEN, W. R. and PASTERNAK, M. 1973: The annual radiation balance of the earth–atmosphere system during 1969–70 from Nimbus 3 measurements. *Journal of Atmospheric Sciences* 30, 341–64.

RIEHL, H. 1965: *Introduction to the Atmosphere*. New York: McGraw-Hill.

ROSSIGNOL-STRICK, M. 1983: African monsoons, an immediate climate response to orbital insolation. *Nature* 304, 46–9.

RUDDIMAN, W. F. and MCINTYRE, A. 1981a: Oceanic mechanisms for amplification of the 23,000-year ice-volume cycle. *Science* 212, 617–27.

RUDDIMAN, W. F. and MCINTYRE, A. 1981b: The North Atlantic during the last deglaciation. *Palaeogeography, Palaeoclimatology, Palaeoecology*, 35, 145–214.

SAGAN, C., TOON, O. B. and POLLACK, J. B. 1979: Anthropogenic albedo changes and the earth's climate. *Science* 206, 1363–8.

SARNTHEIN, M. 1978: Sand deserts during glacial maximum and climatic optimum. *Nature* 272, 43–6.

SHEETS, H. and MORRIS, R. 1974: *Disaster in the Desert*. Washington: Carnegie Endowment for International Peace.

SHUKLA, J. and MINTZ, Y. 1982: Influence of land-surface evapotranspiration on the earth's climate. *Science* 215, 1498–500.

SIESSER, W. G. 1978: Aridification of the Namib desert: evidence from oceanic cores. In Zinderen Bakker, E. M. Van *Antarctic glacial history and world paleoenvironments*. Rotterdam: Balkema, 105–13.

—— 1980: Late Miocene origin of the Benguela upwelling system off northern Zambia. *Nature* 208, 293–85.

SLATYER, R. O. 1968: The use of soil water balance relationships in agroclimatology. In *Agroclimatological Methods. Proceedings of the Reading Symposium*. Paris: UNESCO, 73–87.

SMITH, A. G. and BRIDEN, J. C. 1977: *Mesozoic and Cenozoic paleocontinental maps*. London: Cambridge University Press.

SOLIMAN, K. H. 1953: Rainfall over Egypt. *Quarterly Journal of the Royal Meteorological Society*, 79, 389–97.

STREET, F. A. and GROVE, A. T. 1976: Environmental and climatic implications of late Quaternary lake-level fluctuations in Africa. *Nature* 261, 335–90.

STREET, F. A. and GROVE, A. T. 1979: Global maps of lake-level fluctuations since 30,000 BP *Quaternary Research*, 12, 83–118.

STREET-PERROTT, F. A. and ROBERTS, N. 1983: Fluctuations in closed-basin lakes as an indicator of past atmospheric circulation patterns. In Street-Perrott, A., Beran, M. and Ratcliffe, R. (eds.) *Variations in the Global Water Budget*. Dordrecht: Reidel, 331–45.

SUD, Y. C. and FENNESSY, M. 1982: A study of the influence of surface albedo on July circulation in semi-arid regions using the GLAS GCM. *Journal of Climatology* 2, 105–25.

TETZLAFF, G. and ADAMS, L. J. 1983: Present-day and early-Holocene evaporation of Lake Chad. In Street-Perrott, A. Beran, M. and Ratcliffe, R. (eds.) *Variations in the Global Water Budget*. Dordrecht: Reidel, 347–60.

THORNTHWAITE, C. W. 1933: The climates of the earth. *Geographical Review* 23.

—— 1948: An approach toward a rational classification of climate. *Geographical Review* 38, 55–94.

UNITED NATIONS CONFERENCE ON DESERTIFICATION 1977: *Desertification: its Causes and Consequences*. Oxford: Pergamon.

UN PROTEIN ADVISORY GROUP. 1974: *Recommendations – August 1974*. Geneva: United Nations.

WADE, N. 1974: Sahelian drought: no victory for western aid. *Science* 185, 234–7.

WALKER, J. and ROWNTREE, P. R. 1977: The effect of soil moisture on circulation and rainfall in a tropical model. *Quarterly Journal of the Royal Meteorological Society*, 103, 29–46.

WALLÉN, C. C. 1967: Aridity definitions and their applicability. *Geografiska Annaler* 49, 367–   .

—— 1968: Agroclimatological studies in the Levant. In *Agroclimatological Methods. Proceedings of the Reading Symposium*, Paris: UNESCO, 225–33.

WIESNER, C. J. 1970: *Hydrometeorology*. London: Chapman & Hall.

WINKWORTH, R. E. 1970: The soil water regime of an arid grassland community in central Australia. *Agricultural Meteorology* 7, 387–   .

WINSTANLEY, D. 1970: The North African flood disaster, September 1969. *Weather* 25, 390–403.

—— 1975: The impact of regional climatic fluctuations on man: some global implications. In *Proceedings of the WMO/IAMAP Symposium on Long Term Climatic Fluctuations*. Geneva: World Meteorological Organization, 479–91.

ZINDEREN BAKKER, E. M. VAN 1978: Late Mesozoic and Tertiary paleoenvironments of the Sahara region. In Zinderen Bakker, E. M. Van (ed.). *Antarctic Glacial History and World Palaeoenvironments*. Rotterdam: Balkema, 129–35.

5 Grassland and Vegetated Subsystems

1 Vegetation and water balance

A number of non-dimensional basic parameters exist which are of some importance to theoretical climatology. These concern the radiation, heat and water balances of the surface and atmosphere, and are listed below:

(a) Radiation
surface albedo (short-wave), α;
surface emissivity, ϵ;
Ångström ratio, A (quotient of effective long-wave radiation to that emitted at ground level).

(b) Heat balance.
Bowen ratio, β;
radiational index of dryness, R_N/LP (where R_N is net radiation, L latent heat of condensation, and P is precipitation);
Kelvin temperature at or near the ground surface, T.

(c) Water balance.
run-off ratio, f/P (where f is the run-off);
Exchangeable moisture contained in a vertical column of soil, m.

Relationships between the various non-dimensional parameters may be obtained as follows. For a land surface with a stable climate it follows that the multi-annual means of the soil storage terms of both heat and moisture must vanish. Using annual means (shown by a bar), it is possible to write for the land surface;

$$\bar{R}_N = \bar{C} + LE_t \tag{5.1}$$

where C is the sensible heat transfer and E_t the evapotranspiration,

$$\text{since } \bar{\beta} = \bar{C}/L\bar{E}_t \tag{5.2}$$

$$\text{then } \bar{R}_N = L\bar{E}_t(1 + \beta). \tag{5.3}$$

$$\text{Also } \bar{P} = \bar{f} + \bar{E}_t. \tag{5.4}$$

Combining the last two equations, it is possible to obtain

$$\frac{\bar{R}_N}{L\bar{P}} = (1 + \beta)\left(1 - \frac{\bar{f}}{\bar{P}}\right) \tag{5.5}$$

It also follows that

$$\frac{\bar{E}_t}{\bar{P}} + \frac{\bar{f}}{\bar{P}} = 1 \tag{5.6}$$

Taking into account that the mean dryness of the soil increases with an increase in the radiative income of heat and with a decrease in the amount of precipitation, it can be deduced that

$$\frac{\bar{f}}{\bar{P}} \to 0 \text{ or } \frac{\bar{E_t}}{\bar{P}} \to 1 \text{ as } \frac{\bar{R}_N}{L\bar{P}} \to \infty$$

With a decrease in the ratio $\dfrac{\bar{R}_N}{L\bar{P}}$, which Budyko (1974) called the radiative index of dryness, the values of $\dfrac{\bar{E_t}}{\bar{P}}$ will diminish, the upper layers of the soil will become moist and run-off will appear.

Budyko (1974) has constructed a world map of the radiative index of dryness. Budyko comments that when the map of the radiative index of dryness is compared with the available geobotanic and soil maps, it is seen that the location of the isolines of the dryness index agree well with the location of physical-geographic zones. The smallest values of the dryness index (up to $\frac{1}{3}$) correspond to tundra, index values ranging from one-third to 1 to the forest zone, from 1 to 2 to steppe, more than 2 to semidesert, and more than 3 to the desert zones.

2 Microclimates of grasslands

2.1 Radiation

The atmosphere is nearly transparent to short-wave radiation from the sun, of which large amounts reach the earth's surface, but it readily absorbs infrared radiation emitted by the earth's surface, the principal absorbers being water vapour (5.3 to 7.7 μm and beyond 20 μm), ozone (9.4 to 9.8 μm), carbon dioxide (13.1 to 16.9 μm), and clouds (all wavelengths). Only about 9 per cent of the infrared radiation from the ground surface escapes directly to space, mainly in the 'atmospheric window' (8.5 to 11.0 μm); the rest is absorbed by the atmosphere, which in turn re-radiates the absorbed infrared radiation, partly to space and partly back to the surface. The main receiver of heat from the sun is thus the ground surface, while most of the heat loss to space is from the lower troposphere. This means that there must be a general temperature gradient from the ground surface to the lower troposphere, so as to allow the flow of heat from the surface to the atmosphere and then to space. The earth's surface is thus at a higher average temperature than that which would be observed if the atmosphere were perfectly transparent or completely absent. It has been estimated by Fleagle and Businger (1963) that if the atmosphere were absent, the earth's surface would be 30°–40°C cooler than it is at present. The term 'greenhouse effect' is often applied to this particular phenomenon, because it is analogous to that effect which is supposed to operate in a greenhouse where the glass is transparent to short-wave radiation but nearly opaque to infrared radiation.

Of particular interest to the meteorologist is the so-called net radiation, which is the difference between the total incoming and the total outgoing radiation. The net radiation indicates whether net heating or cooling is taking place; the net radiation will normally be negative at night indicating cooling, but during the day it may be negative or positive depending on the balance of the incoming and the outgoing radiation.

In Figure 5.1, which illustrates the components of the net radiation balance, it is assumed that the atmosphere and soil are dry, that is, there is no evaporation or

Night

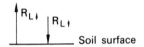

Soil surface

$R_{L\downarrow}$ Long-wave radiation from surface
$R_{L\uparrow}$ Long-wave counter radiation from atmosphere

Day

$R_{L\downarrow}$ $R_{L\uparrow}$ $R_{S\uparrow DIR}$ $R_{S\uparrow DIF}$ $R_{S\downarrow}$ ____ Soil surface

$R_{S\uparrow DIR}$ Direct short-wave radiation from sun
$R_{S\uparrow DIF}$ Diffuse short-wave radiation
$R_{S\downarrow}$ Reflected short-wave radiation

Figure 5.1 Radiation balance of a dry soil surface at night and during the day.

condensation. The simplest case occurs at night, because there is no incoming short-wave radiation, but instead there is a continuous long-wave radiation loss. At night the soil surface emits long-wave radiation ($R_{L\uparrow}$) to the atmosphere, where water vapour and carbon dioxide absorb large amounts of it, which in turn is partly re-radiated downwards ($R_{L\downarrow}$) to be re-absorbed by the soil surface. The difference between $R_{L\uparrow}$ and $R_{L\downarrow}$ is the net radiation loss (R_N), which in this example represents a cooling, and consequently there is a flow of sensible heat from the atmosphere and the lower layers of the soil towards the soil surface.

During the day the situation is more complex because of the incoming short-wave radiation, which can take two forms: direct radiation from the sun and diffuse sky radiation. The term global radiation is used for the sum of the direct and the diffuse short-wave radiation received by a unit horizontal surface. Short-wave radiation in the atmosphere is scattered and reflected by molecules, dust particles, clouds, etc., so that a large percentage of the short-wave radiation reaching the surface will not be coming from the direction of the sun. Indeed in some parts of the world, the Arctic for instance, the diffuse sky radiation can form a major part of the incoming short-wave radiation. Short-wave radiation on reaching the surface can be either absorbed or reflected. The reflected radiation represents a complete energy loss, because it does not heat the surface, and it depends on the albedo of the surface. Of the short-wave energy absorbed, some will be re-radiated as long-wave radiation and some will be transferred as sensible heat into the soil and the atmosphere. The long-wave interactions will be similar to those at night, so the net radiation (Figure 5.1) is given by

$$R_N = (R_{L\downarrow} + R_{S\downarrow DIF} + R_{S\downarrow DIR}) - (R_{L\uparrow} + R_{S\uparrow}) \tag{5.7}$$

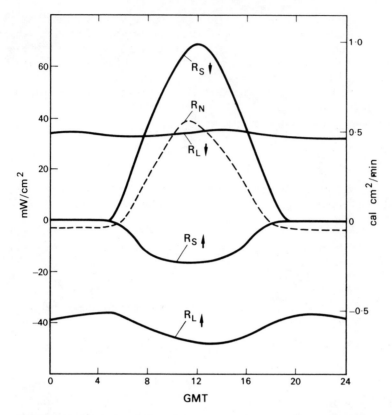

Figure 5.2   Diurnal variation of the components of the radiation balance for a clear August day at Rothamsted (52°N) over a thick stand of grass 40 cm high. (After Monteith and Szeiċz, 1961)

Figure 5.2 from Monteith and Szeicz (1961) illustrates, for temperate latitudes, the relative importance of the various radiative transfers throughout the day in clear weather. The surface consists of grass with an albedo of 26 per cent, a value which has been found to be characteristic of many green crops giving a nearly complete coverage of the soil; therefore approximately 74 per cent of the incident solar radiation is absorbed. The long-wave radiation loss from the surface is greatest around mid-day when the grass is hottest, and least around the time of minimum temperature, which is commonly just before dawn in clear weather. The diurnal variation of the grass temperature is about 20°C, which is about 7 per cent of the absolute temperature. From the Stefan–Boltzmann law a variation of about 28 per cent is to be expected in the long-wave radiation loss from the surface, and this is close to the observed value. The variation in the long-wave radiation from the atmosphere is much less because it depends largely on the temperature of the water vapour and carbon dioxide in the lowest 200 m or so of the atmosphere and this varies but little in comparison with ground surface temperature. The lowest 100 m of the atmosphere contributes strongly to the long-wave radiation reaching the surface and each 100 m above has rapidly diminishing influence so that conditions above 1,000 m are usually of little significance with a clear sky.

The net outgoing long-wave radiation from the ground surface has two basic components: the total long-wave emission from the surface, which is a function of the surface emissivity and temperature, and the counter long-wave radiation from the atmosphere, which is a function of air temperature, water vapour content and the cloud cover. The downwards atmosphere long-wave radiation under cloudless conditions has been found to be closely related to the temperature and the vapour pressure measured in a Stevenson Screen. Various empirical formulas for estimating the counter radiation have been suggested and among these are the following:

$$R_{L\downarrow} = \sigma T^4 (0.805 - 0.235) (10^{-0.052e}) (\text{Ångström, 1916}). \qquad (5.8)$$

$$R_{L\downarrow} = \sigma T^4 (0.52 + 0.065 \sqrt{e}) (\text{Brunt, 1941}). \qquad (5.9)$$

$$R_{L\downarrow} = 5.31 (10^{-14} T^6) (\text{Swinbank, 1963}). \qquad (5.10)$$

where $R_{L\downarrow}$ is the atmospheric counter long-wave radiation, T is the screen temperature, e is the screen vapour pressure, and σ is a constant equal to 5.67×10^{-8} mW cm^{-2} (deg K)^4.

When considering the effect of cloud on atmospheric counter radiation there are two main components, which are cloud amount and the temperature of the cloud base; since high-level clouds have much lower base temperatures than low clouds, it therefore follows that cloud height is also important. Cloud layers are normally considered to radiate as blackbodies, and therefore the effect of cloudiness on the net long-wave radiation ($R_{L\uparrow} - R_{L\downarrow}$) is often taken into account by multiplying the estimate for a clear sky by a factor $(1 - \lambda c)$ in which c is the fractional cloudiness and λ depends on cloud height. For low clouds $\lambda = 0.8$–0.9, for medium cloud 0.6–0.7 and for cirrus cloud about 0.2. Normally dense low clouds are most effective in retarding temperature falls at night, while cirrus clouds have little effect.

It has been found from experiments by Monteith and Szeicz (1961) that changes in the net radiation (R_N) received by bare soil or vegetation during cloudless summer days are related to global radiation (R_s) by the equation

$$R_N = \frac{(1 - \alpha)}{(1 + \beta)} \cdot R_s + L_0 \qquad (5.11)$$

where α is the short-wave albedo; β is a heating coefficient defined as the increase in long-wave radiation loss per unit increase of R_N; and L_0 is the net radiation as $R_s \rightarrow 0$. Values of β vary with climate and depend as much upon atmospheric radiative properties as surface conditions, but for Britain, Monteith and Szeicz (1962) suggest that β is probably close to 0.1 for agricultural crops and natural vegetation which completely cover the ground and are never short of water. If transpiration is physiologically restricted or if ground cover is incomplete, β may lie between 0.1 and 0.2 and for very dry soil between 0.3 and 0.4. Negative values of β have, however, been reported by some authors such as Stanhill, Hofstede, and Kalma (1966). For Britain, Gadd and Keers (1970) suggest values of L_0 equal to $-0.05R_s$. The calculation of net radiation from sunshine duration is considered later.

2.2 Albedo

Radiation falling on a surface may be partly reflected, partly absorbed and partly transmitted. Most natural solid objects are opaque, so light is either reflected or absorbed. In contrast, water is translucent and light penetrates into the surface layers of the oceans, while the atmosphere is nearly transparent to short-wave radiation. Radiation being received from a surface may have resulted from either reflection or radiation by the surface, or indeed both.

Incoming solar radiation to a crop is absorbed, transmitted, or reflected by the plants and absorbed or reflected by the soil. Similarly radiation may be absorbed or reflected by water, desert and ice surfaces. Albedo is a measure of the reflecting power of a surface, being that fraction of the incident radiation which is reflected by a surface. The definition of albedo varies throughout the meteorological literature. In some applications the albedo is restricted to the visible wavelengths while on other occasions the definition is widened to include terrestrial long-wave radiation. Albedo is sometimes called reflectivity, but the latter term properly refers to the reflected–incident ratio for a specific wavelength. Reflection does vary with wavelength, thus grass is green because it reflects much of the green light and absorbs most of the energy in other colours. Plant leaves of different shades of green reflect almost the same fraction of incident radiation and this suggests that most of the reflection takes place outside the visible region of the spectrum and that the dependence of reflectivity on wavelength does not vary greatly among common species of similar type. For the sun, the wavelength of maximum emission is near 0.5 μm, which is in the visible portion of the electromagnetic spectrum, and almost 99 per cent of the sun's radiation is contained in the so-called short wavelengths from 0.15 to 4.0 μm. Observations show that 9 per cent of this short-wave radiation is in the ultraviolet (less than 0.4 μm), 45 per cent in the visible (0.4 to 0.7 μm) and 46 per cent in the infrared (greater than 0.74 μm). Billings and Morris (1951) measured the reflection coefficient of detached leaves with a spectrophotometer in the range 0.4 to 1.1 μm. For the majority of 20 species selected from widely different habitats, reflectivity was relatively low (0.05–0.20) below 0.7 μm but increased abruptly to a constant value between 0.50 and 0.70 from 0.7 to 1.1 μm. This suggests that the visual reflection coefficient of vegetation will normally be smaller than the total reflection coefficient and that the relation between the two depends on the species. The term 'albedo' is often used in meteorological literature to refer to the ratio of the amount of short-wave radiation reflected by a body to the amount incident upon it, short-wave radiation being defined as that radiation approximately between 0.15 and 4.0 μm. Albedo is used in this sense in this book.

At least four factors explain differences in surface albedo. The first is the condition of the surface itself since, in general, higher albedos are associated with dry, light-coloured, smooth surfaces, whereas low albedos are associated with wet, dark-coloured, rough surfaces. In the case of cultivated crops, it depends on the height of the plants, percentage ground cover, angle of the leaves, and leaf area index. The second factor which controls albedo is the zenith angle of the sun, and this leads to marked diurnal variations. The third factor is the state of the sky, with particular reference to the type and amount of cloud. The final important factor is the angle of the surface to the horizontal, and particularly whether it is facing towards or away from the sun. The value of the albedo is normally different from surface to surface and also from one season to another. Albedo also varies throughout clear days, being generally higher in the morning and the evening and much higher later in the afternoon. Some typical albedo values are shown in Table 5.1.

The upward reflected radiation of many plant canopies is complex in nature because it is composed of both reflected and multiple scattered radiation by plant organs and by the ground surface. The diurnal variation of canopy albedo on a sunny day has been shown by a number of authors to be the typical bowl-shaped curve illustrated in Figure 5.3. This has been illustrated for grass and kale by Monteith and Szeicz (1961), for oak forest by Rauner (1976), for spruce forest by Jarvis *et al.* (1976), and for Scots pine forest by Stewart (1971). Graham and King (1961) showed that the albedo increased markedly at low elevations and also that

Table 5.1 Values of albedo for various surfaces (*after Kung et al. 1964 and Kondratyev 1954*)

Surface	Albedo
Black soil, dry	0.14
Black soil, moist	0.08
Ploughed field, moist	0.14
Sand, bright, fine	0.37
Dense snow, dry and clean	0.86–0.95
Sea ice, slightly porous, milky bluish	0.36
Ice sheet, covered by a water layer of 15–20 cm	0.26
Woody farm, covered with snow	0.33–0.40
Deciduous forest	0.17
Tops of oak	0.18
Pine forest	0.14
Desert shrubland	0.20–0.29
Swamp	0.10–0.14
Prairie	0.12–0.13
Winter wheat	0.16–0.23
Heather	0.10
Yuma, Arizona	0.20
Washington, DC (September)	0.12–0.13
Winnipeg, Manitoba (July)	0.13–0.16
Great Salt Lake, Utah	0.30

the diurnal distribution was asymmetrical about noon. Similarly, Munn and Truhlar (1963) reported albedo values for afternoon periods that were higher than the forenoon equivalents, but made no comment regarding the observation. Stanhill *et al.* (1966) showed that albedo increased markedly with decreasing solar altitude, while Impens and Lemeur (1969) also reported a diurnal variation with marked asymmetry about noon.

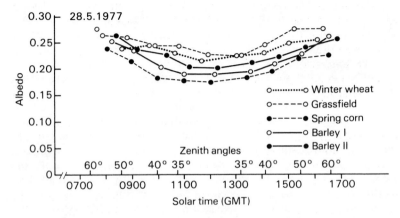

Figure 5.3 Diurnal variation of albedo over various agricultural surfaces.

Most of the observations of diurnal variations in albedo reported above refer to clear sky conditions, but Nkemdirim (1972) has investigated the relationship between albedo and zenith angle under varying sky conditions. He divided the data for each crop type into three ranges of cloudiness. Cloudiness was expressed in terms of transmissivity, which was obtained by dividing the global radiation received at the surface by the global radiation incident at the top of the atmosphere. The three transmissivity ranges selected were less than 0.25, 0.25–0.5, and greater than 0.5, corresponding to overcast conditions, intermittent cloudiness, and cloudless conditions respectively. Nkemdirim found the average albedo over a mixture of surfaces for each zenith angle class under given atmospheric conditions. Ahmad and Lockwood (1979) have also investigated albedo over individual crops. They found that albedo tends to remain almost constant or show least diurnal variation under overcast conditions, but that there is a clear dependence on zenith angle under cloudless skies. Several general comments can be made about both Nkemdirim's, and Ahmad and Lockwood's results.

1 At low zenith angles the albedo tends to remain approximately constant or to decrease as the transmission coefficient increases.
2 At high zenith angles the albedo tends to remain approximately constant or to increase as the transmission coefficient increases.
3 The range of albedo values increases with increasing transmission coefficient, that is the diurnal variation becomes more marked with decreasing cloudiness.
4 The forenoon values of albedo tend to be lower than the afternoon values for similar zenith angles for a significant number of cases.

The albedo of many surfaces depends on the angle of incidence of the direct component of the radiation and also on the proportion of the diffuse sky radiation. Albedo values are often at a minimum when the incoming radiation consists mostly of direct radiation coming from overhead, since under these conditions the trapping of radiation by multiple reflection between leaves is at maximum and radiation will penetrate deep into the canopy and be reflected from the ground. As the angle of approach of the direct radiation falls towards the horizon, the albedo tends to increase. This is because multiple reflection within the canopy decreases, and also because more radiation is reflected off the top of the canopy since the crop as seen by the sun becomes more like a flat surface. So albedo values will tend to be at a minimum under clear skies with the sun nearly overhead, and to increase as the sun sinks towards the horizon. Some of the highest albedo values are observed under clear skies with the sun near the horizon, and this stresses the importance of the angle of incidence of the radiation. Under cloudy or overcast conditions a major proportion of the diffuse radiation will always originate from the clouds overhead, and this accounts for the small variation in albedo values under these conditions. In particular when the sun is low, clouds allow more light to penetrate the vegetation vertically downwards. Thus under cloudy or overcast conditions a greater proportion of the diffuse radiation will originate from overhead when the sun is near the horizon than under cloudless conditions with low sun elevations. Albedo values will therefore tend to be lower under cloudy conditions than under cloudless conditions with the sun near the horizon.

Piggin and Schwerdtfeger (1973) state that for overcast days with a uniform, heavy cloud cover the radiation can be considered practically isotropic all day. They consider that the effective mean solar elevation would thus be about 40°–50°, since the solid angle at 45° contributes maximum radiation to the surface. This arises from a consideration of the combined effects of the cosine law and the solid angle per unit of increment in elevation, 45° being the mean angle of incidence. Piggin and Schwerdtfeger suggest that it might be expected that as the daily mean solar elevation increases beyond 45°, the daily albedo on clear days would have changed from being higher than the daily albedo on overcast days, to being lower. An analysis of their observations at Derrimut, Victoria, Australia appears to indicate a cross-over in the ten-day means of the daily albedo on clear and overcast days for both wheat and barley crops; first clear, then overcast being higher, and the cross-over occurring when the daily mean solar elevation is 40°–45°. They found that surface cover and moisture also affected the cross-over. Piggin and Schwerdtfeger comment that some supporting evidence for this cross-over appears in the measurements of other workers, although they make no mention of a cross-over. They consider that if the daily mean solar elevation has to exceed 45° before the daily cross-over occurs, only at stations equatorwards of 50° latitude could a cross-over possibly be observed. Many European stations lie polewards of 50°N, and this might account for its non-observance.

Under clear skies, the relationship between average albedo for each surface and zenith angle was found by Nkemdirim (1972) to be given by $\alpha = ae^{bz}$ (5.12), where α is the surface albedo, z the zenith angle (in degrees), e the base of natural logarithms, and a and b are constants. Some values of a and b are listed in Table 5.2 along with correlation coefficients. The variation in the AM and PM values of a for each surface are generally small, suggesting that they represent a crop factor. However, Nkemdirim found statistically significant differences between AM and PM

Table 5.2 Dependence of average albedo (α) on zenith angles (z) and period of day for skies with low cloud amounts (a) *after Nkemdirim, 1972*, (b) *after Ahmad and Lockwood, 1979*

Crop	Period	Regression equation	Correlation coefficient
(a)			
Non-irrigated potatoes	AM	$\alpha = 0.613e^{0.181z}$	0.9581
	PM	$\alpha = 0.0460e^{0.0269z}$	0.9794
Irrigated potatoes	AM	$\alpha = 0.0896e^{0.0084z}$	0.9691
	PM	$\alpha = 0.0453e^{0.0271z}$	0.9638
Bare ground	AM	$\alpha = 0.1247e^{-0.0003z}$	−0.0526
	PM	$\alpha = 0.626e^{0.0216z}$	0.9839
Peas	AM	$\alpha = 0.0842e^{0.0134z}$	0.8884
	PM1	$\alpha = 0.559e^{0.0287z}$	0.9800
(b)			
Spring wheat	AM	$\alpha = 0.1765e^{0.0049z}$	0.8871
	PM	$\alpha = 0.1945e^{0.0026z}$	0.5360
Barley	AM	$\alpha = 0.1996e^{0.0040z}$	0.8191
	PM	$\alpha = 0.2262e^{0.0008z}$	0.3230
Grass (rural site)	AM	$\alpha = 0.2540e^{0.0006z}$	0.2181
	PM	$\alpha = 0.2008e^{0.0063z}$	0.9508
Grass (urban park)	AM	$\alpha = 0.2209e^{0.0024z}$	0.9134
	PM	$\alpha = 0.1894e^{0.0061z}$	0.9857
Rough urban surface	AM	$\alpha = 0.1825e^{0.0028z}$	0.9984
	PM	$\alpha = 0.1792e^{0.0037z}$	0.8980
Car park	AM	$\alpha = 0.1495e^{-0.0011z}$	−0.3950
	PM	$\alpha = 0.1852e^{-0.0061z}$	−0.6208

values of the constant *b*. Where there is a reasonable correlation between zenith angle and albedo, it is observed that the slope factor *b* is higher for PM than for AM periods in a significant number of cases, as a result of the more rapid rate of increase of albedo during the afternoon hours.

A number of authors have reported measurements of the seasonal variation in albedo. Typical of the results found are those reported by Idso, Reginato and Jackson (1977), concerning spring wheat growth under irrigation in the USA at Phoenix, Arizona, on Avondale Loam. It has been demonstrated that the changing altitude angle of the sun causes substantial changes in albedo. The effect of these solar zenith angle changes has to be removed if the influence of other factors is to be recognized. So it is possible to plot two albedo curves for any given canopy type, one showing the actual variation of albedo, and the other the variation of albedo which has been normalized to remove the variability caused by the changing altitude angle of the sun during the growing season.

Idso *et al.* found a decline in normalized albedo of their spring wheat crop as the season progressed from values between 0.23 and 0.26 to values between 0.16 and 0.22. They considered that this was due initially to the growth and proliferation of the leaves as they progressively shut out more and more of the high-albedo soil from view, and finally to the growth and proliferation of the plant heads. Albedo showed a sharp rise just before harvesting, and Idso *et al.* suggested that the final grain yield is a linear function of the minimum normalized albedo reached just before the large increases that occur as the heads begin to ripen.

Monteith (1959) measured the seasonal variations of albedo for a number of crops in Britain. The albedo of the Avondale Loam used by Idso *et al.* was high compared with the albedo of plant leaves. The opposite tends to occur in Britain, where soil is inclined to be damp with low albedos. Monteith quotes a value of 0.11 for his soil (clay loam with flints) when the moisture content was at field capacity. The result is that the albedo of British crops increases as the crop develops and reaches a peak when the crop fully covers the ground. Monteith did not correct his albedo values for the changing solar altitude during the growing season, so they do not just reflect changing canopy properties. The albedo of his spring wheat crop increased from 0.11 at the end of April, a value characteristic of the soil rather than the crop, to 0.20–0.21 in May and June. When the crop ripened at the beginning of August there was a sharp drop in albedo to 0.14 followed by a slight increase attributable to vigorous weed growth which had completely covered the ground by the beginning of September. The winter wheat crops showed similar trends, albedo reaching a maximum in May and then decreasing in value. Monteith found that the albedo of a stand of closely cut pure grass was very close to 0.25, showing remarkably constancy throughout the season. He considered that the seasonal changes in albedo were due to changes in crop parameters and particularly to changes in leaf-area index. Leaf-area index is the ratio of the leaf area (i.e. half the total leaf surface) to the area of underlying ground. According to Monteith all his crops achieved a similar maximum value of albedo at or about the time of maximum leaf area. He stated that, provided there is sufficient leaf area and provided leaves are so arranged that they completely shade the ground without shading each other, crop albedos are close to 0.26.

Examples of the seasonal variations in albedo of a number of crops as measured by Ahmad and Lockwood (1979) are shown in Figure 5.4. The raw albedo observations show the effects of cloud cover, crop parameters, soil moisture content and the zenith angle of the sun. In order to remove the effect of the varying zenith angle, the raw albedo data were normalized as suggested by Idso *et al.* The normalizing functions had the approximate effect of reducing the albedo to that of the midsummer value for noon on 21 June. Since the diurnal variation of albedo is complex, the operation of the normalizing function must at best be very crude, but it does make albedo measurements taken at different times approximately comparable. Normalization was carried out by a study of available diurnal albedo curves and a consequent reduction of the observed albedo by a suitable amount. On average on cloudless days the reduction in albedo amounted to 0.01 for every 5° increase in zenith angle over that observed at noon on 21 June. When several observations were available for a particular day, the final normalized albedo was an average of all the normalized albedos for that particular day. Under overcast and heavy broken cloud, the midday value was used as the normalized albedo, since under these conditions the albedo is almost independent of the solar zenith angle.

Typical albedo variations for various British crops during the growing season are shown in Figure 5.4. They were measured on a farm in lower Wharfedale, to the north of Leeds. Soils were mostly of the Dunkeswick, Keswick and Ambergate

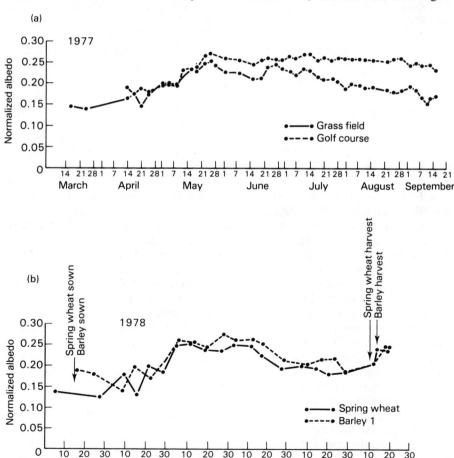

Figure 5.4   Seasonal variations of normalized albedo for selected cultivated crops (after Ahmad and Lockwood, 1979).

series and tended to have low albedo values when wet. The mean daily albedo of Dunkeswick soil was found to vary between 0.22 when dry to 0.17 when wet. This was reflected in the albedo observations of the wheat and barley crops, which started at low values typical of the bare soil, and slowly increased until full ground cover was achieved. Maximum actual albedos of the grain crops measured at full plant cover were around 0.25. After maximum albedo values were reached in June, the normalized albedo fell gradually until harvesting in September, after which it increased sharply. In 1978 a small increase in the albedo of the grain crops was observed before harvesting, corresponding to the observations of Idso *et al.* mentioned earlier.

Piggin and Schwerdtfeger (1973) have measured the albedo of wheat and barley crops during the course of two successive growing seasons, one of which was exceedingly wet and the other abnormally dry, at Derrimut, Victoria, Australia. The first measurements over barley were taken on 30 July when the leaf-area index was less than 1 and the albedo was 0.16. As the crop cover increased so did the albedo, until by 15 September with the leaf-area index at 5 the albedo was 0.26. This value represented the albedo of green barley plants at full crop cover and agrees with the results reported above. From 15 September (LAI5) to 5 October (LAI4) the albedo remained constant at 0.26. Heading occurred during this plateau period but it did not measurably affect the albedo. Fritschen (1957) and Graham and King (1961) have reported similar results, namely, no change in the albedo of cereal crops and maize with earing. After heading, translocation of stored carbohydrates from leaves to ears caused the leaves to yellow, dry and wither, and the leaf-area index dropped sharply. This exposed some of the lower-albedo soil and the measured albedo fell from 0.26 on 5 October to 0.23 on 15 October. Once most of the leaves on the barley plants had withered, the albedo displayed another plateau, since it only decreased from 0.23 on 15 October to 0.22 on 5 November. From the leaf-drying depression, the albedo again rose rather sharply from 0.22 on 5 November to 0.30 on 24 November. The rise was especially marked over the three-day period 19 to 21 November, when the daily albedo values increased from 0.25 to 0.30. The increase in the albedo of the barley cover above its 'green value' of 0.26 resulted as the leaves, stalks and heads dried and yellowed. After harvesting, the albedo of the barley stubble was 0.34, which exceeded the value of the fully ripened barley. Although Piggin and Schwerdtfeger made no attempt to correct their albedo observations for varying solar zenith angles, they do, even so, correspond in general to those made by Ahmad and Lockwood (1979). Clearly the exact nature of the curves will depend to a certain extent on the relative values of the soil and crop albedos. The 'ripening upturn' in albedo at the end of the growing season will probably be much less marked in the damp climate of Britain than in drier climates, and is important in the remote-sensing context.

The normalized albedo values measured over grass by Ahmad and Lockwood (1979), which are illustrated in Figure 5.4, show an increase up to late May when they reached maximum values of around 0.25, but after this the normalized albedo of the pasture field slowly decreased. The golf course grass was kept in good condition and maintained its normalized albedo at around 0.25 for the latter part of the growing season. Albedo observations were also conducted over three types of urban surface. The normalized albedos show only small variations with season, and this is to be expected since the surfaces are relatively constant in nature. This also suggests that the variations in the normalized crop albedo values are a function of the crop canopy.

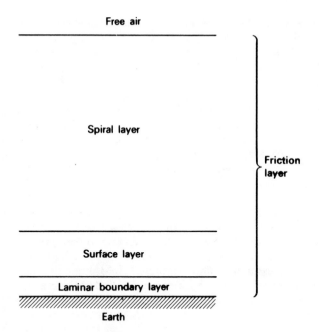

Figure 5.5 Schematic illustration of the sub-layers of the friction layer. The friction layer corresponds to the dynamic part of the planetary boundary layer.

2.3 Atmospheric turbulence

Atmospheric turbulence near the ground is manifested by rapid, and sometimes violent, variations of wind speed and direction. Since turbulence is a very disorganized motion of the air, it is not possible to predict the individual fluctuations from past history. In the wind tunnel, a mean wind can be defined and measured at any point in the flow without too much difficulty, but there have been many arguments concerning what constitutes mean flow and what constitutes turbulence in the atmosphere. A mean wind cannot be defined without specifying the time over which the average flow is defined; therefore the mean wind may be the average for a minute, a day or a year. Fluctuations about the mean are called turbulence. Turbulence cannot extend to the ground, because the vertical velocity vanishes at the surface, but its importance increases rapidly with height in the lowest few centimetres. Figure 5.5 shows a schematic illustration of the sub-layers of the atmosphere near the ground.

Starting from the ground surface, the increase of wind speed with height is first of all in accordance with the logarithmic law in the first 20 m and the direction is at an angle to the upper wind direction. With a further increase in height, the wind increases more gradually and turns into the upper wind direction. Providing the pressure pattern is not changing rapidly, the wind at the top of the planetary boundary layer approximates to the geostrophic wind (V_g), that is to say, the wind blowing parallel to the isobars with the requisite speed for the force due to the horizontal pressure gradient (grad p) to be balanced by the deflecting force caused by the earth's rotation (Coriolis force). The geostrophic wind relationship for latitude ϕ is:

$$V_g = \operatorname{grad} p / 2\rho\Omega \sin \phi \qquad (5.13)$$

where Ω is the angular velocity of the earth. The layer below the level at which the observed wind is in general a close approximation to the geostrophic wind is often known as the friction layer, and the atmosphere above the friction layer is usually called the free air. The friction layer which corresponds to the dynamic part of the planetary boundary layer, can be divided into a number of sub-layers (see Figure 5.5), and these are the spiral layer, the surface layer, and the laminar boundary layer. The laminar boundary layer is the layer where flow is laminar and the viscous stresses dominate the eddy stresses. The surface layer is the layer in which the wind speed varies with height according to the logarithmic law and the eddy stress is an order of magnitude larger than the horizontal pressure force. The spiral layer is a layer in which the eddy stress has approximately the same order of magnitude as the pressure gradient and Coriolis force.

The laminar boundary layer
Because the laminar boundary layer of air adheres strongly to the earth's surface, it prevents the turbulent motions of the atmosphere from extending to the ground. Within the laminar boundary layer, heat, water vapour, and momentum are transferred vertically by molecular processes only, and the rates of transfer are proportional to the vertical gradients of temperature, specific humidity and horizontal air velocity. For example, the vertical transfer of sensible heat (H) is given by

$$ H = - \rho C_p K_h \ \frac{\partial \theta}{\partial z} \tag{5.14} $$

where ρ is the air density;
        C_p is the specific heat of air at constant pressure;
        K_h is the thermal diffusivity of air;
        θ is the potential temperature;
and z is the height.
    If the sun's heat absorbed at the ground were propagated through the air purely by molecular conduction then the diurnal temperature wave would vanish at only 4 m above the surface and at a height of 2 m the maximum temperature would occur at midnight. But the diurnal variation of air temperature extends well above 1,000 m, so clearly a much more effective process than molecular diffusion must operate in the free atmosphere. This more effective process of heat transfer is in fact the general mechanical turbulence of the atmosphere.

The logarithmic layer
Over level ground, away from trees or other obstructions, the wind profile under conditions of neutral stability follows, in the lowest 20 m or so, a logarithmic law of the form:

$$ u_z = A \ln z + B \tag{5.15} $$

where u_z is the wind speed at height z,
and A and B are constants.
    The rate of increase in wind speed with height, i.e. the vertical wind shear ($\partial u / \partial z$), is largest near the ground and decreases uniformly upwards in such a way that when u is plotted against the logarithm of z a straight line can be drawn (Figure 5.6).
    Further studies have improved the above equation so that it may now be written as:

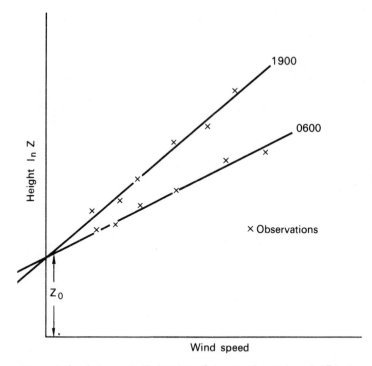

Figure 5.6 Schematic illustration of the roughness length (Z_0). Average profiles of wind speed with height for a number of observations at 0600 and 1900 local time when near neutral stability conditions prevail. It is assumed that the observations are made in the first 20 m over short grass.

$$u_z = \frac{1}{k} \left(\frac{\tau}{\rho} \right)^{1/2} \ln \frac{z}{z_0} \tag{5.16}$$

in which τ is the shear stress, ρ is air density, k is Karman's constant (~ 0.41) and z_0, known as the roughness parameter or roughness length, characterizes the nature of the surface as to aerodynamic roughness. The parameter z_0 is relatively easy to identify in physical terms. It is given by the intercept of the line through the experimental points on Figure 5.6 with the vertical axis. The shearing stress at a height z in the boundary layer can be written in the form

$$\tau(z) = K_M \frac{\partial(\rho u)}{\partial z} \tag{5.17}$$

where K_M is the eddy viscosity.

The logarithmic wind profile applies in the surface layer in which the shearing stress and wind direction are effectively constant with height. If the upper level is taken as that at which τ has fallen to 95 per cent of the surface value, then over grassland this will usually range between 15 and 30 m. The roughness length of natural surfaces is about an order of magnitude less than the height of the roughness elements, e.g. 0.1 cm for a lawn ($h = 1$ cm), 4 cm for barley ($h = 70$ cm) and 20 cm for maize ($h = 230$ cm). The relationship between the roughness length and the vegetation height h can, according to Tanner and Pelton (1960), be expressed by

$$\log z_0 = a + b \log h \tag{5.18}$$

where common (base 10) logarithms are used. A variety of values have been obtained for the constants a and b; according to Tanner and Pelton, $a = 0.883$ and $b = 0.997$, but Kung (1961, 1963) suggests that $a = 1.24$ and $b = 1.19$, and Sellers (1965) that $a = 1.385$ and $b = 1.417$. However the variations in their values is probably not significant and a good approximation is $z_0 = 0.13h$. In practice for natural vegetation that yields to the wind, the roughness length tends to decrease with increasing wind speed, presumably because at high speeds the vegetation bends and presents a relatively flat and smooth surface to the air flow.

For tall vegetation, equation (5.16) becomes

$$u_z = \frac{1}{k} \left(\frac{\tau}{\rho} \right)^{1/2} \ln \frac{z - d}{z_0} \tag{5.19}$$

where d is called the zero-plane displacement, since the theoretical wind speed goes to zero at the height $z_0 + d$, rather than at z_0. The plane $z = d + z_0$ can be regarded as an apparent sink of momentum in the canopy, but the linear relation between u and $\ln (z_0 - d)$ is not valid below the top of the roughness elements and the real windspeed is finite at the height $d + z_0$. Stanhill (1969) reviewed a large number of measurements of d for vegetation ranging in height h from 0.2 to 20 m and established a linear relation between $\log d$ and $\log h$:

$$\log d = 0.9793 \log h - 0.1536 \tag{5.20}$$

Monteith (1973) comments that the determinations of d were too scattered to justify the quotation of constants so precisely and the much simpler relation $d = 0.63h$ fits them just as well.

The spiral layer
Below the top of the friction layer the wind blows obliquely across the isobars with a component towards lower pressure at a speed below that of the geostrophic wind. The angle of obliqueness increases and the speed decreases with increasing depth into the friction layer. This wind spiral with height is analogous to the turning of ocean currents as the effect of wind stress diminishes with increasing depth, and is known as the 'Ekman spiral'. The variation of wind speed and direction in the spiral layer can be explained by the assumption that the eddy viscosity K_M is constant with height. This also indicates that the depth of the friction layer is proportional to $(K_M/\Omega \sin \phi)^{1/2}$, where the eddy viscosity K_M is defined by

$$K_M = \frac{k^2 u_z z}{\ln (z/z_0)} \tag{5.21}$$

where k is Karman's constant, and the other symbols are as previously defined. The above equation indicates that large values of K_M are to be expected with rough surfaces, strong winds and strong instability and that these will create deeper friction layers than smooth surfaces, light winds and stable conditions. The Coriolis parameter is weaker in lower latitudes and this suggests that deeper friction layers may exist in the tropics as compared with higher latitudes.

The influence of the surface on wind speed is shown by the following calculations by Lettau (1959) and Deacon (1969). Assuming a geostrophic wind speed of 25 m sec^{-1} at latitude 43°, Deacon showed that over scrub with a roughness length of 100 cm, the wind at 10 m above ground level would be 7 m sec^{-1}; but if the scrub were cleared and replaced by bare soil with a roughness length of 0.1 cm, the 10 m wind speed would increase to 13 m sec^{-1}. He comments that in a semi-arid climate a wind speed of 13 m sec^{-1} is sufficient to cause soil to start blowing away, and so the vital role that wind-breaks and shelter belts play in such situations becomes clear.

2.4 Convection

Turbulence in general in the atmosphere is partly controlled by the stability, for when the atmosphere is stable, turbulence is on a small scale with no thermals and the transfer of heat and water vapour upwards from the surface is slow. When an inversion is present, that is, when temperature increases with altitude, then the atmosphere is very stable and the transfer of heat and water vapour upwards through the inversion is very slow. In contrast, if the atmosphere is unstable there is a large amount of convective activity, turbulence is marked, and under these conditions heat and water vapour are transferred rapidly upwards. In a city under inversion conditions, it is often very hazy and smoky because there is little vertical exchange of air. If the atmosphere becomes slightly unstable, it often becomes reasonably clear because of the vertical turbulent mixing throughout the lower troposphere. In summary it can be said that atmospheric as well as surface conditions control the rate of supply of energy and water vapour to the lower troposphere.

3 Evapotranspiration

In meteorology, evaporation is the change of liquid water or ice to water vapour, and it proceeds continuously from the earth's free water surfaces, soil, snow and ice-fields. Transpiration is the process by which the liquid contained in the soil is extracted by plant roots, passed upwards through the plant, and discharged as water vapour to the atmosphere; the rate of transpiration is highest during the day and falls almost to zero during night hours. Evapotranspiration is the combined process of evaporation from the earth's surface and transpiration from vegetation. Potential evapotranspiration is the maximum amount of water vapour that can be added to the atmosphere under the given meteorological conditions from a surface covered by green vegetation with no lack of available water.

Evapotranspiration from a land surface depends primarily on the radiant energy supply to the surface, but may be limited by the rate of movement of water to the evaporating surface. Evapotranspiration from natural surfaces is a physical process in which liquid water is vaporized and carried into the atmosphere. According to King (1961), three simultaneous dynamic processes are involved, which are:

1 a flow of water vapour by turbulent and molecular diffusion from the evaporating surface to the atmosphere;
2 a flow of heat by radiation, convection, and conduction to the evaporating surface and the removal therefrom as latent heat of vaporization; and
3 a flow of water through the soil and plants to the evaporating surface.

Turbulent transfer
The water vapour from an evaporating surface first passes through a thin laminar layer next to the surface where only molecular diffusion takes place, and is then transported upwards by turbulent motions. The basic equation for the mean vertical transfer of water vapour (E) is

$$E = - \rho K_v \frac{\partial q}{\partial z} \tag{5.22}$$

where K_v is the eddy diffusivity for water vapour, and q is the specific humidity (mass of moisture per unit mass of moist air). According to Thornthwaite and Holzmann (1939), evapotranspiration can be calculated using the following propositions:

1 the transfer factor for momentum is identical with that for water vapour: $K_M = K_v$;
2 the shearing stress is constant with height;
3 the wind velocity u_z at the height z may be expressed by a logarithmic wind function.

These propositions lead, in a neutral atmosphere, to the following equation for evapotranspiration (E_t):

$$E_t = \frac{\rho k^2 (u_2 - u_1)(q_1 - q_2)}{\left[\ln \dfrac{(z_2 + z_0 - d)}{(z_1 + z_0 - d)} \right]^2} \tag{5.23}$$

The last equation is very similar to the empirical Dalton equation, which relates evapotranspiration to the prevailing wind velocity and the vapour pressure gradient or degree of atmospheric saturation $(e_s - e_a)$. The Dalton equation may be stated as follows:

$$E_t = f(\bar{u})(e_s - e_a) \tag{5.24}$$

where $f(\bar{u})$ is an empirically derived function of the wind velocity, usually given in the form

$$f(\bar{u}) = a(1 + b\bar{u}) \tag{5.25}$$

where $\bar{u}$ is the time averaged wind velocity measured at standard height and a and b are constants.

There are difficulties in making practical use of equations of this type, because some of the terms are unknown or difficult to measure.

Energy supply
Evapotranspiration requires a supply of energy to be used as latent heat of vaporization and this is the second of King's three dynamic processes. Neglecting the amount of energy used for photosynthesis and not taking into account the supply of advective energy, the energy balance can be written as follows:

$$R_N = (1 - \alpha) R_s - R_{NL} = LE_t + C + S + G \tag{5.26}$$

where R_N is the net radiation, α is the albedo of the surface, R_s is the global short-wave radiation, R_{NL} is the net long-wave radiation, L is the latent heat of vaporization, E_t is the evapotranspiration, C is the sensible heat transfer to the atmosphere, S is the sensible heat transfer to the soil, and G is the storage of heat in the crop.

The main difficulty in using the energy balance approach arises from the distribution of energy between latent heat and sensible heat transfer to the atmosphere.

The ratio of the sensible-heat flow to the latent-heat flow is known as the Bowen ratio, and can be written as

$$\beta = \frac{C}{LE_t} = \gamma \left(\frac{K_H}{K_V}\right)\left(\frac{T_s - T_a}{e_s - e_a}\right) \tag{5.27}$$

where γ is the psychrometer constant (about 0.65), K_H is the eddy conductivity, T_s and T_a are respectively the surface and air temperature, and e_s and e_a are the vapour pressure respectively at the surface and in the air.

With the Bowen ratio, the expression for the energy used for evapotranspiration becomes:

$$LE_t = \frac{R_N - S}{1 + \beta} \tag{5.28}$$

Both the aerodynamic and energy balance approaches require detailed and difficult measurements, but Penman (1948) has succeeded in combining both methods in a formula for the calculation of evapotranspiration from standard meteorological data. He neglected the storage of heat in the soil, assumed a saturated vapour pressure at the surface, and derived the following equation:

$$E_t = \frac{\Delta R_N/L + \gamma E_a}{\Delta + \gamma} \tag{5.29}$$

where Δ is the slope of the temperature–vapour pressure curve at air temperature and E_a (aerodynamic evaporation) is calculated from an aerodynamic relationship in the form

$$E_a = (e_s - e_a)(f(\bar{u})) \tag{5.30}$$

Aerodynamic relationships of this type have been discussed earlier, when a variety of forms were suggested. The values of the constants will clearly depend on the nature of the surface, and one particular formulation for E_a, adopted by Penman (1956) is

$$E_a = 0.165(V_s - V_a)(0.8 + u_2/100) \tag{5.31}$$

in which E_a is in mm day^{-1}, V_s and V_a are respectively the saturated and actual vapour pressures at screen level in mbar and u_2 is the wind speed at a height of 2 m in km day^{-1}.

Normally values of the net radiation balance are not available for use in Penman's equation, and therefore under these conditions, the net radiation has to be calculated indirectly. Penman suggested the following empirical formula for calculating the net radiation from climatological observations including the observed hours of bright sunshine:

$$R_N = R_{ss}(1 - \alpha)(a + bn/N) - \sigma T^4(0.56 - 0.08\sqrt{V_a})$$
$$(0.1 + 0.9n/N) \tag{5.32}$$

where R_{ss} is the solar radiation incident on an horizontal surface in the absence of the earth's atmosphere, n/N is the ratio of actual to possible hours of bright sunshine, a and b are constants, and σT^4 is the full blackbody radiation at mean air temperature.

Because of changes in cloud structure, the constants a and b are not universal and vary from region to region; suitable regional values are suggested in Table 5.3, and Glover and McCulloch (1958) consider that while b is effectively constant, a varies with the latitude ϕ up to 60°:

$$a = 0.29\cos\phi \tag{5.33}$$

Table 5.3 Regional regression coefficients for use in Penman's equation (*after Black, Bonython, and Prescott, 1954 and Chia, 1969*)

Locality	Latitude	Regression a	Constants b
Rothamsted (England)	51.8°N	0.18	0.55
Gembloux (Belgium)	50.6°N	0.15	0.54
Versailles (France)	48.8°N	0.23	0.50
Mt Stromlo (ACT Australia)	35.3°S	0.25	0.54
Dry Creek (South Australia)	34.8°S	0.30	0.50
Singapore	1.0°N	0.25	0.476

Flow of water to evaporating surface

The third of the dynamic processes involved in evapotranspiration, suggested by King, is the flow of water through the soil and plants to the evaporating surface, and this flow will be controlled by the apparent diffusion resistance of the soil and crops. According to Rijtema (1968) the apparent diffusion resistance depends on the following factors:

1 soil area covered by crops and leaf area;
2 light intensity in relation to stomatal opening;
3 availability of soil moisture and transport resistances for liquid flow in the plant.

Few data are available concerning the effect of partial plant cover of the soil on transpiration, but Marlatt (1961) considers that row crops covering more than 50 per cent of the soil act as a full cover crop in relation to evapotranspiration. The effect of light intensity on stomatal resistance is well understood in the laboratory, but is difficult to study in the field. Evidence exists (Burgy and Pomeroy, 1958 and Van Bavel *et al.*, 1963) that under conditions of high light intensity, stomatal resistance has no influence and evapotranspiration is an externally controlled process.

The soil water storage capacity is primarily a function of the rooting depth of the vegetation and the water retention characteristics of the soils on which it grows. The amount of water in the soil which is available for plant growth is generally considered to be that retained at soil water potentials below 15 atmospheres. Often, the available water content of the soil is defined as the volumetric difference in water content at soil water potential of 15 atmospheres and 0.3 atmospheres. For most plants, growth ceases at soil water potentials above 15 atmospheres, but evapotranspiration will continue at a greatly reduced rate. Clearly the evapotranspiration rate will partly depend on the available water, which in a soil is normally finite and limited. If evapotranspiration is calculated using Penman's equations, a value will be obtained which assumes that there is always a continuous and unrestricted supply of water. In practice the water supply may be restricted and the actual evapotranspiration rates will be considerably less than the values calculated using Penman's equations. The evapotranspiration resulting under conditions of unlimited and unrestricted water supply is known as the potential evapotranspiration. Because soil water conditions are normally unknown, it is usual in climatological discussions to quote the estimated potential evapotranspiration rather than the actual evapotranspiration.

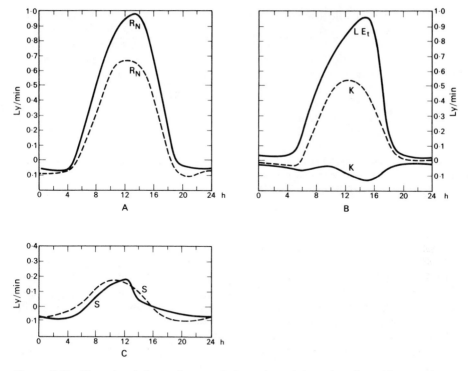

Figure 5.7 Diurnal variations of energy balance in an irrigated oasis and in a semi-desert. Observations from Turkestan semi-desert and an adjacent oasis. Solid line: oasis (irrigated); dashed line: semi-desert.
A: diurnal variation of net radiation.
B: diurnal variation of sensible heat transfer to the atmosphere (K) and latent heat transfer (LE_t).
C: diurnal variation of sensible heat transfer to the soil (after Budyko, 1956)

Oasis effects

Little has been said so far about the size of the evaporating surface, and this particular problem is best illustrated by considering the extreme case of an oasis in a perfectly dry desert. Figure 5.7 illustrates the energy balance of the oasis and of the desert. The net radiation over the desert is lower than that over the basis, because the desert sands have a higher albedo than the vegetation of the oasis, and therefore it might appear that the oasis should have a higher temperature than the desert. In the desert region all the available net radiation is transferred as sensible heat to the atmosphere and ground, there being no water available for evaporation; but in the oasis large amounts of energy are used in evaporating water and therefore the sensible heat fluxes are small. The air arriving over the oasis is dry, and so evaporation rates are high; often they are so high that sensible heat will be drawn from the atmosphere to supply energy for evapotranspiration. Under these conditions the oasis will be cooler than the desert and evapotranspiration rates will be above those calculated from simple radiation balance considerations. If a large oasis is considered, the very high evapotranspiration rates will only be found along the windward edge, because as the air moves deeper into the oasis it will become more moist and slightly cooler and the sensible heat transfer from the atmosphere to the ground will cease. The example quoted is extreme, but clearly the evapotranspiration from a surface will be partly controlled by the size of the surface and partly by the nature of the surrounding conditions. Unusually high evapotranspiration rates from a small surface are often said to arise because of an oasis effect.

3.1 The Bowen ratio

The study of the energy balance of natural surfaces falls into two stages. First, there is the study of the radiation balance, which leads to an estimation of the available net radiation. Secondly, the net radiation has to be divided amongst the sensible heat flows to the soil and the atmosphere, and the latent heat flow to the atmosphere, to produce the full energy balance. The ratio of the sensible heat flow to the atmosphere to the latent heat flow is known as the Bowen ratio, β, and can be written as

$$\beta = \frac{\text{sensible heat loss to the atmosphere } (C)}{\text{latent heat loss to atmosphere } (LE_t)} \tag{5.34}$$

In the absence of atmosphere advection, β can vary between $+\infty$ for a dry surface with no evaporation to zero for an evaporating wet surface with no sensible heat loss. If there is atmospheric heat advection, β may become negative indicating a flow of heat from the atmosphere to the surface.

Both the sensible and latent heat fluxes in the vertical can be written in the convenient, almost symmetrical, forms:

$$C = -\rho\, C_p\, K_H\, \frac{\partial T}{\partial z} \tag{5.35}$$

$$LE_t = -\rho\, \frac{C_p}{\gamma}\, K_V\, \frac{\partial e}{\partial z} \tag{5.36}$$

where

ρ is the density of moist air

C_p is the specific heat of air at constant pressure;

K_V and K_H are the eddy diffusivities for water vapour and heat respectively;

$\frac{\partial e}{\partial z}$ and $\frac{\partial T}{\partial z}$ are the vertical gradients of vapour pressure and temperature respectively;

L is the latent heat of vaporization of liquid water.

The thermodynamic value of the psychometric constant, γ, is given by:

$$\gamma = \frac{C_p p}{0.62L} \tag{5.37}$$

γ becomes 0.66 if the following typical values are introduced into the above equation:

$C_p = 0.240 \text{ cal}°C^{-1}g^{-1}$
$L = 585 \text{ cal g}^{-1}$
$p = 1000 \text{ mbar}$

Therefore, from the above equations it is possible to write for the Bowen ratio:

$$\beta = \gamma \frac{\partial T}{\partial e} \cdot \frac{K_H}{K_V} \tag{5.38}$$

Often the convenient simplification is made that $K_H = K_V$, giving:

$$\beta = \gamma \frac{\partial T}{\partial e} \tag{5.39}$$

Priestley and Taylor (1972) have developed further the idea of the Bowen ratio. To a good approximation the specific humidity q is given by:

$$q = 0.622 \frac{e}{p} \tag{5.40}$$

It is therefore possible to replace e by q in the relationship for the Bowen ratio, and from the equation for γ obtain:

$$\beta = \frac{C_p}{L} \frac{\partial T}{\partial q} = \frac{C_p}{L} S \tag{5.41}$$

This is a useful relationship because the profiles of specific humidity and temperature are often similar if the surface is saturated. Priestley and Taylor (1972) also introduce a term α, where

$$\frac{LE_t}{LE_t + C} = \alpha \frac{S}{S + \gamma} \tag{5.42}$$

For free evaporation from a uniform saturated surface, with no advection effects, they found that $\alpha = 1.26$. If α is taken as 1.26, it follows that:

$$\beta = \frac{C}{LE_t} = \left(1 - 1.26 \frac{S}{S + \gamma}\right) \Big/ \left(1.26 \frac{S}{S + \gamma}\right) \tag{5.43}$$

which has been found to be a function of the surface temperature. Values of β calculated on this basis are given in Figure 5.8. This figure suggests that over a well watered surface the Bowen ratio decreases as the temperature increases, or the proportion of available energy going into latent heat increases. At temperatures above 32°C the sensible heat flow becomes negative implying a flow of heat from the air to the evaporating surface. This would seem to suggest that the highest temperature which can be reached over a freely evaporating surface with the net radiation values experienced on the earth's surface is about 32°C. There is evidence to support this particular view. Priestley (1966) examined the average daily maximum temperature for each month reported by island observing stations and by land stations after periods of heavy rain. His conclusion was that, in the radiation climates that actually exist in nature, air temperatures over a well watered surface do not rise above 32–34°C. Similarly, Linacre (1967) examined the relationship between the temperature of freely evaporating leaves in bright sunshine and of the air and reached the conclusion that leaves are hotter than the air up to about 33°C and, above that, they are cooler.

The Bowen ratio will also rise above that appropriate for a freely evaporating surface if the evaporation is limited for some reason. This may be because the soil becomes dry and the vegetation is under stress, or because the plants themselves limit evaporation.

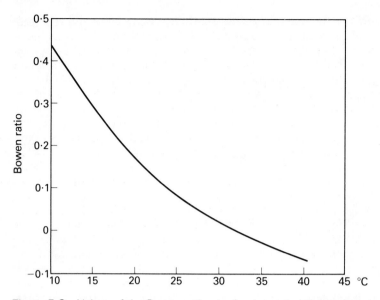

Figure 5.8   Values of the Bowen ratio at a freely evaporating surface for various surface temperatures (T_0). This diagram only applies to a freely evaporating surface, for if the surface becomes dry, the value of the Bowen ratio will increase to above that appropriate for the temperature T_0.

Bowen ratio values can rise to high values if the plant itself restricts transpiration from its leaves and examples of this are found in coniferous forests. According to Jarvis *et al.* (1976), measurements over coniferous forests fall into two groups. For most coniferous forest sites, irrespective of species, the day-time Bowen ratio of a dry canopy varies between 0.1 and 1.5. When the canopy is wet, and evaporation rather than transpiration is occurring, it is reduced to between – 0.7 and + 0.4. However, at two United Kingdom sites, Thetford (52°25′N, 00°39′E) and Fetteresco (56°38′N, 02°24′W), much larger values of β have been found consistently, and similar large values are found occasionally in other extensive records. At Thetford, the Bowen ratio rises to a fairly steady value of between 1 and 4 soon after sunrise; in the afternoon, it may decrease or, in conditions leading to stomatal closure, may climb to even higher values. At Fetteresco, it may reach 2 to 3 in the middle of the day, but declines again in the afternoon.

4 Hydrology of grasslands

When bare soil is thoroughly wetted, the soil surface behaves like water in so far as the relative humidity of air in contact with the surface is 100 per cent. According to Monteith (1981) the rate of evaporation is usually very close to the rate for adjacent short vegetation, despite differences in radiative and aerodynamic properties. Loss of water from the soil surface establishes a gradient of water potential which drives water towards the surface from deeper, wetter layers. This process cannot continue indefinitely because the conductivity of soil for water decreases very rapidly as it dries and it is usually only a few days before the rate of evaporation becomes limited by the upward diffusion of liquid water towards the surface.

In the extratropics, with its large seasonal changes, the soil plays a role analogous to that of the ocean (Shukla and Mintz, 1982). The ocean stores some of the radiational energy it receives in summer and uses it to heat the atmosphere over the ocean in winter. The soil stores some of the precipitation it receives in winter and uses it to humidify the atmosphere in summer. According to Shukla and Mintz, vegetation and clouds play complementary roles. The clouds convert atmospheric water vapour into liquid water, which is transferred to the soil; the vegetation converts soil water into water vapour which is transferred to the atmosphere.

There are three basic requirements for significant land-surface evapotranspiration: moisture in the soil; vegetation, to transfer the moisture from the soil to the interface with the atmosphere; and energy, to convert that moisture to water vapour. During the temperate summer and in warm climates much of the energy for evaporation comes from radiational heating of the surface and therefore depends on surface albedo. In cool seasons and climates much of the energy for evaporation comes from sensible heat advection from the atmosphere. During rain, radiation totals may be small and again much of the energy for evaporation may come from atmospheric heat advection. Therefore evaporation under low radiation conditions is going to depend on the rate of atmospheric heat advection which in turn will depend on the turbulence and stability structure of the lower atmosphere. The amount of turbulence depends partly on the height and roughness length of the vegetation. Indeed, recent work (e.g. Thorn and Oliver, 1977) has indicated that interception loss from wet tall crops can be a large term in the total water balance of a small catchment.

4.1 Resistances

A dense vegetation canopy effectively shields the underlying surface from the effects of solar radiation and wind. This raises the level of the 'active' surface above the level of water concentration and reduces evaporation at the lower level. Nevertheless, in response to a water potential gradient in the liquid phase, water is ordinarily conducted from soil to roots to stems to leaves in living plants, and evaporation occurs at the liquid–atmosphere interface within leaves. This process, and the diffusion of water vapour to the external atmosphere, is called foliar transpiration.

Transpiration from a vegetation canopy covering unit horizontal area A is the sum of individual vapour fluxes from a greater total area of leaf surface A_s. The leaf-area index LAI – A_s/A – is usually expressed in terms of the projected area of a flat leaf (area of one side). In order for LAI to be useful in transpiration studies, however, it is important to stipulate that A_s represents the area of leaf surface through which stomatal transpiration occurs (flat leaves may have stomata on one or both sides). Maximum leaf area indexes are shown in Table 5.4.

The leaves of most land plants consist of cells supplied with water by an elaborate plumbing system and assembled in such a way that evaporation from cell walls keeps the air spaces between them almost saturated with vapour even when the leaf is transpiring. Transpiration is by the loss of water through stomatal pores whose opening and closing is controlled by the hydraulic behaviour of guard cells acting as valves. Monteith (1981) comments that in response to light, stomata open and water vapour escapes from within the leaf, but the degree of opening is reduced when the supply of water from the soil reservoir is restricted or when the temperature of a leaf is too high or low. When stomata are closed, leaves continue to lose water much more slowly through a thin waxy cuticle.

Boundary layer and surface resistances may be defined using Ohm's law in elec-

Table 5.4   Maximum leaf area indexes (*after Thompson et al.*, 1981)

Crop	Green leaf area index
Grass, riparian land	2.0 (Jan), 2.0, 3.0, 4.0, 5.0, 5.0 (June)
	5.0, 5.0, 4.0, 3.0, 2.5, 2.0 (Dec)
Cereals	5.0
Potatoes	4.0
Sugar beet	4.0
Deciduous trees	6.0
Conifers	6.0[1]

[1]   constant throughout year

tricity as a direct analogue. Ohm's law states that the electrical resistance of a wire is equal to the potential difference between its ends divided by the current flowing through it: i.e.

$$\text{electrical resistance} = \frac{\text{potential difference}}{\text{current}} \qquad (5.44)$$

The analogue relationship is obtained by replacing 'potential' by concentration (meaning amount per unit volume) and 'current' by flux (amount per unit area per unit time), so that:

$$\text{boundary layer or surface resistance} = \frac{\text{concentration difference}}{\text{flux}} \qquad (5.45)$$

Monteith (1981) states that for many plant species the spacing between stomata is of the order of 0.1 mm whereas the boundary layer attached to a leaf surface, whether laminar or turbulent, usually has a thickness of several millimetres. The resistances of individual pores therefore behave as if they were wired in parallel with each other and with the cuticle which they perforate and this compound physiological resistance of the leaf surface r_{ss} can be treated as if it were placed in series with the boundary layer resistance r_B. For many plant species the minimum value of r_{ss} for each leaf surface containing stomata is of the order of $100\,\text{s m}^{-1}$, whereas cuticular resistances are between 2,000 and $4,000\,\text{s m}^{-1}$.

The exchange of heat and water vapour between a stand of vegetation and the atmosphere is a much more complex process than the corresponding exchange at the surfaces of individual leaves (Monteith, 1981). Nevertheless Monteith considers that the same basic principles are valid: for every element of foliage throughout the canopy, fluxes of sensible and latent heat are limited by different resistances. A surface resistance r_s for the whole canopy may be defined. Within a canopy there are likely to be systematic vertical differences in the distribution of sources and sinks for heat, water vapour and momentum. It is therefore not evident a priori whether r_s can be regarded as a physiological resistance depending mainly on stomatal components or whether it contains a significant aerodynamic component. In the field, the diurnal and seasonal behaviour of r_s has been determined by measuring the rate of evaporation from vegetation by some independent method. Working with barley, Monteith *et al.* (1965) found that r_s was independent of windspeed (implying a negligible aerodynamic component) and was close to a value estimated for all the component leaves treated as parallel resistors i.e.

Table 5.5 Some stand or crop resistances (r_s s m^{-1}) (*after Perrier, 1982*)

Alfalfa	40	Citrus	250
Barley	70	Coniferous forest	200–300
Cotton	130	Deciduous forest	100–150
Maize	80	Tropical forest	100–300
Potato	70	Grassland (temperate)	100
Rice	80	Grassland (subtropical)	200
Sugarbeet	50	Tundra	400
Sunflower	40		
Wheat	60		

r_{ss}/LAI where LAI is the leaf area index and r_{ss} is the mean stomatal resistance. Similar agreement has now been demonstrated for a range of crops, e.g. for sorghum (Szeicz, *et al.* 1973), for beans (Black, Tanner and Gardner, 1970), for sugar beet (Brown and Rosenburg 1977), and for Douglas Fir (Tan and Black 1976). Monteith (1981) states that further circumstantial evidence for the physiological significance of r_s is provided by its diurnal and seasonal behaviour. Like stomatal resistance, r_s usually increases during the day, particularly when the soil is dry, and increases with the average age of leaves. Typical values of r_s are given in Table 5.5.

4.2 Penman-Monteith evaporation equation
It is often found that it is possible to combine the in-canopy resistances, the stomatal and boundary layer resistances, assuming they act in parallel at a single level in the canopy. The resulting bulk stomatal and boundary layer resistances, r_s and r_B respectively, are then considered to act at an 'effective source height' somewhere close to the apparent sink of momentum ($d + z_0$). In single source models of this type it is no longer relevant to separate the 'boundary layer' and 'eddy diffusive' resistances, since it is more usual to combine them as a single 'aerodynamic' resistance r_A.

Figure 5.9 illustrates a single-source model for the partition of energy into latent and sensible heat fluxes. The latent heat flux differs from the sensible heat flux in that it is subject to the additional stomatal resistance, r_s, in dry canopy conditions. The temperature and humidity are measured at a 'source height' above the vegetation, and have the values T and e respectively; the effective temperature and humidity adjacent to the dry leaf surface are T_0 and e_0. Inside the stomata the air is saturated with a vapour pressure $e_s(T_0)$. From equation (5.45), the latent heat flux is given by (see also equations 5.35 and 5.36)

$$LE = \frac{\rho C_p}{\gamma} \frac{(e_s(T_0) - e)}{r_A + r_s} \qquad (5.46)$$

while the sensible heat flux is given by

$$C = \rho C_p \frac{(T_0 - T)}{r_A} \qquad (5.47)$$

The mean gradient Δ of the saturated vapour pressure curve is defined by

$$\Delta = \frac{e_s(T_2) - e_s(T_1)}{T_2 - T_1} \qquad (5.48)$$

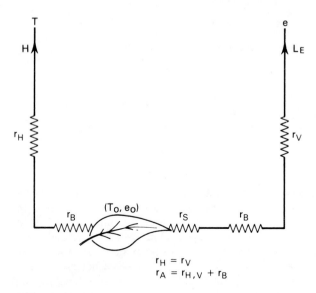

$$r_H = r_V$$
$$r_A = r_{H,V} + r_B$$

Figure 5.9   Schematic diagram of the resistance network used in 'single-source' models of the vegetation/atmosphere interaction. r_s and r_B are the bulk stomatal and boundary layer resistances. r_H and r_V are the bulk transfer resistance for sensible and latent heat, r_H and r_V are usually taken as equal and combined in series with the boundary layer resistance to give the 'aerodynamic resistance' r_A (after Shuttleworth, 1979).

Substituting this last equation into equation (5.47) gives

$$C = \frac{\rho C_p}{\Delta} \frac{(e_s(T_0) - e_s(T))}{r_A} \tag{5.49}$$

Energy conservation requires that

$$C = R_N - LE \tag{5.50}$$

substituting this last equation into equation (5.49), and eliminating $e_s(T_0)$ between the resulting equation and equation (5.46) gives the result

$$LE = \frac{\Delta R_N + \dfrac{\rho C_p}{r_A}(e_s(T) - e)}{\Delta + \gamma(1 + r_s/r_A)} \tag{5.51}$$

This equation, generally called the Penman–Monteith equation, is the fundamental expression used in simple one-dimensional descriptions of the evaporation process, and is the basis of all the more empirical techniques used in estimating evaporation.

4.3 Water balance

Figure 5.10 summarizes the water balance of a small catchment covered by a specified vegetation type over a long time period. The water balance of the catchment may be expressed by the two equations

$$P - I = pP + D + \Delta C \tag{5.52}$$

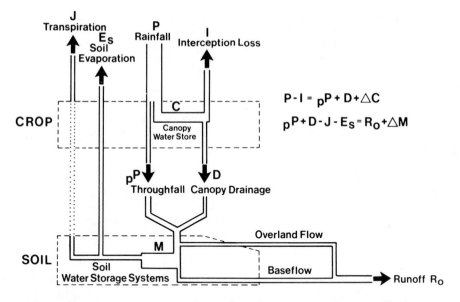

Figure 5.10 The water balance of a small catchment covered by a specified vegetation type over a long time period.

$$pP + D - J - E_s = R_0 + \Delta M \qquad (5.53)$$

where P is the rainfall rate.
  I is the interception loss rate.
  p is the throughfall proportion.
  D is the drainage rate of the whole canopy.
  C is the canopy water storage.
  J is the transpiration loss.
  E_s is the soil evaporation.
  R_0 is the run-off.
  M is the soil moisture storage.

Following the Penman–Monteith equation for evaporation, transpiration from vegetation may be expressed by:

$$LJ = \frac{\Delta R_N + \rho C_p\, \delta e / r_A}{\Delta + \gamma\,(1 + r_s/r_A)} \qquad (5.54)$$

where Δ is the slope of the saturated vapour pressure versus temperature curve.
  R_N is the net radiation;
  δe is the vapour pressure deficit;
  ρ and C_p are the density and specific heat of air respectively;
  γ is the psychrometric constant;
  r_s is the surface resistance;
  r_a is the aerodynamic resistance.

For the evaporation rate I of intercepted water when the canopy is wet, r_s becomes zero. Assuming that the albedo of wet and dry canopies are not materially different, then

$$\frac{J}{I} = \frac{\Delta + \gamma}{\Delta + \gamma\,(1 + r_s/r_A)} \qquad (5.55)$$

When r_s and r_A are of similar size, as they may be in herbaceous communities, then the rate of evaporation of intercepted water will not much exceed the potential transpiration rate in the same conditions. Indeed, for grass it is often assumed that they are the same. In contrast, when r_s is an order of magnitude greater than r_A, as it appears to be in coniferous forest, then intercepted water will be evaporated at 2 to 5 times the current transpiration rate. For well watered arable crops $r_s \approx r_A \approx 50\,\text{s}\,\text{m}^{-1}$.

Thus for grassland areas J, I and E_s are often lumped together and considered as the total evapotranspiration E_t. It will be shown later that this is satisfactory for grassland areas but that it cannot be safely applied to forests.

4.4 Potential evapotranspiration

On the basis of the experimental evidence available, it was for many years believed that the type and form of vegetation cover on the earth's surface (even whether it was there or not) had little effect on the rate of natural evaporation, providing this was limited by the heat supplied to the surface and not by the availability of surface water (Shuttleworth, 1979). Thus in the presence of this belief it is reasonable to conceive an entity, potential evapotranspiration, which might be conservative at a particular location and determined mainly by meteorological conditions. It can be defined (Gangopadhyaya *et al.*, 1966) as, 'the maximum quantity of water capable of being lost, as water vapour, in a given climate, by a continuous, extensive stretch of vegetation covering the whole ground when the soil is kept saturated'. It includes both evaporation from the soil and from the vegetation for a specified region over a given time interval. The concept of potential evapotranspiration is widely used today, and has some validity as a conservative entity serving as a climatological term to which medium-term measurements of actual evaporation in different geographical locations can be related.

Shuttleworth (1979) comments that it has gradually become clear that the initial observation of a lack of dependence on vegetation cover when water is non-limiting, and the idea that potential evapotranspiration represents a maximum rate (which implies that energy advection is a scarce and transient phenomenon), might be related to the fact that many of the original studies took place over short crops. With such crops the control exerted by the atmosphere itself is maximized, since it dictates not only the driving potential in the diffusion process, but it also generates the dominant, possibly controlling, resistance to vapour transfer. This realization, coupled with the continuing desire to create a conservative but surface-independent entity, has given rise to the recreation by Doorenbos and Pruitt (1977) of a more closely defined standard evaporation rate, reference crop evapotranspiration. This entity is defined as 'the rate of evapotranspiration from an extensive surface of 8 to 15 cm tall, green grass cover of uniform height, actively growing, completely shading the ground and not short of water'. Shuttleworth (1979) states that there is considerable overlap between the concept of potential evapotranspiration and reference crop evapotranspiration, particularly in regard to the empirical formulae used to estimate them, but the more exact definition of reference crop evaporation avoids the problem of vegetation control and advection, and therefore increases the probable universality of locally derived empirical equations.

4.5 Soil moisture models and actual evaporation

According to Hounam *et al.* (1975) the following concepts describing the extraction of water from the soil by plants have been used in the last 30 years.

(i)   Evapotranspiration independent of soil moisture. A very simple approach but not reliable.

(ii) Evapotranspiration at potential rate, but reducing in a very dry soil. A variation of this method is used by Penman (1963), who assumes that plants transpire all water from the 'root zone', plus 25 mm of water drawn from below this level, at the potential rate. When this supply is exhausted, transpiration is reduced to one-tenth of the potential rate.

(iii) Evapotranspiration linearly proportional to soil water. This model has been used by a number of workers, notably Thornthwaite and Mather (1955), under conditions of high evaporative demand.

(iv) Evapotranspiration decreasing exponentially. Numerous models of this type have been developed showing a wide variation according to vegetation and soil type and possibly also season as soil density and canopy change. Some models incorporate periods when water content is high, during which evapotranspiration is at the potential rate; thereafter exponential decay takes place.

(v) Multi-level models. In its simplest form water storage is separated into two levels, the 'upper level' which contains most of the plant roots, and the 'lower level' which contains fewer roots. Water in the 'upper' zone is depleted at the potential rate and any deficiency in this zone must be satisfied by rainfall before re-charge of the lower zone commences. Depletion from the lower zone occurs only when there is no water available in the upper zone, the rate of evapotranspiration being proportional to the amount of water available in the lower zone.

Two-stage models of the last type are frequently used in water-balance calculations. Sometimes it is considered that the stages do not necessarily correspond to physical layers in the soil but rather the physical states of the soil during the drying process. Thus evapotranspiration is assumed to take place from the first stage at the potential rate, and it is also assumed that precipitation initially enters the first stage. When the first stage is dry, evapotranspiration (E_a) takes place from the second stage such that:

$$E_a = E_t \frac{dm}{ds} \qquad (5.56)$$

where E_t is the potential evapotranspiration; ds is the maximum available water content of the second stage; and dm is the actual available water content of the second stage. When the first stage becomes saturated, excess water not forming evapotranspiration or run-off is transferred to the second-stage storage. After rainfall the vegetation canopy will normally be wet and the surface acts as a saturated one regardless of the soil-moisture state, so intercepted and surface water may be included in the first stage for the purpose of calculating evapotranspiration. It is assumed that direct run-off originates mostly from the first stage of the soil-moisture store. Water in the second stage of the soil-moisture store does not normally form run-off, since this stage represents water which is tightly bound in the soil as compared with that in the first stage.

Shuttleworth (1983) suggests that actual evaporation may be determined from the equation

$$LE = K_c LE_{RC} \qquad (5.57)$$

where LE_{RC} is the reference crop evapotranspiration, L being the latent heat of vaporization;

and K_c is called the crop coefficient.

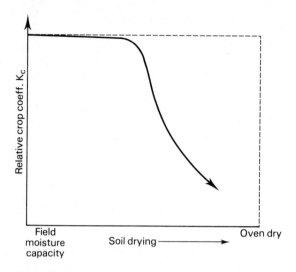

Figure 5.11 Variation in relative crop coefficient K_c in response to changing soil moisture content (after Shuttleworth, 1983).

Shuttleworth comments that studies have been made of the variation in K_c in response to decreasing soil water content. While the results differ considerably in detail, as is to be expected for this type of empirical parameter, the overall behaviour of the crop coefficient during a drying cycle generally follows the pattern shown in Figure 5.11. This behaviour seems to be generally similar for both crops and soil: during the first, fairly constant, stage of the drying cycle, K_c remains quite close to its initial value until a 'break point' is reached, when it begins to decrease in response to declining soil water content. In conditions of prolonged drought, the crop (if present) begins to die and the evaporation rate is then controlled by soil characteristics, especially hydraulic conductivity.

Estimates of evapotranspiration and soil moisture in a typical 8.1 km² grassland Pennine catchment are shown in Figure 5.12. Precipitation and run-off were measured while potential evapotranspiration was estimated by Penman's method. The potential evapotranspiration follows very close by the annual march of global radiation, being at a maximum in summer and a minimum in winter. In winter soil-moisture values are near the upper limit, though the soil is not saturated all the time. During the summer evapotranspiration exceeds rainfall and the soil dries, causing the available soil moisture to reach low values. Under summer conditions the first-stage soil moisture store soon becomes exhausted and actual evaporation values (from equation 5.56) fall below the potential rate.

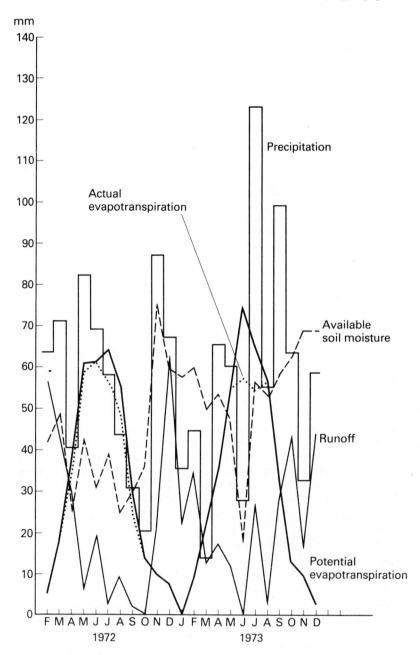

Figure 5.12 Crimple Beck Catchment, North Yorkshire. Monthly values of evapotranspiration, runoff, and soil moisture. All values are monthly totals except soil moisture which is the end of month value. Units: rainfall equivalents in mm (after Lockwood and Venkataswamy, 1975).

References

AHMAD, S. B. and LOCKWOOD, J. G. 1979: Albedo. *Progress in Physical Geography* 3, 510–43.

ÅNGSTRÖM, A. 1916: Über die Gegenstrahlung der Atmosphäre. *Meteorologische Zeitschrift* 33, 529–38.

BILLINGS, W. D. and MORRIS, R. J. 1951: Reflection of visible and infrared radiation from leaves of different ecological groups. *American Journal of Botany* 38, 327–31.

BLACK, J. N., BONYTHON, C. W. and PRESCOTT, J. A. 1954: Solar radiation and the duration of sunshine. *Quarterly Journal of the Royal Meteorological Society* 80, 231–5.

BLACK, T. A., TANNER, C. B. and GARDNER, W. R. 1970: Evapotranspiration from a snap-bean crop. *Agronomy Journal* 62, 66–9.

BROWN, K. W. and ROSENBERG, N. J. 1977: Resistance model to predict evapotranspiration and its application to a sugar-beet field. *Agronomy Journal* 65, 341–7.

BRUNT, D. 1941: *Physical and Dynamical Meteorology*. London: Cambridge University Press.

BRUTSAERT, W. 1982: *Evaporation into the Atmosphere*. Dordrecht: Reidel.

BUDYKO, M. I. 1956: *The Heat Balance of the Earth's Surface*. Translated by N. I. Stepanova. Washington, DC: US Weather Bureau.

—— 1974: *Climate and life*. New York: Academic Press.

BURGY, R. H. and POMEROY, C. R. 1958: Interception losses in grassy vegetation. *Transactions of the American Geophysical Union*, 39, 1095–100.

CHIA, L. S. 1969: Sunshine and solar radiation in Singapore. *Meteorological Magazine* 98, 265–74.

DEACON, E. L. 1969: Physical processes near the surface of the earth. In Flohn, H. (ed.), *General Climatology* 2. Amsterdam: Elsevier, 39–104.

DOORENBOS, J. and PRUITT, W. O. 1977: Crop water requirements. *FAO Irrigation and Drainage Paper* 24. Rome: FAO.

FLEAGLE, R. G. and BUSINGER, J. A. 1963: *An Introduction to Atmospheric Physics*. New York: Academic Press.

FRITSCHEN, L. J. 1967: Net and solar radiation relations over irrigated field crops. *Agricultural Meteorology* 4, 55–62.

GADD, A. J. and KEERS, J. F. 1970: Surface exchanges of sensible and latent heat in a 10-level model atmosphere. *Quarterly Journal of the Royal Meteorological Society* 96, 297–308.

GANGOPADHYAYA, M., URYVAEV, V. A., OMAR, M. H., NORDENSON, T. J. and HARBECK, G. E. 1966: Measurement and estimation of evaporation and evapotranspiration. In *Technical Note 83*, Geneva: World Meteorological Organization.

GLOVER, J. and MCCULLOCH, J. S. G. 1958: The empirical relation between solar radiation and hours of sunshine. *Quarterly Journal of the Royal Meteorological Society* 84, 172–5.

GRAHAM, W. G. and KING, K. M. 1961: Short-wave reflection coefficient for a field of maize. *Quarterly Journal of the Royal Meteorological Society* 87, 425–8.

HENDERSON-SELLERS, A. and HUGHES, N. A. 1982: Albedo and its importance in climate theory. *Progress in Physical Geography* 6, 1–44.

HOUNAM, C. E., BURGOS, J. J., KALIK, M. S., PALMER, W. C. and RODDA, J. 1975: Drought and agriculture. *Technical Note* 38. Geneva: World Meteorological Organization.

IDSO, S. B., REGINATO, R. J. and JACKSON, R. D. 1977: Albedo measurements for remote sensing of crop yields. *Nature* 266, 625–8.

IMPENS, I. and LEMEUR, R. 1969: The radiation balance of several field crops. *Archiv für Meteorologie, Geophysik und Bioklimatologie Series B* 17, 261–8.

JARVIS, P. G., JAMES, E. B. and LANDSBERG, J. J. 1976: Coniferous forest. In Monteith, J. L. (ed.) *Vegetation and the Atmosphere* 2, London: Academic Press, 171–240.

KING, K. M. 1961: Evaporation from land surfaces. In *Proceedings of Hydrology Symposium No. 2: Evaporation*. Ottawa: National Research Council of Canada, 55–82.

KONDRATYEV, K. YA. 1954: *Radiant solar energy*. Leningrad: Gidrometeoizdat.

KONDRATYEV, K. YA., KORZOV, V. I., MUKHENBERG, V. V. and DYACHENKO, L. N. 1982: The shortwave albedo and the surface emissivity. In Eagleson, P. S. (ed.) *Land Surface Pro-*

cesses in Atmospheric General Circulation Models. Cambridge: Cambridge University Press, 463–514.

KUKLA, G. J. 1981: Surface albedo. In Berger, A. (ed.) *Climatic Variations and Variability: Facts and Theories.* Dordrecht: Reidel, 85–109.

KUNG, E. C. 1961: Derivation of roughness parameters from wind profile data above tall vegetation. In *Studies of the Three-Dimensional Structure of the Planetary Boundary Layer*, Annual Report, 1961. Madison: Department of Meteorology, University of Wisconsin.

—— 1963: Climatology of aerodynamic roughness parameter and energy dissipation in the planetary boundary layer of the northern hemisphere. In *Studies of the Effects of Variations in Boundary Conditions in the Atmospheric Boundary Layer*. Annual Report 1963. Madison: Department of Meteorology, University of Wisconsin.

KUNG, E. C., BRYSON, R. A. and LENSCHOW, D. H. 1964: Study of a continental surface albedo on the basis of flight measurements and structure of the earth's surface cover over North America. *Monthly Weather Review* 92, 543–64.

LETTAU, H. H. 1959: Wind profile, surface stress and geostrophic drag coefficients in the atmospheric surface layer. *Advances in Geophysics* 6, 241–57.

LINACRE, E. T. 1967: Further notes on a feature of leaf and air temperatures. *Archiv für Meteorologie Geophysik und Bioklimatologie*, Ser. B. 422–36.

LOCKWOOD, J. G. and VENKATASWAMY, K. 1975: Evapotranspiration and soil moisture in upland grass catchments in the eastern Pennines. *Journal of Hydrology* 26, 79–94.

MARLATT, W. E. 1961: The interactions of microclimate, plant cover and soil moisture content affecting evapotranspiration rates. *Technical Paper 23*. Fort Collins: Department of Atmospheric Science, Colorado State University.

MONTEITH, J. L. 1959: The reflection of short-wave radiation by vegetation. *Quarterly Journal of the Royal Meteorological Society* 85, 386–92.

—— 1973: *Principles of Environmental Physics.* London: Edward Arnold.

—— 1981: Evaporation and surface temperature. *Quarterly Journal of the Royal Meteorological Society* 107, 1–27.

MONTEITH, J. L. and SZEICZ, G. 1961: The radiation balance of base soil and vegetation. *Quarterly Journal of the Royal Meteorological Society* 87, 159–70.

—— 1962: Radiation temperature in the heat balances of natural surfaces. *Quarterly Journal of the Royal Meteorological Society* 88, 496–507.

MONTEITH, J. L., SZEICZ, G. and WAGGONER, P. E. 1965: The measurement and control of stomatal resistance in the field. *Journal of Applied Ecology* 2, 345–55.

MUNN, R. E. and TRUHLAR, E. J. 1963: The energy budget approach to heat transfer over the surface of the earth. *Transactions of the Engineering Institute of Canada* 6. B-7, 1-20.

NKEMDIRIM, L. C. 1972: A note on the albedo of surfaces. *Journal of Applied Meteorology* 11, 867–74.

OKE, T. R. 1978: *Boundary Layer Climates*, London: Methuen.

PENMAN, H. L. 1948: Natural evaporation from open water, base soil, and grass. *Proceedings of the Royal Society* Series A, 193, 120–45.

—— 1956: Evaporation. An introductory survey. *Netherlands Journal of Agricultural Science* 4, 9–29.

—— 1963: *Vegetation and Hydrology.* Commonwealth Bureau of Soils technical communication, 53. Harpenden.

PERRIER, A. 1982: Land surface processes: vegetation. In Eagleson, P. S. (ed.) *Land Surface Processes in Atmospheric General Circulation Models.* Cambridge: Cambridge University Press, 395–448.

PIGGIN, I. and SCHWERDTFEGER, S. 1973: Variations in the albedo of wheat and barley crops. *Archiv für Meteorologie, Geophysik und Bioklimatologie Ser. B21*, 365–91.

PRIESTLEY, C. H. B. 1966: The limitation of temperature by evaporation in hot climates. *Agricultural Meteorology 3*, 241–6.

PRIESTLEY, C. H. B. and TAYLOR, R. J. 1972: On the assessment of surface heat flux and evaporation using large-scale parameters. *Monthly Weather Review* 100, 81–92.

RAUNER, J. L. 1976: Deciduous forests. In Monteith, J. L. (ed.) *Vegetation and the Atmosphere* 2, London: Academic Press, 241–64.

RIJTEMA, P. E. 1968: Derived meteorological data: transpiration. In *agroclimatological methods – Proceedings of Reading Symposium*. Paris: UNESCO, 55–72.

SELLERS, W. D. 1965: *Physical Climatology*. Chicago: University of Chicago Press.

SHUKLA, J. and MINTZ, Y. 1982: Influence of land-surface evapotranspiration on the earth's climate. *Science* 215, 1498–501.

SHUTTLEWORTH, W. J. 1979: Evaporation, *Report No. 56*. Wallingford: Institute of Hydrology.

—— 1983: Evaporation models in the global water budget. In Street-Perrott, A., Beran, M. and Ratcliffe, R. (eds.) *Variations in the Global Water Budget*. Dordrecht: Reidel, 147–71.

STANHILL, G. 1969: A simple instrument for the field measurement of turbulent diffusion flux. *Journal of applied Meteorology* 8, 509– .

STANHILL, G., HOFSTEDE, G.F. and KALMA, J.D. 1966: Radiation balance of natural and agricultural vegetation. *Quarterly Journal of the Royal Meteorological Society* 92, 128–40.

STEWART, J. B. 1971: The albedo of a pine forest. *Quarterly Journal of the Royal Meteorological Society* 97, 561–4.

SWINBANK, W. C. 1963: Long-wave radiation from clear skies. *Quarterly Journal of the Royal Meteorological Society* 89, 339–48.

SZEICZ, G., VAN BAVEL, C. H. B. and TAKAMI, S. 1973: Stomatal factors in the water use and dry matter production of sorghum. *Agricultural Meteorology* 12, 361–89.

TAN, C. S. and BLACK, T. A. 1976: Factors affecting the canopy resistance of a Douglas-fir forest. *Boundary-Layer Meteorology* 10, 475–89.

TANNER, C. B. and PELTON, W. L. 1960: Potential evapotranspiration. Estimates by the approximate energy balance method of Penman. *Journal of Geophysical Research* 65, 3391–413.

THOM, A. S. and OLIVER, H. R. 1977: On Penman's equation for estimating regional evaporation. *Quarterly Journal of the Royal Meteorological Society* 103, 29–46.

THOMPSON, N., BARRIE, I. A. and AYLES, M. 1981: The Meteorological Office rainfall and evaporation calculation system: MORECS (July 1981), *Hydrological Memorandum 45*, Bracknell: Meteorological Office.

THORNTHWAITE, C. W. and MATHER, J. R. 1955: The water budget and its use in irrigation. In *Water – the yearbook of agriculture 1955*. Washington: US Department of Agriculture, 546–58.

THORNTHWAITE, C. W. and HOLZMANN, B. 1939: The determination of evaporation from land and water surfaces. *Monthly Weather Review* 67, 1–8.

VAN BAVEL, C. H. M., FRITSCHEN, L. J. and REEVES, W. E. 1963: Transpiration by Sudan grass as an externally controlled process. *Science* 141, 269–70.

6 Forest Subsystems

Because of the vast extent and distribution of forest, it is very difficult to find a generally valid description. In different countries it includes areas differing markedly in density, height, stratification and diversity of species. Typical distinguishing features of a forest are the height of the crowns above the ground and depth of the roots in the soil. Generally classed as a forest is a biome with trees taller than five metres on an area of at least $100 \, m^2$, with crowns covering a minimum of one-third of this area. The determining factor which establishes a forest as such is whether the trees already form a distinctive environment.

The composition of forests is greatly influenced by their geographic location, or more precisely by their latitude, longitude, altitude, distance from the sea, and by the prevailing currents and winds. The basic geographical division of forests is as follows.

(1) The far northern regions of Eurasia and North America are covered with coniferous forests (taiga). Similar forests (mountain taiga) may be found on all high mountains.

(2) In the temperate regions of Eurasia and North America the forests are composed of deciduous broadleaved trees, giving way to mixed forests towards the taiga.

(3) In regions of both hemispheres having a Mediterranean climate the forests are non-deciduous, sclerophyllous.

(4) In the subtropical and tropical regions where the year is divided into a rainy season and dry season there are continuous savanna woodlands and open tree savannas.

(5) The tropical and equatorial regions with constant precipitation throughout the year are covered with dense, evergreen forests.

The features of grassland vegetation systems outlined in Chapter 5 also apply to forests, and in many respects it is acceptable to view trees as being merely the large-scale counterparts of small plants. According to Oke (1978) forests however do have some special features that have climatic importance. First, the stand architecture is often clearly demarcated. The forest canopy is commonly quite dense, whereas the trunk zone may be devoid of foliage. Second, the increased biomass and the sheer size of the stand volume contribute to the possibility that heat and mass storage may not be negligible over short periods. Third, the height and shape of the trees leads to greater radiation trapping and a very much rougher surface. Fourth, especially in the case of coniferous forests, the stomatal control may be different from that of small plants.

Table 6.1 Mean values of coefficients of albedo, interception and transmission of short-wave radiation according to phenophases and cloud cover (*after Galoux et al., 1981*)

Forest type and author	Cloud cover	Foliated phase			Leafless phase		
		Radiation			Radiation		
		Reflected	Intercepted	Transmitted	Reflected	Intercepted	Transmitted
		Long-term continuous measurements					
Oak forest, Virelles (Grulois, 1968)	Clear sky (10 days)	0.155	0.789	0.056	0.112	0.688	0.200
	Overcast (10 days)	0.194	0.724	0.082	0.165	0.444	0.390
	Rainy (10 days)	0.195			0.167		
Beech forest, Solling (Kiese, 1972)	Clear sky (June–Sept)	0.145	0.786	0.069		0.097 (May)	
	Overcast (June–Sept)	0.125	0.780	0.095			
	Periodic and instantaneous values						
Pinus taeda, N. Carolina (Gay & Knoerr, 1975)	Spring	0.107					
	Autumn	0.128					
Scots pine and beech (Lützke, 1966)		0.132				0.06	
Scots pine, Great Britain (Stewart, 1971)		0.109					
Scots pine, Great Britain (Rutter, 1986)		0.100–0.150					
Picea sitchensis, Picea glauca Great Britain (Jarvis, 1976)		0.150					

1 Forest microclimates

1.1 Radiation

Forest albedo varies with both the state of the sky and also with the zenith angle of the sun. This was discussed for short vegetation types in Chapter 5. Thus Grulois (1968) reported that oak forests during the leafless period have an albedo of 0.112 for clear sky and 0.166 for overcast days; during the foliated period, this changes from 0.155 for clear sky to 0.195 for cloudy skies. Variations in albedo with growth phases are very important in deciduous forests. Thus Grulois (1968) reports that albedo is small (0.105-0.116) during the leafless period in oak forest at Virelles, Belgium. During spring the albedo increases very quickly to 0.20 and remains high during the foliated period in summer. In the autumn during the yellowing of foliage, albedo rises to 0.22 and remains relatively high when the brown foliage falls to the ground, only returning to typical winter values in February. Some typical mean values of forest albedo are given in Table 6.1.

The distribution of radiation in forests depends mainly on the optical properties of individual plant elements and of the canopy as a whole. In forests an exponential function (Lambert–Beer Law) has proved to be a good approximation to the downward flux of short-wave radiation S_t (Rauner, 1976). The average irradiance, $S(z)$, at any level, z, in a canopy may be related to the irradiance above the canopy, $S_t(0)$, an extinction coefficient, κ, and the leaf area index, $L(z)$, calculated from the top of the canopy down to level z,

$$S(z) = S_t(0) \exp(-\kappa L(z)) \tag{6.1}$$

Few extinction coefficients have been reported for forests (see Table 6.2) because of the difficulties in obtaining good estimates of leaf area index and in adequately sampling to determine the mean irradiance (Jarvis *et al.*, 1976). The extinction coefficients listed in Table 6.2 give a general idea of the size of κ in coniferous forest

Table 6.2 Comparative extinction coefficients κ of various stands during the middle of the day (*after Jarvis et al., 1976*)

Species	L^a		Type of radiation	Condition[b]	
Pinus resinosa	2.6	June–July	visible	o	0.35-0.45
Pinus resinosa/strobus	3.1[d]	May–Sept.	short-wave	s	0.28-0.57
				o	0.39-0.45
Pinus sylvestris	4.3	May–Sept.	short-wave	s/o	0.46
Pinus sp.	2.7[d]	—	net	s	0.41
Pseudotsuga menziesii	5.5[d]	—	net	s/o	0.42
			visible	s/o	0.79
Picea abies	8.4[d]	Summer	short-wave	s	0.28
Picea sitchensis	9.8	June–July	visible	s	0.56
			visible	o	0.49
			near infra-red	s	0.53
			near infra-red	o	0.48
		June–August	net	s[c]	0.55
			net	o	0.58

[a] Leaf area index of the stand on a projected leaf area basis.
[b] s, o and s/o are sunny, overcast and mixed conditions respectively.
[c] Sun overhead.
[d] Projectd area taken as total area × 0.389; $(\pi/2 + 1)^{-1} = 0.389$.

canopies but must be regarded as approximate. Under a clear sky the extinction coefficient is a function of the elevation β of the sun above the horizontal. Landsberg *et al.* (1973) found that the extinction coefficient for net radiation in the upper part of the Sitka spruce canopy at Fetteresco increased linearly with increasing (cosec β − 1) from a minimum of about 0.5 in the middle of the day to values of about 2.5 to 3.0 when the sun is near the horizon in the early morning and evening.

Yabuki *et al.* (1978) have studied the microclimate of a warm-temperate evergreen broadleaf forest in eastern Asia. This broadleaf forest formation is a climatic climax almost exclusively confined to the east coasts and adjacent islands of Asia. It is the counterpart of the Mediterranean sclerophyll forest on the opposite side of the Eurasian continent, but differs from the latter in its physiognomy and floristic composition, corresponding to differences in the seasonal distribution of rainfall as well as the continentality of climate. Yabuki *et al.* found that relative solar radiation decreased more or less exponentially with aboveground height, more than 50 per cent of the light being intercepted by the main canopy layer, about 4 m from the canopy surface. That on the forest floor under a cloudy sky tended to be greater in fall and winter seasons than in summer, reaching a maximum of about 10 per cent in March, whereas it was the least in fall and winter (around 3–4 per cent) on fine days, increasing during the growing season. The mean light profiles for two years could be approximated by

$$I_Z = I_H \exp\left[-K(1 - \frac{Z}{H})\right] \tag{6.2}$$

where I stands for solar radiation;
       H for the height of canopy surface;
  and Z for a height level within the forest.
The coefficient of light interception, K, was nearly equal to 2.92. The corresponding value for a Malaysian tropical rain forest was 2.67 (Aoki *et al.*, 1975). In the Malaysian case (Pasoh Forest) Aoki *et al.* (1978) report that 60 per cent of the incoming solar radiation was absorbed by the upper 5 m of the canopy, and only 3 per cent reached the ground surface.

Grulois (1967) obtained for the oak forest of Virelles IBP Site, during overcast days, the constants

$$Y = 12.45 + 170.3e^{-0.7728x} \tag{6.3}$$

where Y is the percentage of global radiation at a certain height and x is the distance, in metres, from the upper canopy surfaces. Kiese (1972) in beech forest at Solling IBP Site found

$$Y = 9.5 + 198.5e^{-0.35x} \tag{6.4}$$

Very few vertical profiles of net radiation have been measured through forest stands (Galoux *et al.*, 1981). Baumgartner (1956, 1957) has measured a radiation profile (Table 6.3) in a young spruce forest in Bavaria. In the Solling beech forest (IBP Site) at 26.8 m height with the first branches at 12 m, Kiese (1972) observed a larger balance value at a small distance below the highest living branches, than above the total canopy. This comes from an infrared emission of these branches at high temperature in the middle of the day. Values of net radiation above or under the herbaceous layer under the tree canopy depend strongly on the radiation exchange at the level of the canopy, and are usually very low (see Table 6.3). In autumn the net radiation above and under the canopy in deciduous forests

Table 6.3 Profile of net radiation balance (spruce forest, Bavaria, after Baumgartner 1956, 7)

Height (m)		Day		Night	
		(cal cm^{-2})	(%)	(cal cm^{-2})	(%)
10.5	Above canopy	565	100	− 22	100
5.0	Upper canopy level	555	98	− 21	96
4.1	Sun crowns	223	39	− 4	18
3.3	Shaded crowns	36	6	− 2	9
0.2	Trunks	35	6	− 1	5

approach each other. In deciduous forests the main active exchange layer during winter is the soil surface or herbaceous layer.

1.2 Temperature and humidity

During the day with clear skies, forest leaf surfaces are warmer than the surrounding air, but at night it is generally the reverse. Gay and Knoerr (1975) found in 1965 over a *Pinus taeda* forest a maximum difference of + 9.5°C at 1200 h in May. The largest differences found at the Virelles IBP site (oak forest) was 4.7°C, 1 May 1966, in the leafless period (Galoux and Grulois, 1968) and + 6.1°C in July in the foliated period (Galoux, 1973). Yabuki *et al.* (1978) found for evergreen oak forest in April that the upper layer of the forest became warmer as the sun rose at a greater rate than did lower strata; the temperature difference between the two parts reached 4°C at 10 h. From around 15 h in the afternoon, the temperature at the canopy surface began to drop, resulting in maximum temperatures at 4–8 m above ground. Aoki *et al.* (1978) found for tropical rainforest at Pasoh (West Malaysia) IBP site that the maximum air temperature appeared at the top of the canopy surface at 47 m, while the night temperatures are fairly evenly distributed inside the stand space. During the day the temperature gradient between the soil surface and the top of the canopy at 47 m reached up to 8°C, but most of the temperature gradient was concentrated in the top 7 m of the forest canopy.

Yabuki *et al.* (1978) found that the relative humidity changes in evergreen oak forest in April corresponded to those in temperature. Relative humidity started to drop after sunrise until a minimum of about 40 per cent was reached on the canopy surface at 14 h. Similarly Aoki *et al.* (1978) found for the Pasoh IBP site that relative humidity was about 100 per cent throughout the tropical rainforest at night but decreased to round 50 per cent in the top few metres of the canopy during the day.

Table 6.4 shows the properties of the boundary layer above oak forest at Virelles IBP site. During the day there is a marked gradient of temperature and vapour density above the canopy, the direction of the gradients being reversed at night. During the day the canopy is warmer than the air above while at night it is colder.

The exact level at which the maximum temperatures occur in the forest canopy during the day depend on the canopy structure of the tree species. It will vary between well layered species and those in which the foliage is highly concentrated at one level.

Table 6.4 Boundary layer above the forest canopy (between + 16 and + 25 m (°C/100 m) (oak forest, Virelles, after Galoux, 1973)*

Hour	2	4	6	8	10	12	14	16	18	20	22	24
a. Temperature gradient in air (1967) (°C/100 m)												
May	−3.6	−3.9	−3.7	0.4	2.4	3.1	3.3	3.6	0.8	−1.3	−3.7	−3.9
June	−3.3	−3.3	−1.5	2.3	4.5	4.6	4.1	3.6	1.7	−0.4	−3.7	−3.5
July	−5.6	−5.3	−4.7	1.7	5.8	7.3	6.3	5.0	2.0	−1.7	−4.8	−6.5
August	−4.5	−4.5	−3.0	0.3	2.6	4.1	3.4	1.6	1.5	−3.2	−4.7	−5.1
September	−3.2	−3.1	−2.8	−0.7	2.0	3.4	2.4	0.4	−1.0	−3.2	−3.4	−3.3
December	−0.3	−1.4	−1.4	−1.0	−1.4	−2.8	−1.4	−0.9	−0.7	−1.1	−1.5	−1.2
b. Vapour density gradient in air (1967) (g m^{-3}/100 m)												
May	−0.1	−0.1	0.2	1.1	0.3	0.0	0.6	−0.7	−0.4	0.1	0.6	0.4
June	0.7	0.6	0.9	2.3	2.8	3.8	4.0	3.9	3.8	2.2	0.9	1.1
July	−0.9	−0.9	0.2	2.8	3.2	3.7	3.6	4.3	2.0	0.8	0.0	0.1
August	−0.1	−0.5	−0.2	2.1	3.2	4.0	3.8	3.6	2.9	1.5	0.3	0.3
September	−0.8	−0.7	−0.3	1.1	2.1	3.0	3.9	3.4	2.2	0.8	0.1	0.4
December	−1.8	−2.0	−1.8	−1.5	−1.7	−1.0	−1.1	−0.9	−1.9	−1.8	−1.7	−2.0

* Oak forest Virelles-Blaimont, Belgium.

1.3 Vertical exchanges

The exchanges of water vapour, carbon dioxide, heat and momentum between a canopy and the bulk air above it, depend upon the turbulent exchange properties of the wind profile generated in the boundary layer above the canopy (Jarvis *et al.*, 1976). These are determined by wind speed and by the properties of the canopy. Under near-neutral conditions the wind-speed, u, at a height z is given by the well known logarithmic equation described in Chapter 5.

The term r_a in the Penman–Monteith equation (equation 5.51) expresses the 'aerodynamic' resistance to the diffusion of water vapour from the surface itself, where vapour pressure has an unknown value e_0, to some reference level above, where the vapour pressure is e. In ideal conditions of neutral atmospheric stability and large fetch, the resistance to the transfer of momentum to the effective surface of vegetation from a distance z above its zero plane displacement level d (see e.g. Thom, 1975) is

$$r_a = \{\ln(z/z_0)\}^2/(k^2u) \tag{6.5}$$

in which z_0 is the aerodynamic roughness parameter of the vegetation (order $h/10$ where h is vegetation height) u is wind speed, and k is von Karman's constant. However, according to Thom and Oliver (1977); with non-neutral conditions and for water vapour exchange, the theoretically rigorous expression

$$r_a = \{\ln(z/z_0) - \psi\}\{\ln(z/z_0') - \psi'\}/(k^2u) \tag{6.6}$$

applies, in which each ψ (see Paulson, 1970) is a specified function of the stability of the surface boundary layer most conveniently expressed in terms of the gradient Richardson number R_i, and where a prime distinguishes a parameter of water vapour exchange from one of momentum exchange. Because of the enhancement of momentum transfer to a rough surface by the action of bluff-body forces (pressure drag) on its individual elements, z_0 generally exceeds z_0' (Thom and Oliver, 1977). In particular, for vegetation z_0' is usually about one-fifth as large as z_0 (Thom, 1972, 1975). Nevertheless, in view of the many sources of error in estimating regional evaporation, Thom and Oliver consider that the simplified expression

$$r_a = \{\ln(z/z_0) - \psi\}\{\ln(z/z_0) - \psi'\}/(k^2u) \tag{6.7}$$

and equation (6.5) itself, adequately specify resistances to water vapour exchange. Clearly the theory could be developed further, but the important point that emerges is that both the height and the surface wetness of the vegetation will influence evaporation rates. Indeed, Thom and Oliver consider that where tall vegetation predominates, attempts to unravel the water balance of a vegetated region will be unrewarding unless there is adequate recognition of the effects of surface roughness and surface wetness.

2 Forest hydrology

Vegetation may intercept water falling in the form of snow, hail or rain, as well as droplets of water in low cloud and fog which would not otherwise be precipitated. In the initial stage of interception by a dry canopy, much of the water is retained. There appears, however, to be a fairly well defined storage capacity for any given vegetation canopy, and when this is succeeded, further intercepted water either drips from the canopy or runs down the stems. The combination of water which

Table 6.5 Interception storage capacities (*after Rutter, 1975*)

Vegetation		C(mm)
Coniferous forest		
Pinus sylvestris		1.6
Picea abies		1.5
Pseudotsuga menziesii		2.1
Pinus nigra		1.0
Deciduous forest		
Carpinus betulus	summer	1.0
	winter	0.6
Old *Quercus robur* coppice	summer	1.0
	winter	0.4
Ericaceous		
Calluna vulgaris		2.0
Herbaceous		
Zea mais		0.4–0.7
Mixed grasses and legumes		1.0–1.2
Lolium perenne, 10 cm high		0.5
48 cm high		2.8
Molinia caerulea		0.7
Pteridium aquilinum		0.9

drips from the canopy and falls directly through gaps is usually called throughfall and the sum of throughfall and stemflow is net precipitation. The difference between gross precipitation and net precipitation is often called interception loss, the loss representing water which is evaporated from plant surfaces.

The storage capacity of a canopy can be estimated approximately as the constant C in the regression equations for individual storms:

$$\text{net precipitation} = b(\text{gross precipitation}) - C \qquad (6.8)$$

The coefficient b will be unity only if the storms are large enough to wet the canopy completely, that is, consisting of continuous falls with negligible evaporation, separated by periods long enough for all the surface stored water to be evaporated, some typical storage values are given in Table 6.5.

Precipitation that is intercepted by a forest canopy constitutes a source of water for evaporation that is basically unlike any other source. Intercepted water is not only directly exposed to the atmosphere but also spread over a much larger surface area than it might otherwise occupy. This suggests that the rate of evaporation of intercepted water I could exceed the rates from other sources under the same atmospheric conditions. The distinction between aerodynamic and physiological resistances has proved to be particularly useful in predicting the loss of water from foliage fully wetted by rain. Transpiration from vegetation may be expressed by equation (5.54). For the evaporation of intercepted water r_s is zero and this equation becomes

$$LI = \frac{\Delta R_N + \rho C_p \delta e / r_a}{\Delta + \gamma} \qquad (6.9)$$

Assuming that the albedos of wet and dry canopies are not materially different, then the ratio of J to I is given by equation (5.55). When r_s and r_a are of similar size, as they may be for many well-watered arable crops, then the rate of evaporation of intercepted water will not much exceed the potential transpiration rate in the same conditions. Indeed, for grass it is often assumed that they are the same. In contrast, when r_s is an order of magnitude greater than r_a, as it appears to be in coniferous forest, then intercepted water will evaporate at 2 to 5 times the current transpiration rate.

Summaries of the results from many traditional interception studies were made by Helvey and Patric (1965) for mature hardwood forests of the eastern United States and by Helvey (1971) for conifers in North America. The general form of prediction for throughfall, stemflow, and interception loss for individual storms is a linear regression with gross precipitation as the independent variable. A composite of the summaries is given in Table 6.6 (from Waring *et al.*, 1981), with estimated interception loss assuming 1 cm of precipitation. Annual or periodic throughfall and stemflow can be computed by solving the equations for annual (or periodic) gross rainfall and multiplying the constant term by the number of storms in which gross rainfall equals or exceeds the constant term. Total rainfall delivered in storms which are smaller than the constant terms in the equation, is added to the difference between precipitation – (throughfall + stemflow) to obtain interception loss.

Table 6.6 Summary of equations for computing throughfall and stemflow for coniferous[a] and hardwood[b] forests from measurements of gross rainfall (*after Waring et al., 1981*)

Species	Average equations (gross rainfall in cm)		Interception loss (cm)
	Throughfall	Stemflow	
Red pine	0.87p–0.04	0.02p	0.15
Loblolly pine	0.80p–0.01	0.08p–0.02	0.15
Shortleaf pine	0.88p–0.05	0.03p	0.14
Ponderosa pine	0.89p–0.05	0.04p–0.01	0.13
Eastern white pine	0.85p–0.04	0.06p–0.01	0.14
Average (pines)	0.86p–0.04	0.05p–0.01	0.14
Spruce-fir-hemlock	0.77p–0.05	0.02	0.26
Mature mixed hardwoods			
Growing season	0.90p–0.03	0.041p–0.005	0.10
Dormant season	0.91p–0.015	0.062p–0.005	0.05

[a] Conifer equations from Helvey (1971).
[b] Hardwood equations from Helvey and Patric (1965).

Table 6.7 shows for a variety of forest types interception loss expressed as percentage of precipitation. It is seen that actual values vary widely with both forest type and annual rainfall.

From observations in a stand of Corsican pine (*Pinus nigra*) Rutter *et al.* (1971) developed a model (Figure 6.1) of the evaporation of intercepted rainfall. The canopy is regarded as having a surface storage capacity, S, which is charged by rainfall and discharged by evaporation and drainage. The rate of input of water to canopy storage is $(1 - p)P$, where P is the rainfall intensity and p is the fraction which falls through gaps in the canopy.

When the amount of water (C) on the canopy equals or exceeds S, the evaporation rate is given by the standard equation (6.9).

Table 6.7A Interception loss in various equatorial rain forest types

Forest type	Site (* denotes IBP site)	Year	Precipitation (P, mm)	Interception (% P)	Throughfall (% P)	Stem flow (% P)
Sub-equatorial forest (Huttel, 1975)	Yapo (*) (Côte d'Ivoire)	1969–71	1950.0	22.0		→78.0←
	Banco (*) (Côte d'Ivoire)	—		—		
	Plateau	1969–71	1800.0	10–12		→88–90←
	Talweg	1969–71	1800.0	10–12		→88–90←
Tropical rain forest (Odum, Moore & Burns 1972)	El Verde (Puerto-Rico)	—	3759	26.4	55.6	18.0
Alstonia scholaris plantation (Banerjee, 1973)	Arabari Range (Bengal)	10/1/70-9/30/71	1622.8	21.3	—	—
Alstonia scholaris plantation (Ray, 1970)	Arabari Range (Bengal)	—	—	21.6 to 36.3	48.7–59.1	13.4–22.9
Shorea robusta plantation (Ray, 1970)	Arabari Range (Bengal)	—	—	16.5 to 35.4	59.3–74.2	5.3–10.1
Shorea curtisii forest (Low, 1972)	Sungai Lui Catch (W. Malaysia)	9 months	—	36.0	—	—
Tropical moist forest (McGinnis et al., 1969)	(Panama)	—	—	17.0	—	—
Tropical wet forest (McColl, 1970)	(Costa Rica)	—	—	—	94.5	—

[1] →values← are sums of colums 6 and 7.

Table 6.7B Interception loss in various tropical forest types

Forest type	Site	Year	Precipitation (P, mm)	Interception (% P)	Throughfall (% P)	Stem flow (% P)
Pine Forest (Smith, 1974)	Lat. 33° 24'S Long. 150°E (Australia)	10/17/68-4/28/71	861.6	18.7	78.3	<3
Eucalypt Forest (Smith, 1974)	Lat. 33° 24'S Long. 150°E (Australia)		861.6	10.6	86.4	<3
Pinus longifolia plantation (Dabral et al. 1968)	Dehra Dun (India)	Summer monsoon, 1960	—	22.1	74.3	3.6
Tectona grandis plantation (Dabral & Rao, 1968)	Dehra Dun (India)	Summer monsoon, 1960	—	27.0	69.7	3.3
Shorea robusta plantation (Dabral & Rao, 1969)	Dehra Dun (India)	Wet period, 1961	2159	38.2	54.6	7.2
Acacia catechu plantation (Dabral et al. 1969)	Dehra Dun (India)	Wet period, 1961	2159	28.5	67.3	4.2
Eucalyptus globulus plantations Thomas, Chanara Sekhar & Haldorai, 1972)	Nilgiris water sh. (India)	—	1300	—	—	—
Tropical forest (Hopkins, 1960)	Mpanga Res. For. (Uganda)	Two wet periods	1130	33.6	66.4	—

Table 6.7C Interception loss in various mediterranean forest types

Forest type	Site (* denotes IBP site)	Year	Precipitation (P, mm)	Interception (% P)	Throughfall (% P)	Stem flow (% P)
Quercus ilex forest (Ettehad et al., 1973)	Le Rouquet (*) France La Madeleine (*) (France)	1967–70	987.7	30.7	65.1	4.2
		1966–70	636.6	33.7	61.2	5.0
Foothill woodland (Major, 1967)	Rocklin (California)	–	569.0	–	–	–
Chaparral (Rowe, Storey & Hamilton 1951)	North Fork (California)	1934–8	1149.6	5.0	80.0	15.0
Chaparral (Rowe, Storey & Hamilton 1951)	Tanbark Flat (California)	1942–5	528.6	11.1	80.5	8.4

Table 6.7D Interception loss in various temperate deciduous forest types

Forest type	Site (* denotes IBP site)	Year	Precipitation (P, mm)	Interception (% P)	Throughfall (% P)	Stem flow (% P)
Mixed oak forest (Schnock, 1970)	Virelles (*) (Belgium)	1964–8	965.9	17.0	76.8	6.7
Oak-hornbeam forest (G. Schnock & A. Galoux (unpublished))	Ferage (*) (Belgium)	1964–8	898.0	30.7	65.6	3.7
Oak-hornbeam forest (G. Schnock, R. Dalebroux & A. Galoux (unpublished))	Virelles (*) (Belgium)	1964–8	994.0	16.6	80.3	3.1
Oak-hazel forest (G. Schnock & A. Galoux (unpublished))	Ferage (*) (Belgium)	1964–8	898.0	28.3	68.5	3.2

Forest (study)	Location	Period				
Oak-hornbeam forest (Intribus, 1975)	Bab (*) (Czechoslovakia)	11/71–10/72	771.9	17.8	78.6	3.6
Oak-hornbeam forest (Ukrecky, Smolik & Lanar, 1974)	Brno (*) (Czechoslovakia)	4/71–10/71	38.5	24.5	75.5	—
Mixed deciduous forest (White & Carlisle, 1968)	Meathop (*) (United Kingdom)	12/66–11/67	1554.4	14.2	77.3	8.5
Oak-ash-lime forest (Molchanov, 1971)	Tellermanovsky (*) (USSR)	1952–61	513.0	14.4	—	—
Oak (*Q. daschorochensis*) forest (Cepel, 1967)	Near Istamboul (Turkey)	5 yr	1032.7	20.0	69.1	10.9
Oak-pine forest (Bodeux, 1954)	Campine (Belgium)	1950	849.6	26.8.	65.5	7.7
Old birch coppice (Noirfalise, 1959)	Ottignies (Belgium)	1945–6 (one year)	876.9	29.8	63.9	6.3
Beech forest (Benecke & van der Ploeg, 1975)	Solling (Germany)	10/31/69	896.3	11.8	88.2	
Beech forest (Lemee, 1974)	Fontainebleau* (France)	8/19/70-8/16/73	554.0	—	74.4	
Beech-hornbeam forest (Aussenac, 1970)	Nancy (France)	5/66–6/67	724.2	17.0	76.0	7.0
Beech forest (Eidmann, 1959)	(Germany)	1952–7	1216.0	7.6	75.9	16.5
Beech (*F. orientalis*) forest (Cepel, 1967)	Near Istamboul (Turkey)	5 yr	1032.7	17.4	67.1	15.5
Beech forest (Delfs, 1967)	Sauerland (Germany)	4 yr	1150.0	7.6	75.8	16.6
Oak forest on plateau (Galoux, 1963)	Virelles* (Belgium)	1961	921.9	—	84.2	—
Thermophilous oak forest on south slope (Galoux 1963)	Virelles* (Belgium)	1961	921.9	—	74.6	—
Maple-ash-elm forest on north slope (Galoux, 1963)	Virelles* (Belgium)	1961	921.9	—	86.8	—

Table 6.7E Interception loss in coniferous artificial stands in temperate deciduous forest areas

Forest type	Site (* denotes IBP site)	Year	Precipitation (P, mm)	Interception (% P)	Throughfall (% P)	Stem flow (% P)
Spruce forest (Benecke et al., 1975)	Solling* (Germany)	1/1/69–10/31–69	896.3	26.7	—	73.3
Spruce forest (Strauss, 1971a,b)	(München*) (Germany)	1969	841.0	—	—	—
Spruce forest (Aussenac, 1970)	Nancy (France)	5/65–6/67	750.5	34.4	63.8	1.8
Spruce forest (Eidmann, 1959)	Germany	1953–8	1216.0	25.9	73.3	0.7
Spruce forest Mature spruce	Sauerland	4 yr	1150.0	25.9	73.4	0.7
forest (Delfs 1967)	Hartz	4 yr	1356.0	36.0	63.2	0.8
Pine forest (Aussenac, 1970)	Nancy (France)	5/65–6/67	750.5	30.9	65.7	1.6
Pinus sylvestris plantation (Rutter, 1967)	— (England)	4/58–3/59	710.0	33.8	52.1	14.1
Pinus sylvestris planation pool stage High Wood (Lützke & Simon, 1975)	Berlin (DDR)	10/1/67–9/30–71	626.2 / 627.0	33.1 / 28.5	64.2 / 71.0	2.7 / 0.5
Grand Fir plantation (Aussenac, 1970)	Nancy (France)	5/65–6/67	750.5	42.3	56.7	1.0

Table 6.7F Interception loss in various boreal coniferous forest types (*after Galoux et al., 1981*)

Forest type	Site (* denotes IBP site)	Year	Precipitation (P, mm)	Interception (% P)	Throughfall (% P)	Stem flow (% P)
Northern taiga (Molchanov, 1971) Pine forest and mixed pine-spruce forest of various ages	(*) Lat. 63° 30'N Long. 40° 7'E	1948–9	525	6.3–24.4	—	—
Middle taiga (Molchanov, 1971) Spruce forests	(*) Lat. 60° 30'N Long. 34° 35'E	—	600	26.3–27.7	—	—
Mixed birch-spruce forests	Lat. 60° 30'N Long. 34° 35'E	—	600	29.7–30.5	—	—
Birch forest	Lat. 60° 30'N Long. 34° 35'E	—	600	5.5–17.7	—	—
Southern taiga (Molchanov, 1971) Spruce forests	(*) Lat. 59°N Long. 39°E	—	730	17.5–20.3	—	—
Birch forests	Lat. 59°N Long. 39°E	—	730	10.8–11.0	—	—

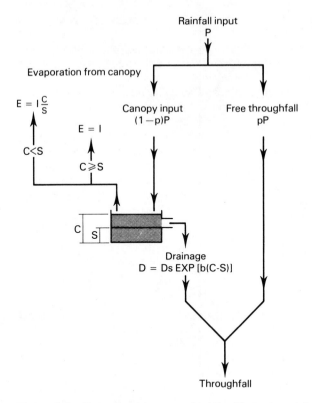

Figure 6.1   Conceptual framework of the 'Rutter' model of rainfall interception by forest vegetation (after Shuttleworth, 1983).

When $C < S$, the evaporation rate is

$$E = I\frac{C}{S} \tag{6.10}$$

The rate of drainage, D, from the canopy is given by

$$\ln D = a + bC \tag{6.11}$$

where b is a drainage coefficient for the canopy. Rutter *et al.* found that when C was equal to S in the Corsican pine stand, D was 0.002 mm min^{-1} and therefore

$$a = \ln(0.002) - bS \tag{6.12}$$

Alternatively, the drainage rate may be given by the algorithm:

$$\begin{aligned} D &= Ds \exp[b(C - S)] && C \geq S \\ D &= 0 && C < S \end{aligned} \tag{6.13}$$

where Ds is the drainage rate from the canopy when the canopy storage is at capacity (typically 0.002 mm min^{-1}).

The empiricism in the drainage calculation only influences the calculation of the evaporation through its effect on the amount of time taken to dry out the canopy after a rain storm: the drainage rate of a supersaturated canopy is commonly high, and the fall to canopy capacity quick, so that the time to drying out is mainly a function of the calculated evaporation rate.

Given the inflow rate $(1 - p)P$ and the outflow rate described by the above equations, it is possible to calculate a running water-balance of the canopy and hence the interception loss.

Stem flow was negligible in the Corsican pine stand since little water was diverted from the canopy to the trunks, but it is a recorded factor of some other types of forest. Rutter *et al.* consider that the diversion of some of the rainfall to the trunks would place it in a situation where evaporation rates would be lower. They suggest that under these conditions separate water balance calculations be made for the canopy and the trunks as follows. If p is the proportion of rain falling through gaps in the canopy to the ground and p_t is the proportion intercepted by the trunks, then the input to the canopy is $(1 - p - p_t)P$ and the calculations of canopy water balance are modified accordingly.

Robins (1974) showed that, for storms large enough to cause stem-flow in Douglas fir and preceded by rainless days so that the trunk was initially dry, stem-flow is described by

$$F = 0.15P - 0.89 \text{ mm} \tag{6.14}$$

where the constant 0.89 represents the storage capacity (S_t) of the trunks and the branches which drain to them, and the coefficient 0.15 is an estimate of p_t. Rutter *et al.* comment that similar expressions have been published for other tree stands. The expression neglects losses by evaporation, but Rutter *et al.* consider that expected differences in boundary layer thickness, ventilation and exposed areas of trunks and leaves suggest that the potential evaporation rate per unit land area from the trunks will be less than 10 per cent of the potential rate from the canopy. The main evaporation from the trunks would therefore be the slow evaporation of the stored water during the hours or days following a storm.

Implicit in equations (5.51) and (6.5) is the notion that the canopy is in effect a fixed single-layer source or sink of water vapour and sensible heat, and sink of momentum, all situated at a height $z_0 + d$ (Hancock *et al.*, 1983). Under conditions of neutral stability the same mechanism of turbulent eddy diffusion is envisaged as transporting these quantities and hence a similarity hypothesis is invoked between the diffusivities K_M, K_H and K_V of momentum, sensible heat and water vapour respectively such that

$$K_M = K_H = K_V \tag{6.15}$$

Since the reference level within the vegetation and the measurement level above it are regarded as the same for each transported quantity the corresponding resistances will also be equal, i.e.

$$r_{aH} = r_{aV} = r_a \tag{6.16}$$

This assumption is implicit in equation (5.51) above, but it is usually argued that r_{aM} will be lower than r_a owing to bluff-body effects (Thom, 1975).

Hancock *et al.* (1983) comment that in recent meticulous experimental work, major discrepancies have been found between estimates of latent and sensible heat transport by different but theoretically equivalent methods. Notably Thom *et al.* (1975) found aerodynamic (profile gradient) estimates which were two or three times less than independent energy budget estimates, and Grip *et al.* (1979), using energy balance (Bowen ratio) and water balance methods estimated evapo-

Table 6.8   Aerodynamic resistance values (referred to top of canopy) derived for the 14 evaporation events with relevant meteorological data (*after Hancock et al., 1983*)

Event no	Mean canopy wetness fraction C/S	Mean canopy evaporation $\bar{E}$ (mm h^{-1})	Canopy top windspeed u (ms^{-1})	Momentum aerodynamic resistance r_{aM} (sm^{-1})	Effective aerodynamic resistance r_{a1} (sm^{-1})
1	0.29	0.04	2.5	2.0	0.5 ± 0.1
2	0.14	0.03	2.4	2.1	0.3 ± 0.1
3	0.75	0.09	2.4	2.1	1.8 ± 0.5
4	0.61	0.18	2.2	2.3	2.6 ± 0.8
5	0.61	0.04	2.2	2.3	3.6 ± 2.0
6	0.86	0.12	2.3	2.2	1.5 ± 0.5
7	0.85	0.04	2.2	2.3	3.8 ± 3.0
8	0.78	0.12	2.5	2.0	1.1 ± 0.5
9	0.68	0.26	2.8	1.8	1.2 ± 0.3
10	0.43	0.37	2.9	1.7	1.6 ± 0.3
11	0.93	0.04	0.7	7.1	2.8 ± 2.0
12	0.92	0.16	2.1	2.4	2.2 ± 0.5
13	0.58	0.32	2.4	2.1	1.5 ± 0.3
14	0.08	0.14	2.6	1.9	0.8 ± 0.2

transpiration to be either 350 mm or 150 mm respectively. In both cases the authors have suggested that discrepancies may have arisen because the similarity hypothesis (6.15) does not hold for aerodynamically rough vegetation.

A field experiment to collect canopy storage data by monitoring the cantilever deflection of wet branches is described by Hancock and Crowther (1979). Table 6.8 sets out the results derived from the canopy storage and meteorological data (Hancock *et al.*, 1983). Hancock *et al.* calculate mean evaporation $\bar{E}$ from

$$E = -\frac{dC}{dt} \tag{6.17}$$

where C is measured directly every 5 minutes in the field. The effective aerodynamic resistance r_{a1} is calculated from

$$r_{a1} = \frac{\rho\, C_p\, \delta e}{(\Delta + \gamma)L\frac{S}{\tau} - \Delta R_N} \tag{6.18}$$

where $\tau = \dfrac{S}{E_p}$,

and E_p is the evaporation rate of intercepted water obtained by setting the bulk surface resistance r_S to zero in the Penman–Monteith equation. Values of r_{a1} may be compared with resistances r_{aM} calculated from equation (6.5) using the values of windspeed u at the top of the canopy and the generally accepted values of $d = 0.75\,h$ and $z_0 = 0.1\,h$, when h is the crop height.

Hancock *et al.* (1983) comment that in Table 6.8, notwithstanding the deficiencies of the experimental site and limitations of the data, a considerable range of effective aerodynamic resistances has been derived and this requires explanation. Most of the more precise values lie between 1 s m^{-1} and 2 s m^{-1}, which is considerably less than resistances derived for momentum flux r_{aM}. However these values are also much smaller than the average value of 3.5 s m^{-1} derived by Calder (1977) by optimization and supported by Gash *et al.* (1980) both for broadly similar forest canopies.

Hancock *et al.* (1983) offer five possible explanations for the variations in resistances.

(i) The experimental site is not ideal as regards fetch and slope and it may be argued that there may have been enhanced turbulence and horizontal flux divergence leading to apparently lower transport resistances.

(ii) The Thom (1971) values of $d = 0.75$ and $z_0 = 0.1$ h used in the standard calculation of r_{aM} may be inappropriate for this experimental site and if so perhaps inappropriate for the entire upland region.

(iii) A more plausible explanation might be that the similarity hypothesis fails for tall vegetation (Thom *et al.*, 1975; Grip *et al.*, 1979).

(iv) Another possible explanation could be errors in experimental values.

(v) Finally, movement of sensible heat and vapour sources could provide the most likely explanation.

Excepting the effects of wind deformation and non-neutrality, both of which are reported by Jarvis *et al.* (1976) to be slight in coniferous forest, momentum characteristics of crop canopies may be expected to vary little over a wide range of environmental conditions. The same cannot be assumed for the location of sensible heat and vapour sources.

Sellers and Lockwood (1981a) have used a four-layer numerical simulation to show that a saturated pine canopy should dry progressively from the top of crop downwards (Figure 6.2), and that the energy required to satisfy the evaporative demand of wetted surfaces should be such that the strongest vapour source corresponds to the main heat sink, resulting in a downward transfer of sensible heat into the most actively evaporating leaf layers. The simulation therefore predicted a progressive lowering of the apparent vapour source and sensible heat sink throughout the drying period.

Hancock *et al.* (1983) therefore suggest that the vertical movement of the apparent sources of sensible heat and water vapour might explain the differences in r_{al} and r_{aM} in Table 6.8. In further analysis they suggest that the implied water vapour source position for a fully wetted canopy is approximately 0.1 m above the effective sink for momentum but drops to 0.2 m below as the canopy dries preferentially in the upper levels. The Sellers–Lockwood (1981a) model predicts for one year of real data a 20 per cent increase in interception loss from coniferous forest over the Rutter model estimate.

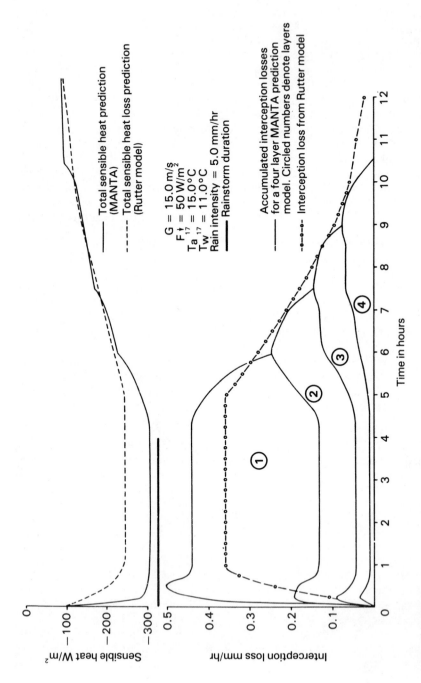

Figure 6.2 Predictions by Sellers/Lockwood and Rutter models of interception loss during and after a simulated high-intensity rainstorm (after Sellers and Lockwood, 1981a).

3 Forest clearance

That vegetation influences climate – and, especially, that the clearing of forest reduces rainfall – is an old idea (Shukla and Mintz, 1982). Approximately 10 per cent of the globe is covered by forests, or 33 per cent of the continental area. Roughly 53 per cent of the world forest area is situated within the tropics, and the tropical land surfaces are about 42 per cent covered by forests. South America is the main forest continent with 55 per cent forest cover. Since tree surfaces are in very good contact with the air, they absorb and intercept radiation, precipitation, aerosols and the momentum of the air extremely well. It is well established that the principal effects of forest on the hydrological cycle are in the reception and disposal of precipitation (Pereira, 1972). Forests provide the greatest surface area for the interception and re-evaporation of water and are effective traps for the absorption of solar radiation. Tree root depths are characteristically large, so trees are frequently not under moisture stress at times when grasslands have dried up. Sagan *et al.* (1979) comment that simply altering ground cover affects surface albedo and run-off, changes the ratio of sensible to latent heat transport, and greatly modifies the surface winds. These variations in turn cause soil moisture, temperature, and erosion rates to change.

The most widespread modification of the climate by man in the past has been achieved inadvertently by the conversion of the natural vegetation into arable land and pastures (SMIC Report, 1971, p. 62). According to Flohn (1973), over the past 8,000 years, about 11 per cent of the land area of the world has been converted into arable land, and 31 per cent of the forest land is no longer in its natural state. Preagricultural peoples not only caused numerous accidental fires but also intentionally set fire to bushes and trees. Thus evidence from European peat bogs reveals that there was a decline in tree pollens after about 3000 BC, associated with the creation of shepherd-farmer culture and a consequent reduction in forest cover (Seddon, 1967). Agricultural peoples have laid waste natural vegetation, primarily to expand farm and grazing land. According to Darby (1956), about 60 per cent of central Europe has been converted from forest to farmland during the last 1,000 years. Grainger (1980) comments that if the present rate of clearance continues then the figures indicate that there will be no tropical moist forests in 60 years time.

3.1 The multilayer crop model

According to Lettau and Baradas (1973), evapotranspiration climatonomy is a numerical approach to the determination of moisture storage, run-off, and evapotranspiration resulting from gravitation and the sun's work on precipitation intercepted by a natural watershed. The concept, originally developed by Lettau (1969) using eastern North America for illustration, has been applied to a number of climatic regions. The approach is well illustrated by the study of the $46 \, \text{km}^{-2}$ Mabacan River basin at Laguna, Philippines, by Lettau and Baradas (1973). They used a numerical model to produce estimates of monthly run-off, soil moisture storage, and evapotranspiration. The input or forcing function to their model consists of monthly values of precipitation and global radiation, they also take into account variations in basin parameters such as albedo. Unfortunately the models of this type are too general to produce any useful conclusions about variations in basin vegetation type.

Figure 5.10 summarizes the water balance of a small basin covered by a specific vegetation type over a long time period. The effect of changing a crop type within a basin is to alter the interception loss, transpiration and, to a lesser extent, soil evaporation rates. Changes in any of these will have feedback effects on the soil moisture content and the run-off. In view of the close relationship between basin behaviour and crop type, the simulation part of this study is directed towards the building of physically realistic models capable of simulating all the exchanges shown in Figure 5.10.

For a short crop, such as grass, the simulation techniques are fairly simple and well proven. Soer (1977) demonstrated that a single layer model with a finite difference soil heat flux routine could perform accurately under non-precipitating conditions. The basic model known as TERGRA, has been adapted to include a simple description of rainfall interception. For a short crop the unilayer treatment does not diverge too much from reality as the system can be effectively described as a single surface in contact with the atmosphere. This is not true for tall crops such as pine and oak because here the leaf area is distributed over a relatively large vertical distance.

The MANTA multilayer crop model described in this chapter has the advantage of simulating all the exchanges shown in Figure 5.10 in a physically realistic manner. It has a distinct advantage over the general models of the Lettau type in that is uses hourly meteorological observations and detailed descriptions of the crop parameters. Summary descriptions of the model may be found in Sellers and Lockwood (1981a and b). A complete detailed description of the model is contained in Sellers (1981).

A flow diagram for the basic computer model is outlined in Figure 6.3. The inputs to the model consisted of hourly observations of air temperature, wet bulb temperature, wind speed at 10 m, low cloud cover, and total cloud cover at standard meteorological sites on airfields in the United Kingdom. Since the airfields were mainly grass-covered the data are not directly applicable to the conditions above another vegetation type, and therefore have to be converted. The energy and water balances of the standard sites were computed and used to estimate air temperature and vapour pressure at 20 m, and also geostrophic wind speed. These data then formed the inputs to the alternative vegetation types. Four vegetation types were chosen for investigation, all typical of temperate maritime climates: grassland, pine which is typical of coniferous forest, oak which is typical of deciduous forest, and winter wheat which is typical of arable cultivation. Allowance was made for the changing leaf area index, albedo, roughness length, moisture tension, and stomatal resistance in each crop as the seasons progressed. A unilayer model was used for grassland, otherwise crop models of four or more layers were used for the other vegetation types. The temperature and vapour pressure at 20 m and the geostrophic wind speed from the standard meteorological site, together with radiation data estimated from cloud amount, and precipitation values, were used to simulate rainfall interception, evaporation and transpiration in each vegetative canopy. The effects of atmospheric non-neutrality were incorporated into the description of turbulent transfer conditions above both the airfield model and the four crop models. The hydrological model used to estimate run-off represents the complete catchment as a vertical series of cascading soil moisture stores, the contents of which determine the run-off. Rainfall, run-off and actual evaporation data from the 18.6 km^{-2} Grendon Underwood river basin in Buckinghamshire, UK, were used to optimize the parameters for the hydrological model. During wetting and drying of the vegetation canopies the model used 12-minute time steps, when the canopy was dry the time step was extended to one hour.

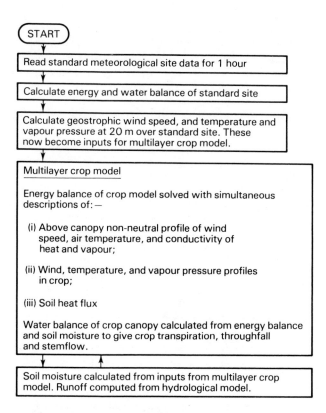

Figure 6.3 Schematic diagram of the multilayer crop model.

3.2 Some results of a multilayer crop model

Because of the extensive computing services required it has so far only been possible to run MANTA with one year of data, which was for 1977 from Benson in Oxfordshire, UK. The hydrological run-off model applies to the nearby Grendon Underwood river basin, and this should be considered to be completely covered by each hypothetical vegetation type in turn. Monthly mean climatological data for 1977 at Benson are shown in Table 6.9. The simulated development of run-off interception loss, and transpiration for the four crop types shown in Figure 6.4. It

Table 6.9 Monthly averages of climatological observations for Benson, for 1977 (*after Sellers and Lockwood, 1981b*)

	J	F	M	A	M	J
Daily air temperature (°C)	3.2	6.1	7.4	7.6	10.9	12.3
Monthly total precipitation (mm)	71.1	78.9	54.9	28.6	39.5	77.7
Daily mean hours of bright sunshine	1.30	3.21	3.36	5.37	7.32	4.60

	J	A	S	O	N	D
Daily air temperature (°C)	16.3	15.7	13.7	12.3	6.5	6.5
Monthly total precipitation (mm)	20.8	177.6	20.1	44.3	39.9	70.7
Daily mean hours of bright sunshine	7.07	4.51	3.75	3.57	2.99	1.53

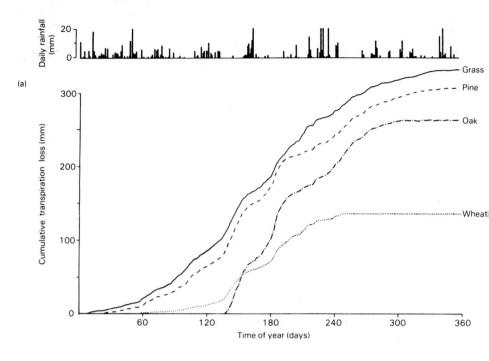

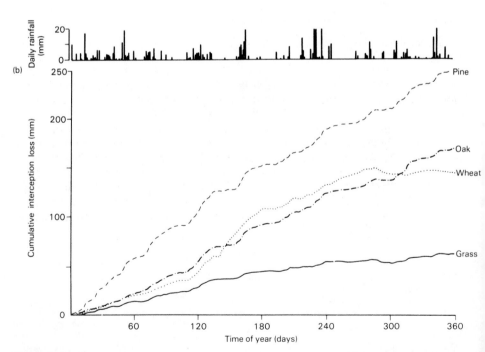

Figure 6.4 Simulated values over one year, for four crop types of (a) accumulated transpiration loss; (b) accumulated interception loss;

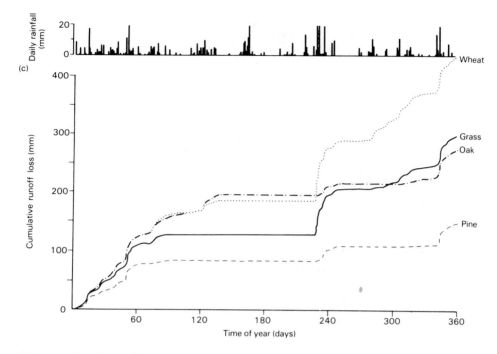

(c) accumulated run-off.
Meteorological data set from Benson, Oxfordshire, for 1977 (after Lockwood and Sellers, 1983a).

should be noted that interception loss includes canopy loss plus soil evaporation. The lowest run-off values throughout the year are associated with pine, and the highest for much of the year with wheat. Run-off values from wheat and oak are roughly equal until July, after which they diverge rapidly. The causes of these changes in run-off become clear from a study of the interception loss and transpiration curves.

Simulated annual values of the hydrological components are shown in Table 6.10 where slight variations in run-off and evaporative totals are due to small variations in soil moisture at the beginning and end of the year. It is seen that changes in vegetation type bring about profound changes in the operation of the surface hydrological system illustrated in Figure 5.10. In particular there are fundamental changes in the partitioning of water between run-off and evaporation, and also between evaporation from intercepted water and transpiration from plants.

Table 6.10   Annual totals of simulated run-off, interception loss and transpiration for four vegetation types (in mm rainfall equivalent) (*after Sellers and Lockwood, 1981b*)

	Run-off	Interception loss (+ soil evaporation)	Transpiration
Wheat	417	158	134
Pine	152	253	309
Oak	279	172	263
Grass	318	62	335

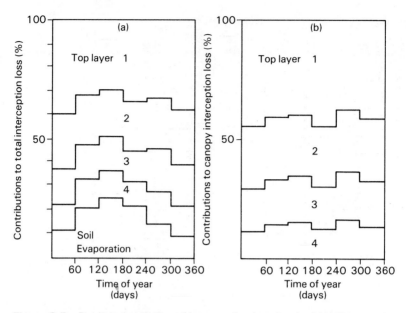

Figure 6.5   Predicted variation of interception loss for the MANTA pine model for 1977: (a) canopy interception and soil evaporation; (b) canopy interception only. Circled numbers denote leaf layer position. Meteorological data set from Benson, Oxfordshire, for 1977 (after Lockwood and Sellers, 1983a).

Changing forest to a non-forest vegetation type leads to an increase in run-off. This is particularly marked with a change from coniferous forest (pine) to any other type of vegetation, or from any other type of vegetation to arable land (wheat). The change in run-off is less significant with a change from deciduous forest (oak) to grass.

The vertical distribution of interception loss contribution from a pine canopy as predicted by the MANTA pine model is shown in Figure 6.5. The most important prediction is that the topmost leaf layer is by far the most dominant source of interception loss, providing 42 per cent of the canopy's total annual loss, in spite of being wet for the shortest time. Simulations using synthetic data (Sellers 1981; Sellers and Lockwood, 1981a) indicate that the dominance of the top leaf layer is likely to vary according to weather conditions. Figure 6.5 demonstrates that soil evaporation is a relatively large loss during the summer months but is suppressed during the winter by an actively evaporating canopy which is able to consume a larger proportion of the available energy. Similarly, if canopy interception loss only is considered, it is evident that the upper layers are slightly more dominant during the winter months. This is because under conditions of low evaporative potential the topmost leaf layer uses up a relatively large proportion of the available energy, and in the process suppresses the evaporation from lower layers.

An inspection of Figure 6.6 reveals that the annual course of interception loss from oak is closely related to the state of the crop's foliage. The leafless crop gives rise to a much reduced rate of interception loss due to its low canopy water storage content, high throughfall coefficient, small maximum exposed wetter area, all of which give rise to a low bulk vapour source. Over the year as a whole, the canopy contribution to interception loss amounted to 53 per cent of the total, but as this

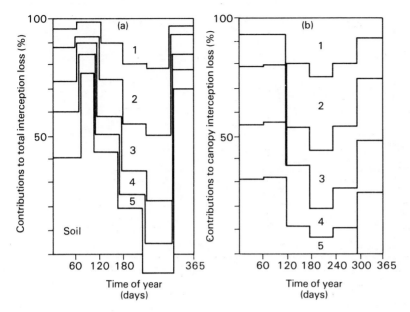

Figure 6.6 Predicted variation of interception loss for the MANTA oak model for 1977: (a) canopy interception and soil evaporation; (b) canopy interception only. Circled numbers denote leaf layer position. Meterological data set from Benson, Oxfordshire, for 1977 (after Lockwood and Sellers, 1983a).

incorporates predictions for the leafless and fully foliated canopy, it reveals little specifically about the physical interactions between a structured crop and the atmosphere. A comparison of the period of day 1–60, when the crop is completely leafless, to that of day 181–240, when the crop is fully foliated, is more relevant. Over these two periods, the canopy contribution to interception loss amounted to 59 per cent and 80 per cent of the total interception losses respectively. As the leaf area index increases and the rainfall income decreases during the summer, the canopy contributes a large proportion of the total interception loss. With full foliage, the upper part of the canopy contributes relatively more to the total interception loss while the leafless tree has a distribution of interception loss almost directly related to the twig area index of each leaf layer.

The predicted cumulative transpiration loss from the MANTA wheat model is shown in Figure 6.4. The model predicts the lowest annual total of transpiration loss of the four crop types and Sellers (1981) considers that the total annual evapotranspiration loss is underestimated by some 20 to 50 mm. This is due to the fixed root resistance sub-model. Figure 6.7 shows the contributions of leaf layers and soil to interception loss over the year. It is apparent that soil evaporation accounts for over 77 per cent of the total interception loss, and the soil is the most important intercepting surface. During the summer, from day 181–240 when the canopy is fully developed with a leaf area index of 5.5, the soil evaporation loss still makes up 65 per cent of the total. The predicted annual total of interception loss from the wheat foliage was just over 5 per cent (37 mm) of the annual precipitation. This low value (compared to the equivalent value of 29 per cent for pine) is the result of the relatively short period of full plant cover in the summer and the complete absence

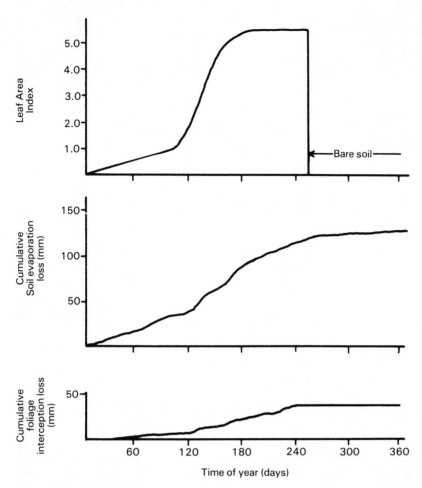

Figure 6.7 Interception losses from soil and foliage as predicted by the MANTA wheat model for 1977. Meteorological data set from Benson, Oxfordshire, for 1977 (after Lockwood and Sellers, 1983a).

of foliage during other wetter months. Loss from the soil surface, which is predicted to be 17 per cent (123 mm) of the annual precipitation, is related to be the availability of soil moisture and the seasonal variation of evaporative potential. The period of greatest soil evaporation is from day 121 to 180 when the moisture supply is ample, the radiation income high and the plant canopy not fully developed. Later in the summer, from day 181 to 240, the soil moisture content drops, reducing, both transpiration and interception loss rates. From day 241 to 365, the evaporative potential is steadily reduced and so, although the soil moisture supply is more than adequate, the soil evaporative loss is small.

The MANTA prediction of the total of interception loss from grass is 8.7 per cent (63 mm) of the annual precipitation. The low value of the water storage capacity chosen for the MANTA grass model makes it a suitable representation of a grazed pasture.

The same hydrological model and maximum available soil moisture store were

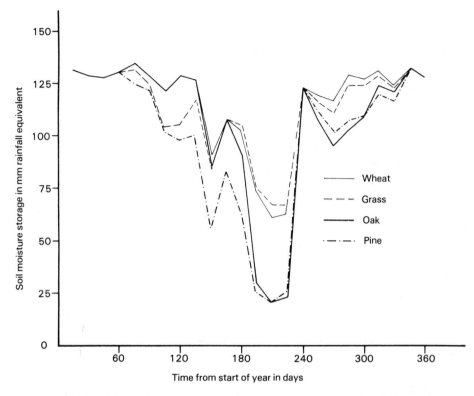

Figure 6.8 Simulated value over one year for each crop type of soil moisture storage (after Sellers and Lockwood, 1981b).

used for each vegetation type, since this allows for a direct comparison of the effects of rainfall interception on the water balance of each plant community. In the real world differing root depths will also influence evaporation rates and soil moisture contents. Though the transpiration rate from grass is relatively large, the interception loss is small compared with forest. This means that considerably more of the rainfall penetrates through the vegetation canopy and into the soil moisture store as compared with forest areas. Therefore the predicted soil moisture levels (Figure 6.8) are generally higher under grass than under forest, and particularly than under coniferous forest.

The oak model predicts the most extreme run-off regime of the four models as there is no transpiration and little interception loss during the winter months and, by contrast, the highest predicted transpiration loss of the four crop models during the summer. As a result, the oak model yields a soil moisture content roughly equivalent to that of wheat and slightly higher than that for grass over the first 90 days of the year. Thereafter, the soil moisture content drops rapidly as the canopy develops and falls below that of pine at day 189. The oak model is then predicted to maintain the lowest level of soil moisture until day 302 when the effects of leaf fall reduce interception loss and halt transpiration. From this period until the end of the year, the soil moisture content converges onto the equivalent prediction for pine. As a result of the unbalanced annual distribution of evapotranspiration loss,

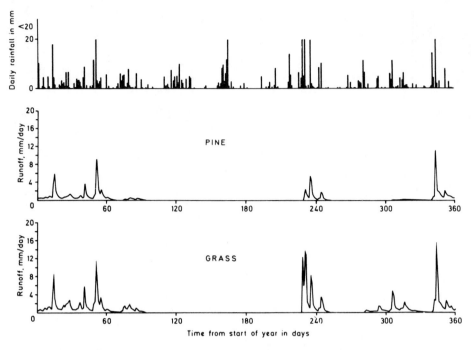

Figure 6.9   Predicted daily run-off totals for pine- and grass-covered areas. Meteorological data set from Benson, Oxfordshire, for 1977 (after Lockwood and Sellers, 1983a).

the oak model's run-off regime is characterized by high peak flows and high levels of soil moisture in winter and an extended period of low flows during the summer.

For much of the year the lowest soil moisture totals are associated with pine and this is reflected in the low run-off totals predicted by MANTA. Again during the summer there is an extended dry period with negligible run-off associated with large soil moisture deficits. The numerical model predicts the highest run-off loss of all four crops from wheat. This is because of the long periods of the year during which the soil is bare or nearly bare. Thus the soil moisture content is maintained at a high level during the whole year and only drops below that of grass for a brief period during the late summer. From day 250 to day 365 the bare soil model used to represent the post-harvest state of the crop predicts the wettest soil moisture conditions. The probable error in the calculations of transpiration losses from the wheat model has been estimated at 20 mm to 50 mm. It is likely that the run-off losses from day 120 to day 250 should be reduced from their currently predicted values and correspondingly the soil moisture deficit of the late summer increased. Nevertheless, the bare soil still ensures very high run-off losses from day 250 to the end of the year. The predicted soil moisture record for grass shows that a generally high level is maintained throughout the year. The cumulative run-off prediction is characterized by an even distribution of losses throughout the year.

Predicted daily run-off totals for the four vegetation types are shown in Figures 6.9 and 6.10. Statistical analysis of the run-off values indicates that an increase in the mean discharge over a period leads to a proportionately greater increase in the frequency of the high flows. Also it is found that the run-off regime generated by

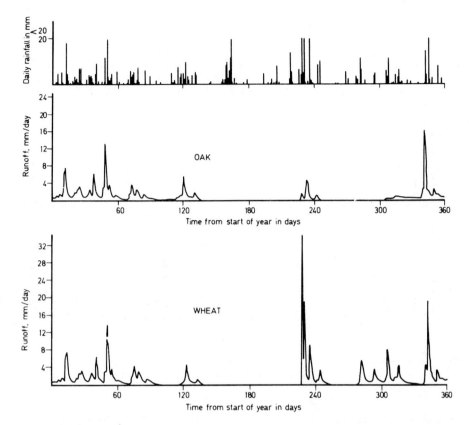

Figure 6.10 Predicted daily run-off totals for oak- and wheat-covered areas. Meteorological data set from Benson, Oxfordshire, for 1977 (after Lockwood and Sellers, 1983a).

each crop model is idiosyncratic in terms of the characteristics of its frequency distribution. Grass maintains the least leptokurtic and least skewed run-off regime throughout the year which implies that its run-off frequency distribution is the most even of the four crops. Oak and wheat show highly skewed and leptokurtic run-off frequency distributions with a much greater preponderance of high run-off rates during the winter when neither crop is able to regulate soil moisture levels via evapotranspiration to the extent predicted for grass and pine. Pine, with the highest predicted annual evapotranspiration loss, is predicted to have the highest frequency of low flows, as can be seen in Figure 6.11, but not necessarily the lowest peak flows.

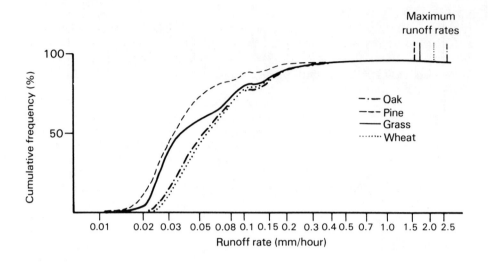

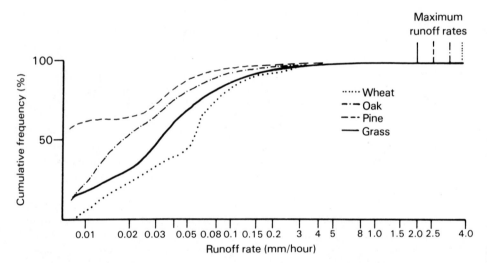

Figure 6.11 Predicted run-off rate frequency distribution for the four crop types: (a) over the period day 1 to 60; (b) over the period day 301 to 365 (after Lockwood and Sellers, 1983a).

3.3 Experimental studies

Table 6.11 compares the interception losses measured in experimental catchments with coniferous forest cover in the United Kingdom with the predicted MANTA interception losses. It is seen that the MANTA prediction of interception loss from a pine cover corresponds with the results obtained from coniferous forest sites in the dryer parts of the United Kingdom.

The large interception loss from trees has been illustrated in the field by Clarke and Newson (1978) who compared the water yields from the upper Wye and Severn river basins, which are under hill pasture and largely coniferous forest, respectively. For the upper Wye basin, the mean annual loss (precipitation minus streamflow) for the years 1970 to 1975 inclusive was 18 per cent of the mean annual precipitation of 2,415 mm, while for the upper Severn, the mean annual loss over the same period was 30 per cent of its mean annual precipitation, 2,388 mm. Moreover, the latter made no allowance for the fact that about one-third of the Severn is unforested; when Clarke and Newson made a statistical adjustment to allow for this, mean annual loss from the forested area of the upper Severn rises to 38 per cent of mean annual precipitation. Clarke and Newson claim that the reality of these results has been confirmed by plot studies within the Severn forest on 'natural' lysimeters underlain with impermeable boulder clay.

A number of experimental studies have been made of the response of streamflow to a change in vegetation type. Hibbert (1967) reports some results from two river basins at Coweeta, North Carolina, USA. Two small forested river basins near Coweeta were calibrated together for 3 years at the beginning of the study period. The natural vegetation consisted mainly of oak–hickory forest. In 1940, one of the basins was clearcut. In the first year after clearcutting, annual streamflow from the treated basin increased some 373 mm over the value expected from the regression equation with the control basin. As the deciduous forest regrew during the following 23 years, streamflow from the treated basin decreased logarithmically until at the end of this period the flow was only 75 mm greater than the value predicted for a full forest cover. Clearcutting again at that time resulted in a second marked increase in streamflow from the treated basin almost exactly equivalent to the response after the first cutting.

Swift *et al.* (1975) have developed a phenomenological model of water exchange between soil, plant and atmosphere which was used to simulate evapotranspiration and annual drainage for 2 years from a mature oak–hickory forest in the southern Appalachians. In a year of unusually high precipitation the simulated annual drainage was within 1–5 per cent of measured streamflow. Simulations were also performed by using the same 2 years of meteorologic data, but vegetation parameters were changed to represent a young white pine plantation and a regrowing hardwood forest one year after clearcutting. These results were comparable to changes of – 200 and + 300 mm observed in corresponding basin experiments near Coweeta. Simulated evapotranspiration during the summer was nearly identical for hardwood and pine forests, while winter and early spring water loss was greater for pine. Simulation suggests that the greater evapotranspiration by pine was due to increased interception in all seasons and increased transpiration in the dormant season. These results are in agreement with those from the MANTA model.

Afforestation reducing run-off has been reported from a number of sites in varying climatic regimes. Lill *et al.* (1980) report the results of a basin experiment on the Eastern Transvaal escarpment, South Africa. Gauging of flow from the

Table 6.11 Interception losses from experimental catchments with coniferous forest cover in the UK (*after Lockwood and Sellers, 1983*)

Site	Species	Duration of study	Rainfall, P (mm)	Interception loss, I (mm)	I/P × 100 (%)	Source
Hodder forest (Yorkshire Pennines)	Sitka spruce	4 June 1955–8 June 1956	984	371	37.7	Law 1956
Crawthorne, Berkshire	Scots pine	Oct. 1959–March 1960	430	161	37.4	Rutter 1963
		April–Sept. 1958	366	128	34.8	
		958–1960	≈ 800 p.a.	—	31.7	
Thetford Chase (Norfolk)	Scots pine	365 days in 1977 (continuous)	595	213	35.8	Gash & Stewart 1977
Upper Severn, Plynlimon (Wales)	Mixed mature conifers	6 Feb. 1974–1 Oct. 1976	5444	1580	29.0.	Calder & Newson 1979
Hafren forest (Plynlimon)	Sitka spruce	351 (continuous) days	1757	479	27.2	Gash *et al.* 1980
Roseisle forest, Moray Firth (Scotland)	Scots pine	281 (continuous) days	493	209	42.4	Gash *et al.* 1980
Kielder forest (Cheviot Hills)	Sitka spruce	302 (continuous) days	802	254	31.7	Gash *et al.* 1980
River Ray (Buckinghamshire)	Pine model	1977	724	252	34.8	MANTA prediction (Sellers 1981)

catchment under natural grass cover began in 1956. One of the catchments was planted to *Eucalyptus grandis* in 1969. Simple regression analysis showed that this afforestation exerted an observable influence from the third year after planting, with a maximum apparent reduction in flow, expressed as rainfall equivalent of between 300 and 380 mm yr $^{-1}$ and with maximum reductions in seasonal flow of about 200–260 mm yr $^{-1}$ in summer and 100–130 mm yr $^{-1}$ in winter. Langford *et al.* (1980) have described a series of experiments near Melbourne, Australia. These investigated the effects on long-term streamflow of reducing the density of *Eucalyptus regnans*. They found that the decrease in streamflow which would result from the conversion of an open old-growth forest to a dense regrowth forest is of the order of 100–200 mm yr $^{-1}$ rainfall equivalent.

A considerable amount of evidence that forest cover reduces run-off has been summarized by Bosch and Hewlett (1982). They consider that pine and eucalypt forest cause on average about 40 mm change in annual water yield per 10 per cent change in forest cover. Deciduous hardwoods are associated with about 25 mm change in yield per 10 per cent change in cover, while 10 per cent changes in brush or grasslands seem to result in about 10 mm change in annual yield.

4 Interception, evaporation and run-off in tropical forests

In tropical rainforest, the vegetation is a key intermediary between the soil and the large volume of water regularly cycling through or being stored in the ecosystem. Removal of the forest upsets the local cycling of water through the ecosystem and can cause changes in both evaporation and run-off. Tree canopies intercept large amounts of rainfall which can be readily evaporated back to the atmosphere and this is illustrated for tropical trees in Table 6.7.

The Amazon is the world's greatest river and although Amazonia remains by far the largest area of tropical forest on earth, it is being subjected to the same rapid ravages of deforestation as the rest of the world's tropical forests. Gentry and Lopez-Parodi (1980) have posed the question – has deforestation in upper Amazonian Ecuador and Peru already resulted in significant changes in the water level for the Amazon at Iquitos, Peru? The records show a pronounced and statistically highly significant increase in the height of the annual crest of the Amazon at Iquitos during the last decade. Before 1970 the annual flood crest at Iquitos had never reached a depth of 26 m; after 1970 it has never been lower than 26 m. The height of the annual low water mark has remained virtually unchanged during this time. Gentry and Lopez-Parodi (1980) conclude that run-off water from upper Amazonia has increased during the last decade, and that this is largely independent of precipitation and thus more likely to be related to the changes in drainage and run-off associated with deforestation. Lettau *et al.* (1979) have modelled the atmospheric and hydrologic phases of the water cycle in Amazonia. In a numerical experiment they simulated the effect of a decrease in the forest cover of central Amazonia (between 57.5° amd 67.5°W) to one-half of its original value. Their computations show that, due to recycling, the precipitation increases at all longitudes between 57.5° and 77.5°W; however, the primary augmentation of evaporation outweighs the precipitation increase. Consequently total run-off from the entire basin decreases from the original 1,075 mm yr $^{-1}$ to 1,018 mm yr $^{-1}$. This conclusion by Lettau *et al.*, regarding the effects of deforestation on run-off, seems to be at variance with actual experience within the tropics, and therefore requires further investigation.

The interception loss I (see equations 5.52 and 5.53) from a wet canopy is a primary loss from the system and will influence the amount of water available for the soil moisture store and eventually run-off. According to Rutter (1975), the interception storage capacities of vegetation canopies are relatively small and represent between 1 and 3 mm of rainfall. Canopies therefore soon become fully saturated after significant rainfall, and often evaporation from the canopy represents a large water loss. The atmosphere is rarely fully saturated when it is raining (Rutter, 1975), so significant evaporation can take place under these conditions. Rain forms in saturated air aloft and normally falls through drier air below. This air takes time to become saturated, as is shown by the lack of surface fog during rainfall. Unsaturated air during rainfall is often assumed in the analysis of interception data, e.g. see Rutter *et al.* (1971) and Gash *et al.* (1980). Shuttleworth and Calder (1979) comment that the simple experimental observation of a finite atmospheric humidity deficit near the surface of forest vegetation in wet canopy conditions very often implies the presence of sensible heat advection, since the incident radiation is often low at such times.

Sellers (1981) and Sellers and Lockwood (1981a and b) have used a multilayer crop model (MANTA) to simulate the interception loss from oak, pine, wheat and grass canopies. Figures 6.12–6.14 show the predicted evapotranspiration rates from the four canopies under various meteorological conditions. The height of the oak forest was fixed at 15 m with a foliage area index of 6.0 and a stem area index of 0.47. The pine forest was represented as having a normally distributed canopy extending from 5 m to 15 m with a total leaf area index of 6.0. A leaf area index of 5.5 and a height of 1 m was used for the wheat model. The grass model was based on an existing unilayer model, the TERGRA model of Soer (1977).

The geostrophic wind speed refers to the wind speed just above the friction layer at an altitude of about 1,000 m and is defined as being independent of surface conditions. The use of this value as an input to hydrometeorological models allows

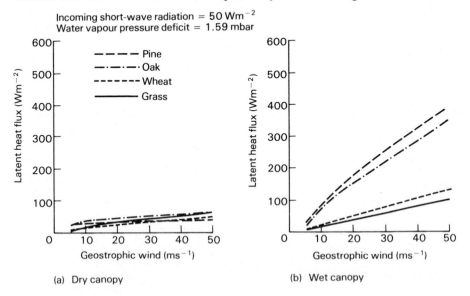

Figure 6.12 Predicted evapotranspiration rates for canopies under conditions of low evaporative potential (after Lockwood and Sellers, 1982b).

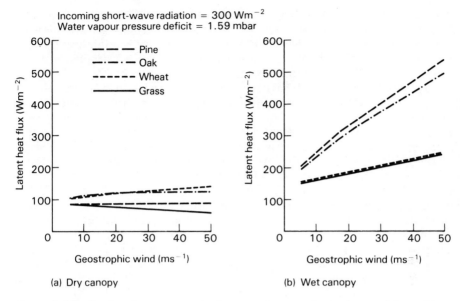

Figure 6.13 Predicted evapotranspiration rates for canopies under conditions of low vapour pressure deficit (after Lockwood and Sellers, 1982b).

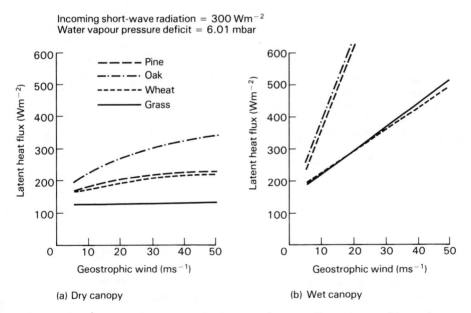

Figure 6.14 Predicted evapotranspiration rates for canopies under conditions of moderate evaporative potential (after Lockwood and Sellers, 1982b).

differing vegetation types to be compared under the same free atmosphere wind conditions. The wind speed just above the canopy is related to the geostrophic wind speed by a number of factors which include air temperature, sensible heat flux and the roughness length of the surface. The wind profile from geostrophic height to the top of the canopy was defined by means of the functions of Ayra (1975) and Paulson (1970). Deacon (1969) has considered the ratio of the wind speeds at 13 m to that at geostrophic height over a downland site near Salisbury, England. At high geostrophic wind speeds, 20 m s^{-1} or more, it made little difference whether it was a clear summer noon or a clear night – the dynamical turbulence is so strong that thermal (non-neutral) effects are swamped and the transfer conditions are effectively neutral. Under these conditions Deacon (1969) found a ratio of wind speed at 13 m to the geostrophic wind speed of about 0.4. For a variety of conditions with wind speeds above about 5 m s^{-1} the ratio varied between about 0.35 and about 0.5. The numerical model predicted that, during January, 1977, over a grass-covered area in southern England the ratio between the wind at 10 m and the geostrophic wind speed was 0.38. Deacon (1969) found however that as wind speed decreases, stability has a rapidly increasing effect until for low wind speeds and a possible positive sensible heat flux, vertical mixing becomes so vigorous that the difference in wind speed between 13 m and higher levels becomes only about half as much as with neutral conditions. Near the equator the concept of geostrophic wind does not strictly apply and here it may be taken as being the wind just above the friction layer. Data given by Snow (1976) suggests that at Cayenne, French Guiana (04°50′N), the three-monthly ratio between the surface wind and the 900 m wind varies between 0.49 and 0.76. The input values of vapour pressure deficit used in the model refer to a height of 17 m and indicate the general atmospheric conditions.

In Figures 6.12–6.14, the upper diagrams refer to simulated steady-state evapotranspiration from dry crops supplied with ample soil moisture. The lower diagrams refer to evapotranspiration loss from a saturated crop and soil surface. The predicted latent heat fluxes above the wet canopies were modelled assuming that the foliage was saturated with water to the appropriate interception capacity. Water draining out of the canopy to the soil surface does not affect the water balance of the plant/soil system, whereas water evaporated from the canopy and soil surface represents a loss which may be reflected in changes in run-off from the system. Here a comparison is made of the interception losses from various canopies – clearly the higher the interception loss, the less water there is to be transferred to the soil to form run-off. It is apparent that an increase in the dominance of the aerodynamic contribution to the evaporative potential has a far greater effect on interception loss rates from forest vegetation as compared to the two shorter crops. The reason for this effect is straightforward; an increase in wind speed leads to a greater reduction, in absolute terms, of the aerodynamic resistance above the aerodynamically rough canopies of tall crops for which, under normal conditions, evaporative potential is more dependent upon the drying power of the air than on radiative income. Consequently, an increase in wind speed leads to a far greater rise in the evaporative potential over forests than over shorter crops. The ratio of the gradients of the rate of interception loss against wind speed for saturated crops demonstrates this effect clearly since, in general, these gradients for the tall crops are 3 to 4 times higher than those of wheat and grass. Increasing the vapour pressure deficit increases the dependence on wind speed and also increases the transpiration loss from oak as compared with grass. In a simulation using actual British lowland observations, Sellers (1981) found that the transpiration during summer from oak was 263 mm as compared with 219 mm from grass over the same period. Increasing the incoming short-wave radiation increases the absolute values of the transpiration and interception losses without markedly changing the dependence on wind speed.

Rutter *et al.* (1971) give the mean vapour pressure deficit for all weather conditions in southern England as about 2.2 mbar over the year, ranging from about 0.9 mbar in December to February and 3.4-4.1 mbar in June to August. Mean monthly wind speeds at London Airport (Heathrow) range from about 4 m s^{-1} to 5 m s^{-1}. Rainfall in Britain is often associated with fronts, with well mixed deep boundary layers and large transverse wind speeds, typically 20 m s^{-1} (Shuttleworth and Calder, 1979). Under these conditions, Shuttleworth and Calder (1979) consider that air movement over forests with dimensions of the order of 100 km would be necessary before the cooling and humification of the boundary layer was sufficient to significantly reduce evaporation rates. Figure 6.12 is probably typical of conditions in middle latitudes during and just after rain, and it is seen that with geostrophic wind speeds of say 15 m s^{-1} the evaporation from a wet tree canopy could be up to 4 times that from wet grass. Sellers and Lockwood (1981b) found in a simulation of water balance using one year's data from lowland Britain that the interception loss (including soil evaporation) from pine trees exceeded that from grass by 4.1 times.

In the equatorial trough zone, vapour pressure deficits tend to be on average, in all weather conditions, about 6 mbar and the short-wave radiation intensity is relatively high (average annual global radiation around 200 W m^{-2}). Richards (1964, figure 26) produces data on the daily march of saturation deficit in the undergrowth and tree tops, in tropical forest at Shasha Forest Reserve, Nigeria. During the night in the wet season, average all weather tree-top saturation deficits are around 1.0-1.5 mbar but during the day they reach around 13 mbar. Saturation deficits in the undergrowth are around 0.5-2 mbar throughout the day. Dry-season values are 2-3 mbar higher in the tree tops than wet-season values. Similar saturation deficits are reported in rubber trees in Malaysia by Moraes (1977, Figure 3). Robinson (1966, p. 138) quotes for December at Kinshasa (04°S) midday peak radiation values of 700 W m^{-2} (slightly greater than 2/3 of the maximum June value in Britain) with a 24-hour mean value of 200 W m^{-2}.

Figures 6.12-6.14 suggest that the enhanced interception loss by trees over that of grass is critically dependent on wind speed. In this context oak can be considered similar to tropical evergreen forest. In the equatorial trough zone wind speeds are usually very low. For example, at Manaus (03°08'S, 60°01'W) in the Amazon Basin, monthly mean wind speeds are below 2 m s^{-1}, suggesting a wind speed above the friction layer of about 5 m s^{-1} and this is typical of much of the region (Ratisbona, 1976). It would seem likely therefore that changing the vegetation type from forest to grassland in equatorial regions will have little influence on the interception loss rate and therefore on the amount of run-off. This conclusion depends critically on wind speeds and particularly on wind speeds during rainfall, since if these are relatively high they will increase interception loss. Figures 6.13 and 6.14 suggest that with the average wind speeds prevailing in the equatorial trough zone, evaporation from a wet tree canopy could exceed that from wet grass by some 15-32 per cent. This is small compared to the 400 per cent increase in interception loss from pine trees as compared to short grass suggested for middle latitudes. It is therefore seen that the relative changes in interception loss from differing wet canopies in the equatorial zone are small.

Nevertheless some evidence has been produced, and discussed earlier, of increasing run-off following deforestation within the tropics, and this needs further consideration. Changes in tropical run-off following deforestation could be due to a variety of causes, some of which are listed below.

(a) If there is a complete clearance of vegetation, and rain falls on bare soil, then the surface hydrology will be completely changed and dramatic increases in run-off could take place under equatorial conditions. UNESCO (1978) comments that, from the admittedly inadequate information available, under dense forest, surface flow is negligible or absent and seepage quick, and that therefore storm-flow in streams will be moderate, except after extreme cloudbursts. Temporary or permanent exposure of the soil surface leads to rapid run-off and dramatic degradational processes, which are difficult, costly and slow to reverse.

(b) Interception loss is very sensitive to wind speed, which could be relatively high in hilly and coastal locations. Wind speeds during tropical showers can be surprisingly high due to a highly turbulent boundary layer. For example, at Malacca, Malaysia, Watts (1954) found from anemograms that during the decade 1931–40, there were gusts of $13.4 \, \text{m s}^{-1}$ (30 mph) or more on 619 days, and of $17.9 \, \text{m s}^{-1}$ (40 mph) on 146 days of which the highest gust was $29.5 \, \text{m s}^{-1}$ (66 mph). Watts found that most gusts occurred in marked squalls of duration ranging from a few minutes to half an hour. Malacca, being a coastal station, may be unusually windy and not typical of inland stations. If a mean geostrophic wind of $10 \, \text{m s}^{-1}$ is assumed during rain then from Figure 6.13 this implies a decrease of interception loss of about 40 per cent on changing from forest to grassland. A similar value is also found from Figure 6.12. This is still low compared to values predicted for changes from pine to grass in middle latitudes, but could lead to changes in run-off.

(c) Figure 6.14 suggests that transpiration rates from broad-leaf forest are higher (by at least 50 per cent) than those from grassland, even at low wind speeds. Thus soils will be drier under forest than under grassland and this in turn will lead to lower run-off from forested environments. This is clearly demonstrated for temperate latitudes in Sellers and Lockwood (1981b), and illustrated in Table 6.12. Sellers and Lockwood (1981b) carried out a water–balance simulation using a numerical model running on actual British lowland observations (the results shown in Table 6.12 are based on this study). In the latter part of August, 1977, about 145 mm of rain fell, but prior to this the weather had been comparatively dry. Table 6.12 shows the soil moisture values predicted by the numerical model prior to the period of heavy rainfall. The lowest values are shown under the trees because of both their high transpiration and interception losses. The soil moisture values predicted under both wheat and grass are considerably higher though they are exposed to exactly the same weather conditions. The period of heavy rain towards the end of August saturated the soil and caused some run-off. The highest run-off values are associated with the wetter soils under the short vegetation types. Thus it is seen that differences in soil moisture under various plant communities exposed to the same weather conditions could lead to differences in run-off. It also indicates that heavy showers under equatorial conditions could give greater run-off from grassland than from forest.

Table 6.12 Simulated water balances following heavy rain after a dry period (units are rainfall equivalents in mm) (*after Lockwood and Sellers, 1982*)

	Grass	Wheat	Oak	Pine
Soil moisture, day 225 (end of dry period)	67	64	24	26
Soil moisture, day 240 (after heavy rain)	127	126	124	125
Runoff, days 225–240	68.0	67.6	15.8	19.1

4.1 Deforestation of the Amazon basin

The main channel of the Amazon runs almost parallel to the equator, with the eastward streamflow being opposite to the easterly winds prevailing over equatorial South America during the year (Lettau *et al.*, 1979). Lettau *et al.* estimate that at 75°W about 88 per cent of the total amount of rain falls at least a second time as the result of recycling of vapour evaporated from land regions. This high value is a direct result of the low wind speeds and large size of the Amazon basin. Deforestation in the upper Amazon would produce the increased run-off observed by Gentry and Lopez-Parodi (1980), if the uplands were relatively windy, thus causing an enhanced interception loss rate from trees as compared to other types of vegetation. If the whole of the Amazon basin were deforested and replaced mainly with grassland then the effects could be different. Figures 6.12–6.14 suggest that replacing forest by grassland will lead to both lower transpiration and interception loss rates, which could well lead to increases in run-off in the lower Amazon. This implies that water, which would normally be evaporated back to the atmosphere to form further rainfall inland, could be carried rapidly back to the sea as increased run-off. Therefore in a deforested Amazonia less water vapour would be advected inland leading to slightly lower rainfall and correspondingly slightly lower run-off in the upper basin. As a result further increases in run-off at Iquitos, as reported by Gentry and Lopez-Parodi (1980), may well depend on the progress of deforestation lower down the river basin. Lettau *et al.* (1979) may well be right in stating that deforestation of the middle Amazon basin (between 57.5 and 67.5°W) will lead to a slight decrease in total run-off from the entire basin, but are probably wrong in predicting a precipitation increase. The assumption in the reasoning above is that wind speeds over Amazonia are low. Higher wind speeds over the basin could lead to greater oceanic moisture advection, less water vapour recycling, and therefore tend to nullify changes in rainfall caused by deforestation. Predictions about the effects of deforestation in Amazonia will depend critically on assumed wind speeds and moisture advection values.

Shukla and Mintz (1982) have carried out numerical simulations of global July precipitation for wet and dry soil cases. In the first case the evapotranspiration is always set equal to the potential evapotranspiration calculated by the model (this is the evapotranspiration when the soil is moist and completely covered by vegetation); in the second case no evapotranspiration is allowed to take place. According to Shukla and Mintz the first of these conditions would be physically realizable on an earth that is completely covered with vegetation and is irrigated where necessary, whereas the second would be approached on an earth that is completely and permanently devoid of vegetation. Shukla and Mintz found over South America that the rainfall near the equator in the wet soil case is about 6 mm day^{-1}, which is about 2 mm day^{-1} larger than the evapotranspiration. In the dry soil case, which could correspond to a deforested Amazonia, the rainfall is almost as large, all of it

(a) Streamline vertmn DJF 63-73

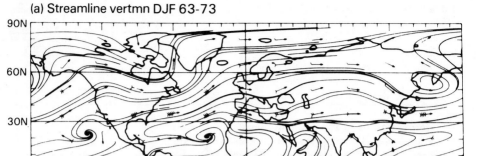

(b) Streamline vertmn JJA 63-73

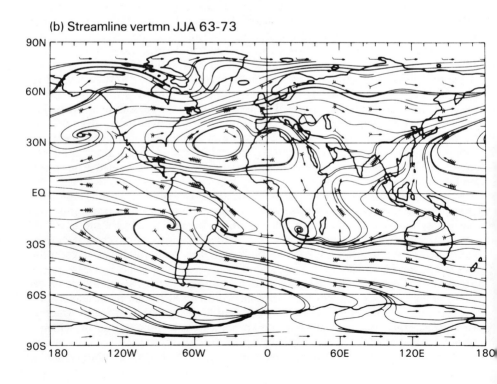

(c) Streamline vertmn YR 63-73

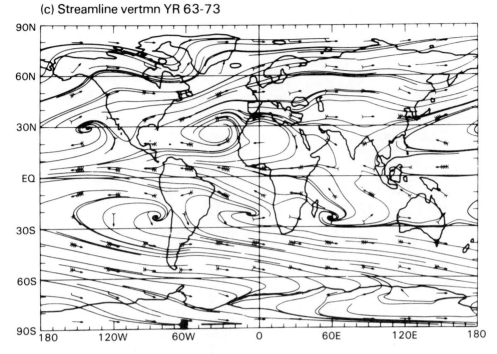

Figure 6.15 Planetary distributions of the total vectors transport fields of water vapour in the atmosphere, and some corresponding streamlines; each barb represents $2 \, \text{m s}^{-1} \, \text{g kg}^{-1}$. (a) DJF; (b) JJA; (c) year (after Peixoto and Oort, 1983).

being water transported from the ocean. In this case the rainfall over Amazonia is reduced by about 2 mm day^{-1} with the greatest falls in the east.

Peixoto and Oort (1983) have mapped the vector transport field of water vapour in the atmosphere (Figure 6.15). In addition to the vectors, streamlines are also shown on these maps. In a hypothetical steady state, the streamlines show the prevailing paths of water vapour in the atmosphere after its release from the various regions at the earth's surface. Figure 6.15 shows that the main sources of water vapour for the atmosphere are localized over the subtropical oceans. The maps also clearly show that the water vapour necessary for precipitation over the continents comes from the oceans. In steady-state conditions, this net inflow of moisture to the continents is compensated by the run-off of the rivers into the oceans. The maps are also interesting in showing a relatively strong flow of moisture from the east into the Amazon basin, thus supporting the numerical calculations of Shukla and Mintz (1982). The effects of the deforestation of Amazonia on rainfall will depend on exactly how much recycling there is of water between the forest and the atmosphere. If this really is large then deforestation could greatly reduce rainfall in the upper regions of the basin. In contrast, if it is small compared with atmospheric water advection, then deforestation will have little effect on rainfall. The calculations by Shukla and Mintz (1982) suggest that in the extreme case deforestation could reduce rainfall over Amazonia to about 70 per cent of its present value.

References

ALDRIDGE, R. and JACKSON, R. J. 1973: Interception of rainfall by hard beech (Nothofagus truncata) at Taita, New Zealand. *New Zealand Journal of Science* 16, 185–98.

AOKI, M., YABUKI, K. and KOYAMA, H. 1975: Micrometeorological environment and assessment of primary production of a tropical rainforest in West Malaysia. *Journal Agricultural Meteorology* (Tokyo) 31, 115–24.

—— 1978: Micrometeorology of Pasoh Forest. *The Malayan Nature Journal*, 30, 149–59.

ARYA, S. P. S. 1975: Geostrophic drag and heat transfer relations for the atmospheric boundary layer. *Quarterly Journal of the Royal Meteorological Society*, 101, 147–61.

AUSSENAC, G. 1970: Action du couvert sur la distribution au sol des precipitations. *Annales des Sciences Forestieres*, 27, 383–99

BANERJEE, A. K. 1973: Computing transpiration and soil evaporation from periodic soil moisture measurements and other physical data. *Indian Forester*, 99, 82–91.

BAUMGARTNER, A. 1956: Untersuchungen uber den Wärme- und Wasserhaushalt eines jungen Waldes. *Ber. deuts. Wetterdienstes*, 5, 4–53.

—— 1957: Beobachtungswerte und weitere Studien zum Wärmehaushalt eines jungen Waldes. *Wissenchaftliche Mitteilungen Meteorologischen Institut, Universität München* 1, No. 4

BAUMGARTNER, A. and KIRCHNER, M. 1980: Impacts due to deforestation. In Bach, W., Pankrath, J. and Williams, J. (eds.) *Interactions of Energy and Climate*. Dordrecht: Reidel, 305–16.

BENECKE, P. and VAN DER PLOEG, R. R. 1975: Nachhaltige Beeinflüssung des Landschaftwasserhaushaltes durch die Baumartenwahl. *Forstarchiv* 46, 97–102.

BEVEN, K. J. and KIRKBY, M. J. 1979: A physical based variable contributing area model of basin hydrology. *Hydrological Sciences Bulletin* 24, 46–69.

BODEUX, A. 1954: Recherches écologiques sur le bilan de l'eau sous la forêt et la lande de Haute Campine. *Agricultura*, II, 1–80.

BOSCH, J. M. and HEWLETT, J. D. 1982: A review of catchment experiments to determine the effect of vegetation changes on water yield and evapotranspiration. *Journal of Hydrology* 55, 2–23.

CALDER, I. R. 1977: A model of transpiration and interception loss from a spruce forest in Plynlimon, central Wales. *Journal of Hydrology*, 33, 247–65.

CALDER, I. R. and NEWSON, M. D. 1979: Land use and water resources in Britain – a strategic look. *Water Resources Bulletin, US* 15, 1628–39.

CARLISLE, A., BROWN, A. H. F. and WHITE, F. J. 1966: Litter-fall production and the effects of defoliation by Tortrix Viridara in a sessile oak canopy. *Journal of Ecology* 54, 65–81.

CEPEL, V. N. 1967: Interzeption (= Niederschlagsverdünstung im Kronensaum) in einem Buchen-, einem Eichen- und einem Kiefernbestand des Belgrader Waldes bei Istanbul. *Forstwissenschaftliches Centralblatt*, 5, 301–14.

CLARK, R. T. and NEWSON, M. D. 1978: Some detailed water balance studies of research catchments. *Proceedings of the Royal Society London*, A363, 21–42.

DABRAL, B. G. and RAO, B. K. S. 1968: Interception studies in chir and teak plantations, New Forest. *Indian Forester*, 94, 541–51.

DABRAL, B. G. and RAO, B. K. S. 1969: Interception studies in Sal (Shorea robusta) and Khair (Acacia catechu) plantations. *Indian Forester*, 95, 314–23.

DARBY, H. C. 1956: The clearing of the woodland in Europe. In Thomas, W. C. (ed.) *Man's Role in Changing the Face of the Earth*. Chicago: University of Chicago Press, 183–6.

DEACON, E. L. 1969: Physical processes near the surface of the earth. In Flohn, H. (ed.) *General Climatology* 2. Amsterdam: Elsevier, 39–102.

DELFS, J. 1965: Interception and stem flow in stands of Norway spruce and beech in West Germany. In *International Symposium on Forest Hydrology*. Oxford: Pergamon Press, 179–85.

EIDMANN, F. E. 1959: *Die Interception in Buchen- und Fichtenbeständen. Ergebnis mehrjähriger Untersuchungen im Rothaargebirge (Sauerland)*. Gentbrugge: Publication de l'Association Internationale d'Hydrologie Scientifique.

ETTEHAD, R., LOSSAINT, P. and RAPP, M. 1973: Recherches sur la dynamique et le bilan de l'eau des sols de deux écosystèmes Mediterranéens a chêne vert. *Editions du Centre National de Recherches Scientifiques, (Paris)*, 40, 199–289.

FLOHN, H. 1973: Natürliche und anthropogene Klimamodifikationen. *Annalen der Meteoologie* NF 6, 59–66.

GADD, A. J. and KEERS, J. F. 1970: Surface exchanges in a 10-level model. *Quarterly Journal of the Royal Meteorological Society* 90, 297–308.

GALOUX, A. 1963: Budgets et bilans dans l'écosystème forêt. *Lejeunia, Revue de Botanique*, 21.

—— 1973: La chênaie mêlangée calcicole de Virelles-Blaimont. Flux d'energie radiante, conversions et transferts dans l'écosystème (1964–7). *Travaux Station de Recherches des Eaux et Forêts, Groenendaal-Hoeilaart, A14.*

GALOUX, A., BENECKE, P., GIETH, G., HAGER, H., KAYSER, C., KIESE, O., KNOERR, K. R., MURPHY, C. E., SCHNOCK, G. and SINCLAIR, T. R. 1981: Radiation, heat, water and carbon dioxide balances. In Reichle, D. E. (ed.) *Dynamic Properties of Forest Ecosystems*. Cambridge: Cambridge University Press, 87–204.

GALOUX, A. and GRULOIS, J. 1968: La chênaie de Virelles-Blaimont. Echanges radiatifs et convectifs en phase vernale. *Travaux Station de Recherches des Eaux et Forêts, Groenendaal-Hoeilaart, A13.*

GASH, J. H. C. 1979: An analytical model of rainfall interception by forests. *Quarterly Journal of the Royal Meteorological Society*, 105, 43–55.

GASH, J. H. C. and STEWART, J. B. 1977: The evaporation from Thetford forest during 1973. *Journal of Hydrology* 35, 385–95.

GASH, J. H. C., WRIGHT, I. R. and LLOYD, C. R. 1980: Comparative estimates of interception loss from three coniferous forests in Great Britain. *Journal of Hydrology* 48, 89–105.

GAY, L. W. and KNOERR, K. R. 1975: *The Forest Radiation Budget*. Bulletin 19. Durham, North Carolina: Duke University School of Forestry and Environmental Studies.

GENTRY A. H. and LOPEZ-PARODI, J. 1980: Deforestation and increased flooding of the upper Amazon. *Science* 210, 1354–56.

GOUDRIAAN, J. 1977: *Crop micrometeorology: a simulation study*. Pudoc: Wageningen Centre for Agricultural Publishing and Documentation.

GOUDRIAAN, J. and WAGGONER, P. E. 1972: Simulating both aerial microclimate and soil temperature from observations above the foliar canopy. *Netherlands Journal of Agricultural Science*, 20, 104–24.

GRAINGER, A. 1980: The state of the world's tropical forests. *The Ecologist*, 10, 6–54.

GRIP, H., HALLDIN, S., JANSSON, P. E., LINDROTH, A., NOREN, B. and PERTTU, K. 1979: Discrepancy between energy and water balance estimates of evapotranspiration. In Halldin, S. (ed.) *Comparison of Forest Water and Energy Exchange Models*. Copenhagen: International Society for Ecological Modelling. 237–55.

GRULOIS, J. 1968: La chênaie de Virelles-Blaimont. Reflexion, interception et transmission du rayonnement de courtes longueurs d'oude variations au cours d'une année. *Bull. Soc. Roy. Bot. Belg.* 102, 13–25.

HANCOCK, N. A. and CROWTHER, J. M. 1979: A technique for the direct measurement of water storage on a forest canopy. *Journal of Hydrology* 41, 105–22.

HANCOCK, N. H. SELLERS, P. J. and CROWTHER, J. M. 1983: Evaporation from a partially wet forest canopy. *Annales Geophysical*, 1, 139–46.

HELVEY, J. D. 1971: A summary of rainfall interception by certain conifers of North America. In *Proceedings of the International Symposium for Hydrology Professors, Biological Effects in the Hydrological Cycle*. Lafayette, Indiana: Purdue University, 103–13.

HELVEY, J. D. and PATRIC, J. H. 1965: Canopy and litter interception by hardwoods of Eastern United States. *Water Resources Research* 1, 193–206.

HIBBERT, A. R. 1967: Forest treatment effects on water yield. In Sopper, W. E. and Lull, H. W. (eds.) *Forest Hydrology*. Oxford: Pergamon Press, 536–8.

HOPKINS, B. 1960: Rainfall interception by a tropical forest in Uganda. *East Afr. Agric. Journal*, 25, 255–8.

HUTTEL, C. 1975: Recherches sur l'Ecosystème de la forêt subequatoriale de basse Côte d'Ivoire. *La Terre et al Vie, Rev. Ecol. Appl.* 29, 192–202.

INTRIBUS, I. 1975: Water balance factors in the ecosystem of an oak–hornbeam stand at the object in Bab. *Research Project IBP Progress Report*, 2. Bratislava: Slovak Academy of Sciences, 337–51.

JARVIS, P. G. 1976: Exchange properties of coniferous forest canopies. In *XVI Congress of International Union of Forest Research Organizations, Norway, Division II*, 90–8.

JARVIS, P. G., JAMES, G. B. and LANDSBERG, J. J. 1976: Coniferous forest. In Monteith, J. C. (ed.) *Vegetation and the Atmosphere* 11. Oxford: Academic Press, 171–240.

KIESE, O. 1972: Bestandsmeteorologische Untersuchungen zur Bestimmung des Wärmehaushalts eines Buchenwaldes. *Berichte des Instituts für Meteorologie und Klimatologie der Technischen Universität Hannover.*

LANDSBERG, J. J., JARVIS, P. G. and SLATER, M. B. 1973: The radiation regime of a spruce forest. In *Plant Response to Climatic Factors, Proceedings of the Uppsala Symposium.* Paris: UNESCO, 411–18.

LANGFORD, K. J., MORAN, R. J. and O'SHAUGHNESSY, P. J. 1980: The North Maroondah experiment pre-treatment phase comparison of catchment water balances. *Journal of Hydrology* 46, 123–45.

LAW, F. 1956: The effect of afforestation upon the yield of water catchment areas. *Journal of the British Waterworks Association*, 38, 484–94.

—— 1958: Measurement of rainfall, interception and evaporation losses in a plantation of Sitka spruce trees. *International Association of Scientific Hydrology General Assembly Toronto*, 2, 397.

LEGG, B. J. and LONG, I. F. 1975: Turbulent diffusion within a wheat canopy II. *Quarterly Journal of the Royal Meteorological Society*, 80, 198–212.

LETTAU, H. H. 1969: Evapotranspiration Climatonomy I: A new approach to numerical prediction of monthly evapotranspiration, run-off, and soil moisture storage. *Monthly Weather Review*, 47, 691–9.

LETTAU, H. H. and BARADAS, M. W. 1973: Evapotranspiration climatonomy II: Refinement of paramaterization exemplified by application to the Mabacan River Watershed, *Monthly Weather Review*, 101, 636–49.

LETTAU, H., LETTAU, K. and MOLION, L. C. B. 1979: Amazonia hydrologic cycle and the role of atmosphere recycling in assessing deforestation effects. *Monthly Weather Review*, 107, 227–38.

LILL, V. W. S., KRUGER, F. J. and WYK, V. D. B. 1980: The effect of afforestation with Eucalyptus Grandis Hill Ex. Maiden and Pinus Patula Schlecht – Et Cham on streamflow from experimental catchments at Mokobulaan, Transvaal. *Journal of Hydrology* 48, 107–18.

LOCKWOOD, J. G. 1979: *Causes of Climate.* London: Edward Arnold.

LOCKWOOD, J. G. and SELLERS, P. J. 1982: Comparisons of interception loss from tropical and temperate vegetation canopies. *Journal of Applied Meteorology*, 21, 1405–12.

—— 1983: Some simulation model results of the effect of vegetation change on the near-surface hydroclimate. In Street-Perrot, A., Beran, M. and Ratcliffe, R. (eds.) *Variations in the Global Water Budget.* Dordrecht: Reidel, 463–77.

LOW, K. S. 1972: Interception loss in the humid forested areas (with special reference to Sungai Lui catchment, West Malaysia). *Malayan Nature Journal*, 25, 104–11.

LÜTZKE, R. 1966: Vergleichende Energieumsatzmessungen im Walde und auf einer Wiese. *Arch. Forswes.* 15, 995–1015.

LÜTZKE, R. and SIMON, K. H. Zur Bilanzierung des Wasserhaushalts von Waldestanden auf Sandstandorten der Deutschen Demokratischen Republik. In *Beiträge fur die Forstwirtschaft*, 1. Akademie-Verlag, Berlin; in *Allgemeine Forstschrift*, 37, Munchen, 806–7.

MCCOLL, J. G. 1970: Properties of some natural waters in a tropical wet forest of Costa Rica. *Bioscience*, 20, 1096–100.

MCGINNIS, J. T., GOLLEY, F. B., CLEMENTS, R. G., CHILD, G.I. and DUEVER, M. J. 1969: Elemental and hydrologic budgets of the Panamian tropical moist forest. *Bioscience* 19, 697–700.

MCNAUGHTON, K. G. and JARVIS, P. G. 1983: Predicting effects of vegetation changes on transpiration and evaporation. In Kozlowski, T. T. (ed.) *Water Deficits and Plant Growth, Vol. VII. Additional Woody Crop Plants.* New York: Academic Press, 1–47.

MOLCHANOV, A. A. 1971: Cycles of atmospheric precipitation in different types of forests of natural zones of the USSR. Ecology Conservation 4. In *Productivity of forest ecosystems.* Proceedings of the Brussels Symposium. Paris: UNESCO.

MONSI, M. and SAEKI, J. 1953: Über den Lichtfaktor in der Pflanzengesellschaft und seine Bedeutung fur die Stoffproduktion. *Jan J. Bot*, 14, 22–52.

MORAES, V. H. F. 1977: Rubber. In Alvim, P. de T., and Kozlowski, T. T. *Ecophysiology of Tropical Crops.* Oxford: Academic Press, 315–31.

MURPHY, C. E. and KNOERR, K. R. 1975: Evaporation of intercepted rainfall from a forest stand. An analysis by simulation. *Water Resources Research*, 11, 273–380.

NOIRFALISE, A. 1959: Sur l'interception de la pluie par le couvert dans quelques forets belges. *Bull. Soc. R. For. Belg. (Bruxelles)* 10, 433–9.

ODUM, T. H., MOORE, A. M. and BURNS, L. A. 1970: Hydrogen budget and compartments in the rain Forest. In *A Tropical Rain Forest* 3. Washington, DC: US Atomic Energy Commission, 105–22.

OKE, T. R. 1978: *Boundary Layer Climates.* London: Methuen.

OLIVER, H. R. 1978: Ventilation in a forest. *Agricultural Meteorology* 14, 347–55.

PAULSON, C. A. 1970: The mathematical representation of wind speed and temperature profiles in the unstable atmospheric surface layer. *Journal of Applied Meteorology* 9, 857–61.

PEARCE, A. J., GASH, J. H. C. and STEWART, J. B. 1980: Rainfall interception in a forest stand estimated from grassland data. *Journal of Hydrology*, 46, 147–63.

PEARCE, A. J. and ROWE, L. K. 1981: Rainfall interception in a multi-storied evergreen mixed forest: estimates using Gash's analytical model. *Journal of Hydrology*, 49, 341–53.

PEARCE, A. J., ROWE, L. K. and STEWART, J. B. 1980: Nighttime, wet canopy evaporation rates and the water balance of an evergreen mixed forest. *Water Resources Research*, 16, 955–9.

PEIXOTO, J. P. and OORT, A. H. 1983: The atmosphere branch of the hydrological cycle and climate. In Street-Perrott, A., Beran, M. and Ratcliffe, R. *Variations in the Global Water Budget.* Dordrecht: Reidel, 5–65.

PEIXOTO, J. P., SALSTEIN, D. A. and ROSEN, R. D. 1981: Intra-annual variation in large-scale moisture fields. *Journal of Geophysical Research*, 86, 1255–64.

PEREIRA, H. C. 1972: The influence of man on the hydrological cycle. In IASH *World Water Balance* 3, Gentbrugge, 533–69.

RATISBONA, L. R. 1976: The climate of Brazil. In Schwerdtfeger, W. (ed.) *Climates of Central and South America.* Amsterdam: Elsevier, 219–93.

RAUNER, YU L. 1976: Deciduous forest. In Monteith, J. L. (ed.) *Vegetation and the Atmosphere* 11. Oxford: Academic Press, 241–64.

RAY, M. P. 1970: Preliminary observations on stem flow, etc., in *Abstonia scholans* and *Shorea robusta* plantations at Arabari, West Bengal. *Indian Forester*, 96, 482–93.

RICHARDS, P. W. 1964: *Tropical Rain Forest.* Cambridge: Cambridge University Press.

ROBERTS, J., PYMAR, C. F., WALLACE, J. S. and PITMAN, R. M. 1980: Seasonal changes in leaf area, stomatal and canopy conductancies and transpiration from bracken below a forest canopy. *Journal of Applied Ecology*, 17, 409–22.

ROBINS, P. C. 1974: A method of measuring the aerodynamic resistance to the transport of water vapour from forest canopies. *Journal of Applied Ecology* 11, 315–25.

ROBINSON, N. 1966: *Solar Radiation.* Amsterdam: Elsevier.

ROWE, P. B., STOREY, H. C. and HAMILTON, E. L. 1951: Some results of hydrologic research. *US Department of Agriculture Forest Service, Californian Forest and Range Experiment Station, Miscellaneous Publications*, 1.

RUTTER, A. J. 1963: Studies in the water relations of *Pinus sylvestris* in plantation conditions 1. Measurement of rainfall and interception. *Journal of Ecology*, 51, 191–203.

—— 1967: Evaporation in forests. *Endeavour*, 26, 39–43.

—— 1968: Water consumption by forests. In Kozlowski, T. T. (ed.) *Water Deficits and Plant Growth* 2. New York: Academic Press, 23–84.

—— 1975: The hydrological cycle in vegetation. In Monteith, J. C. (ed.) *Vegetation and the Atmosphere*. Oxford: Academic Press, 111-54.

RUTTER, A. J., KERSHAW, K. A., ROBINS, P. C. and MORTON, A. J. 1971: A predictive model of rainfall interception in forests. I: Derivation of the model from observations in a plantation of Corsican Pine. *Agricultural Meteorology* 9, 367-84.

RUTTER, A. J., MORTON, A. J. and ROBINS, P. C. 1975: A predictive model of rainfall interception in forests. II. Generalization of the model and comparison with observations in some coniferous and hardwood stands. *Journal of Applied Ecology*, 12, 367-82.

SAGAN, C., TOON, O. B. and POLLACK, J. B. 1979: Anthropogenic albedo changes and the earth's climate. *Science*, 206, 1363-8.

SCHNOCK, G. 1970: *Le bilan de l'eau et ses principales composantes dans une chênaie melangée calcicole de Haute-Belgique (Bois de Virelles-Blaimont)*. Thèse Université Libre de Bruxelles, Bruxelles.

SEDDON, B. 1967: Prehistoric climate and agriculture review of recent paleoecological investigations. In Taylor, J. S. (ed.) *Weather and Agriculture*. Oxford: Pergamon Press, 173-85.

SELLERS, P. J. 1981: *Vegetation Type and Catchment Water Balance: a Simulation Study*. Ph.D. thesis, University of Leeds.

SELLERS, P. J. and LOCKWOOD J. G. 1981a: A computer simulation of the effects of differing crop types on the water balance of small catchments over long time periods. *Quarterly Journal of the Royal Meteorological Society* 107, 395-414.

—— 1981b: A numerical simulation of the effects of changing vegetation type on surface hydroclimatology. *Climatic Change*, 3, 121-36.

SHUKLA, J. and MINTZ, Y. 1982: Influence of land-surface evapotranspiration on the earth's climate. *Science*, 215, 1498-500.

SHUTTLEWORTH, W. J.. 1983: Evaporation models in the global water budget. In Street-Perrott, A., Beran, M. and Ratcliffe, R. (eds.) *Variations in the Global Water Budget*. Dordrecht: Reidel, 147-71.

SHUTTLEWORTH, W. J. and CALDER, I. R. 1979: Has the Priestly-Taylor equation any relevance to forest evaporation? *Journal of Applied Meteorology*, 18, 639-46.

SINGH, B. and SZEICZ, G. 1979: The effect of intercepted rainfall on the water balance of a hardwood forest. *Water Resources Research*, 15, 131-8.

SMITH, M. K. 1974: Throughfall, stemflow and interception in pine and eucalyptus forest. *Aust. For.* 36, 190-7.

SNOW, J. W. 1976: The climate of northern South America. In Schwerdtfeger, W. (ed.) *Climates of Central and South America*. Amsterdam: Elsevier, 295-403.

SOER, G. J. R. 1977: The Tergra model, *NOTA 1014*. Wageningen: Instituut voor Cultuurtechniek en Waterhuishouding.

STEWART, J. B. 1971: The albedo of a pine forest. *Quarterly Journal of the Royal Meteorlogical Society*, 97, 561-4.

—— 1977: Evaporation from the wet canopy of a pine forest. *Water Resources Research*, 13, 915-21.

STRAUSS, R. 1971a: *Energiebilanz and Verdünstung eines Fichtenwaldes im Jahre 1969*. München: Universität München, Meteorologisches Institut.

—— 1971b: Energiehaushalt eines Fichtenwaldes, 1969. Meteorologisches Institut der Universität München. *Wiss. Mitt.* 21, 17-19.

STUDY OF MAN'S IMPACT ON CLIMATE (SMIC) REPORT 1971: *Inadvertent Climate Modification*. Cambridge, Mass: MIT Press.

SWIFT, L. W., SWANK, W. T., MANKIN, J. B., LUXMOORE, R. J. and GOLDSTEIN, R. A. 1975: Simulation of evapotranspiration and drainage from mature and clearcut deciduous forests and young pine plantation. *Water Resource Research*, 11, 667-73.

TAJCHMAN, S. J. 1972: The radiation and energy balances of coniferous and deciduous forests. *Journal of Applied Ecology*, 9, 359-75.

THOM, A. S. 1971: Momentum absorption by vegetation. *Quarterly Journal of the Royal Meteorological Society*, 97, 414-28.

—— 1972: Momentum, mass and heat exchange of vegetation. *Quarterly Journal of the Royal Meteorological* Society, 98, 124-34.

—— 1975: Momentum, mass and heat exchange of plant communities. In Monteith, J.C. (ed.) *Vegetation and the Atmosphere Vol. 1.* London: Academic Press, 57-109.

THOM, A. S. and OLIVER, H. R. 1977: On Penman's equation for estimating regional evaporation. *Quarterly Journal of the Royal Meteorological Society*, 103, 345-57.

THOM, A. S., STEWART, J. B., OLIVER, H. R. and GASH, J. H. C. 1975: Comparison of aerodynamic and energy budget estimates of fluxes over pine forest. *Quarterly Journal of the Royal Meteorological Society*, 101, 93-105.

THOMPSON, F.B. 1972: Rainfall interception by oak coppice (*Quercus robur*, L.) In Taylor, J.A. (ed.) *Research Papers in Forest Meteorology.* Aberystwyth: Cambrian News, 59-74.

UKRECKY, I., SMOLIK, Z. and LANAR, M. 1975: Preliminary evaluation of the precipitation balance in the floodplain forest near Lednice in Moravia, Ecosystem study on floodplain forest in South Moravia. *PT-PP-IBP*, Report 4. Brno, Czechoslovakia: Institute of the IBP, 287-96.

UNESCO 1978: *Tropical Forest Ecosystems.* Paris: UNESCO.

WAGGONER, P. E. and REIFSNYDER, W. E. 1968: Simulation of temperature, humidity and evaporation profiles in a leaf canopy. *Journal of Applied Meteorology* 7, 400-9.

WARING, R. H., ROGERS, J. J. and SWANK, W. T. 1981: Water relations and hydrologic cycles. In Reichle, D. E. (ed.) *Dynamic Properties of Forest Ecosystems.* Cambridge: Cambridge University Press, 205-64.

WATTS, I. E. M. 1954: Line squalls of Malaya. *Journal of Tropical Geography* 3, 1-14.

WHITE, E. J. and CARLISLE, A. 1968: The interception of rainfall by mixed deciduous woodland. *Q. J. For.* 310-20.

YABUKI, K., AOKI, M. and HAMONTANI, K. 1978: Characteristics of the forest microclimate. In Kira, T., Ono, Y. and Hosokama, T. (eds.) *Biological Production in a Warm-Temperate Evergreen Oak Forest of Japan.* Tokyo: University of Tokyo Press, 55-64.

ZINKE, P. J. 1967: Forest interception studies in the United States. In Sopper, W. E. and Lull, H. W. (eds.) *Forest Hydrology.* Oxford: Pergamon Press, 137-61.

7 **Climate and energy**

World-wide energy use has increased at an average annual rate of about 3.5 per cent over the last 50 years, accelerating to about 5.5 per cent during the latter half of this period, but with a perceptible slowing in the late 1960s. By 1971, the per capita consumption of energy among the developed countries had reached around 5×10^3 W, having tripled since the beginning of the Industrial Revolution (Cooke, 1971).

Llewellyn and Washington (1977) have reviewed the regional and global aspects of energy use. They comment that the combined result of population increases, changing patterns of population densities, and escalating per capita energy consumption has been to concentrate in some regions of the world very large energy fluxes per unit area (energy flux density) when compared with the world average. According to Landsberg and Perry (1977) annual energy consumption for the world as a whole is about 0.016 W m^{-2}, but for the United States it is about 0.26 W m^{-2}, while for Africa it is only 0.004 W m^{-2}, or around 1.5 per cent of the level of the United States. The present world energy flux density distribution has been mapped (Figure 7.1) by Llewellyn and Washington (1977). They show that in 1970 the average flux density of the eastern USA, Germany and Japan was equivalent to between 0.8 and 1.4 W m^{-2}. In contrast much of Southeast Asia had an average energy flux density in the range $0\text{--}0.008$ W m^{-2}, with Thailand reaching around 0.04 W m^{-2} and only eastern China, the Philippines and southeast Australia reaching between 0.16 and 0.8 W m^{-2}.

According to Llewellyn and Washington (1977) the net population growth rate for the developing countries has risen sharply since 1950, largely as a result of declining infant death rates. This, coupled with an increasing concern and ability on the part of these nations for improving the living conditions of their people, has resulted in rising energy consumption to the extent that the ratio of per capita energy consumption for the developed to the developing countries has remained nearly constant at about 8 since 1960. Thus even though the population of the developing world is increasing more rapidly than that of the developed world, the difference in per capita energy consumption has not changed markedly, indicating that the aggregate energy consumption by the developing countries is becoming a larger fraction of the world total as time passes. Llewellyn and Washington (1977) suggest that if this trend continues it will lead to a more rapid increase of energy flux densities in the developing countries than in the developed nations. Indeed, aggregate consumption of all fuels by the developing nations will equal that among the present developed nations in less than 120 years, if both groups maintain the average increase in energy use typical of the past 15 years. The upward trend of energy use in the developing countries comes partly from a need to feed an increasing population and partly from an increase in industrial activity.

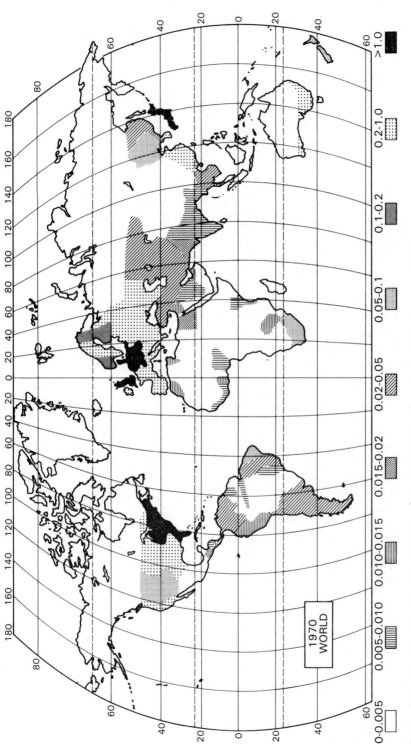

Figure 7.1 World energy flux density (W m^{-2}) AD 1970 (after Llewellyn and Washington, 1977).

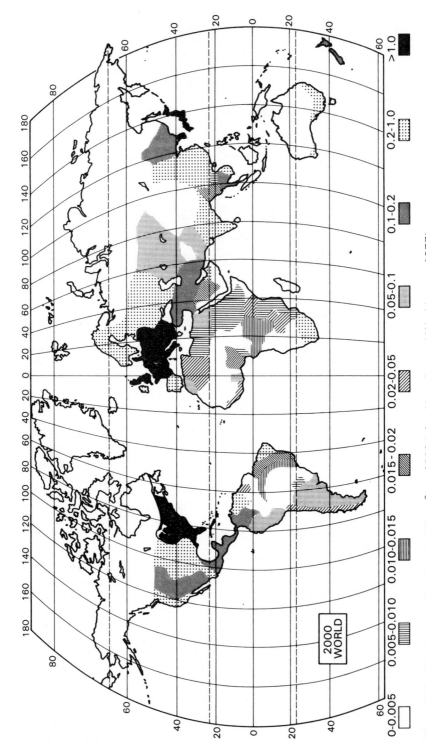

Figure 7.2 World energy flux density (W m^{-2}) AD 2000 (after Llewellyn and Washington, 1977):

0-0.005   0.005-0.010   0.010-0.015   0.015-0.02   0.02-0.05   0.05-0.1   0.1-0.2   0.2-1.0   >1.0

Llewellyn and Washington (1977) have estimated the world energy flux densities for 2000, 2025, 2050 and 2075 (Figure 7.2), on the basis of present values and recent trends in per capita energy consumptions and population densities. Calculations are on a country-by-country basis, except where available data made possible smaller geographical computations. In their predicted analysis, Southeast Asia shows a rapid energy growth, with eastern China and India reaching an energy flux density equal to that of western Europe and the eastern USA by 2050. Much of mainland Southeast Asia is forecast to have an energy flux density of at least 0.16 W m^{-2} by the year 2000, which is a substantial increase over 1970 values. In 2000 AD much of China, India, and parts of Malaysia and Thailand are forecast to have energy flux densities equivalent to those now found in southeast Australia. Nevertheless there are a few countries which are forecast to have low energy flux densities throughout the period.

Climate is related to this increasing energy use in three major areas:

(i) Climate as an energy source;
(ii) The pollution of the atmosphere by energy use;
(iii) Climatic controls on energy use and the effects of climatic variations.

1 Climate as an energy source

The principal growth in energy use will come from coal, oil and gas with perhaps some nuclear energy, but climatic energy sources such as solar, wind and water power will also play some part. Brown and Howe (1978) comment that most of the people in the developing countries of Asia, Africa and Latin America live in rural areas. The prohibitive costs of large central generators and massive transmission and distribution systems, as well as the slow pace of the spread of rural electrification programmes, discourage hopes that rural energy needs can be met with a national electric grid. Hence it is useful to inquire into the potential of small-scale technologies that use renewable energy sources coming from the sun. Current solar energy comes from four major systems: (i) photosynthesis, which is the basis of all life, both plant and animal; (ii) rainfall, which comes from the hydrological cycle, which in turn is driven by the sun; (iii) wind, caused by the atmosphere pressure differences due to changing amounts of solar energy falling on different places; and (iv) direct sunshine.

2 The pollution of the atmosphere by energy use

Among the important pollutants produced by energy use are carbon dioxide, dust and waste heat. Over 97 per cent of the energy demand of the industrial world is met today by the burning of conventional fuels. Most of the by-products of fossil-fuel combustion used at present are injected into the atmosphere where they may interfere with the natural radiative processes. The most important gaseous by-product is carbon dioxide, and if the present fossil-fuel consumption growth rate continues, this will lead in the year 2000 to atmospheric carbon dioxide levels about 30 per cent above the pre-industrial base, with levels perhaps doubling around 2050–2075.

2.1 Carbon dioxide
Rotty and Weinberg (1977) comment that most official forecasts of future energy use give to coal a large and increasing role during the next 25 years. Thus nearly all scenarios for future energy supply systems show heavy dependence on coal, the

exact magnitude depending on assumptions as to the reliance on nuclear fission, degree of electrification, and rate of GNP growth. The carbon in the carbon dioxide produced from fossil fuels each year is about 1/10 the net primary production by terrestrial plants, but the fossil fuel production has been growing exponentially at 4.3 per cent per year. Observed atmospheric CO_2 concentrations have increased from 315 ppm in 1958 to 330 ppm in 1974 – in 1900, before much fossil fuel was burned, it was about 290–295 ppm. With current rates of increase in fossil fuel use, the atmospheric concentration should rise to twice its pre-1900 value towards the end of the next century. A shift to coal as a replacement for oil and gas gives more carbon dioxide per unit of energy; thus Rotty and Weinberg (1977) suggest that if energy growth continues with a concurrent shift towards coal, high concentrations can be reached somewhat earlier.

The stores of carbon that are interacting in the short term of a few years consist of fossil fuels, plants and humus, the ocean and the atmosphere. The fluxes that connect these pools are a continuous exchange with the biota, a constant exchange between the atmosphere and the oceans, and the emission of carbon dioxide through combustion of fossil fuels. Woodwell *et al.* (1978) comment that the fact that the terrestrial biota and humus together contain a pool of carbon that is probably between two and three times the total in the atmosphere means that any appreciable change in the biota has the potential to cause a short-term change in the carbon dioxide content of the air. This is seen in the seasonal variations in carbon dioxide concentrations observed at Mauna Loa (Hawaiian Islands). (Figure 7.3) Here carbon dioxide is at a maximum in spring and a minimum in early autumn, the difference between maxima and minima being about 5 ppm or 1.5 per cent of the average concentration.

The carbon dioxide content of the atmosphere (Figure 7.3) is increasing at a rate that has ranged over the past 20 years between about 0.5 and 1.5 ppm yr^{-1}, with an average of about 1.0 ppm yr^{-1}. The increase in atmospheric carbon dioxide has been commonly assumed to be due to the combustion of fossil fuels which release about 5×10^{15} g carbon annually into the atmosphere. When the amount of carbon dioxide released by burning fossil fuels is compared with the total amount accumulating in the atmosphere, it is found that only about half has remained in the atmosphere. It has been thought that some of the 'lost' carbon dioxide went into the oceans, which are a potentially large sink, and that some went into the biosphere, on the assumption that many kinds of photosynthetic systems use it faster as the concentration increases. Kellogg (1979) comments that our current understanding of these sinks is clouded by the realization that we do not know whether the mass of the biosphere is actually growing or shrinking, and there is some evidence that is difficult to refute indicating that deforestation, especially in the tropics, is causing a decrease of this mass. The biosphere may be another source of carbon dioxide as harvested organic material is burnt or allowed to decay, and according to some estimates it could be an even larger source than that of fossil-fuel burning. The entire ocean contains some 60 times more carbon dioxide than resides in the atmosphere, so it is potentially an enormous sink for new carbon dioxide; however, only the upper, well mixed part of the ocean is in contact with the atmosphere, representing a layer with an average depth of several hundred metres which is only about 10 per cent of the total ocean volume. The rate at which the newly added carbon dioxide can be transported downwards in the oceans by eddy diffusion or large-scale over-turning is extremely slow, hence the total exchange time to go half way to a new equilibrium is estimated to be 1,000 to 1,500 years. Therefore for the next several decades at least it is reasonable to believe that a little over one-

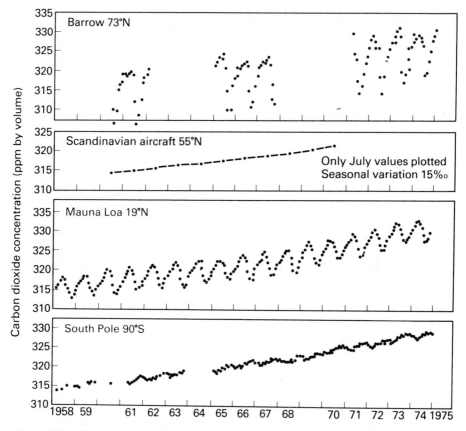

Figure 7.3   Changes in the atmospheric concentration of carbon dioxide. Except for Scandinavian aircraft data, the points denote monthly average concentrations (after Machta *et al.*, 1977).

half of our added carbon dioxide will continue to remain in the atmosphere. On this basis the atmospheric carbon dioxide content will reach twice its pre-1900 level by the middle or end of the next century.

Future changes in the atmospheric CO_2 concentration will depend on the future demand for energy and the proportion of this demand that will be satisfied by fossil fuels. The four major factors determining energy demand are: population growth, economic growth, technological progress in the processes and machines involved in energy conversion, and structural changes within national or regional economies. The global demand for primary energy in 1975 was 8.21 TW yr yr^{-1}. The IIASA Energy Systems Programme (1981) has made two projections for primary energy demand in 2030, amounting to 35.65 TW yr yr^{-1} and 22.39 TW yr yr^{-1} respectively. Rotty and Marland (1980) project a global demand for primary energy of 26.88 TW yr yr^{-1} in 2025. Lovins (1980) in contrast, suggests that a global western European material standard of living could be maintained for 8 billion people (twice the present global population) with today's rate of global energy use. In 1975 fossil fuels supplied 93 per cent of the total primary energy demand. The IIASA scenarios project that about 70 per cent of the total demand in 2030 will be supplied by fossil fuels.

Numerous projections of the increase in atmospheric CO_2 concentration as a result of fossil-fuel consumption have been made. The IIASA Energy Systems Programme (1981) scenarios result in an atmospheric CO_2 concentration in 2030 of 430 pp mv and 500 pp mv respectively. The World Climate Programme (1981) projects an atmospheric CO_2 concentration in 2025 of 410–490 pp mv with 450 pp mv a most likely value. This value may be compared with estimates of atmospheric CO_2 concentrations of as low as 200 pp mv at the end of the last glaciation (Oeschzer *et al.* (1980), Delmas *et al.* (1980)).

Carbon dioxide is important because it is one of the gases which exerts a so-called 'greenhouse effect' in the atmosphere. The radiative equilibrium or planetary temperature of the earth is determined by the balance between solar radiation received by the earth and outgoing infrared radiation at the top of the atmosphere, and is found to be about 257 K ($-16°C$). The average surface temperature, however, is about 15°C or 288 K, which is some 31 K warmer than the planetary temperature. The difference is due to the fact that there are a number of gases in the atmosphere that absorb infrared radiation, which would otherwise escape to space from the surface. This phenomenon is often called 'the greenhouse effect' though the analogy to a greenhouse is actually not a very good one. An important greenhouse gas is carbon dioxide, and any changes in its concentration will affect surface temperature.

The zero-order, purely optical, effect of doubling atmospheric CO_2 content is an increase of about 0.2°C in surface temperature. Depending on what is assumed regarding various feedback mechanisms (atmospheric temperature, humidity, lapse rate, evaporation, clouds, etc.) this zero-order sensitivity may be multiplied by a factor ranging from 0.1 to 50. The problem is best approached by using general circulation models (GCMs), which have the ability to portray many of the non-linear feedback processes which serve to regulate atmospheric (and hence climatic) changes. The first application of a GCM to the CO_2 climate problem was made by Manabe and Wetherall (1975). Using idealized geography with a swamp-like ocean and annually-averaged insolation, they found that a doubling of atmospheric CO_2 resulted in a 2–3°C increase of globally averaged surface air temperature. Their results also showed the warming to be several times larger in higher latitudes than in low, and the globally averaged precipitation to be increased by a few per cent relative to the control. In subsequent simulations with GCMs including more realistic geography and seasonal variations, generally similar results (global increases of about 2°C for double CO_2) have been found; such GCMs indicate that the largest warming due to increased CO_2 will occur in the winter in high latitudes, with the summer precipitation shifted slightly poleward along with increased surface dryness in mid-latitudes.

Manabe and Wetherald (1975) investigated the climatic effects of an increase of atmospheric carbon dioxide using a highly simplified model of the atmospheric general circulation. The simplified characteristics of the model included a limited computational domain with idealized geography, no seasonal variations, no heat transport by ocean currents and fixed cloudiness. The model used in Manabe and Wetherald (1980) is similar to their earlier one, but is used to study the geographical character of the carbon dioxide-induced climatic change.

Manabe and Wetherald (1980) found a number of important changes in global climate accompanying an increase in atmospheric carbon dioxide. For instance the meridional temperature gradient in the lower troposphere markedly reduces in response to an increase in atmospheric carbon dioxide (Figure 7.4). As discussed by Manabe and Wetherald (1975), one of the important reasons for this reduction is

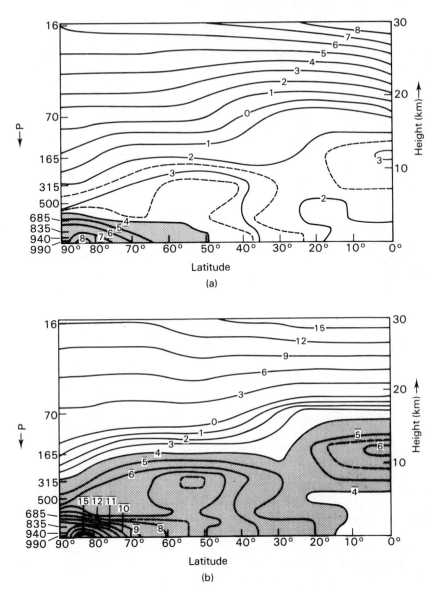

Figure 7.4 Latitude–height distribution of the change of the zonal-mean temperature (K) in response to (a) a doubling of CO_2 content and (b) a quadrupling of CO_2 content (after Manabe and Wetherald, 1980).

the enhancement of the warming in high latitudes by the poleward retreat of the highly reflective snow cover and the confinement of the additional heat in the lowest layer of the model troposphere by the stable stratification. Another important reason for the general reduction of the meridional temperature gradient is the large increase in the poleward transport of latent heat due to the general warming of the model atmosphere. The two authors now appear to be convinced that the effectiveness of the atmosphere as a whole for transporting latent heat increases with increasing atmospheric temperature and results in the reduction of the meridional temperature gradient in the lower model atmosphere. In other words, the meridional temperature gradient in the lower troposphere is reduced irrespective of the existence of the snow-albedo feedback mechanism.

Manabe and Wetherald found that the climatic effect of a carbon dioxide increase may be far from uniform and may have significant geographical variation. Thus the increase in the poleward moisture transport has important hydrologic implications. They found that the enhanced penetration of the moisture to high latitudes markedly increases precipitation there and shifts the mid-latitude rainbelt poleward. Thus, the rate of run-off from the model continent increases significantly in high latitudes. On the other hand, the poleward shift of the mid-latitude rain-belt results in the absence of a significant increase in the precipitation rate and the reduction of soil moisture over the model continent in a zonal belt centred around 42° latitude (Figure 7.5). In the model subtropics, an increase of carbon dioxide in the model atmosphere intensifies the monsoonal precipitation along the east coast of the model continent and this results from the increase of the northward moisture transport along the periphery of the oceanic anticyclones.

Manabe and Stouffer (1979) have described some further results from an improved general circulation model. The mathematical model consisting of both a general circulation model of the atmosphere and a simple mixed-layer ocean model with uniform thickness, has a global computational domain and realistic geography. The ocean model is a static isothermal water layer of uniform thickness with provision for a sea-ice layer. 68 m is chosen as the thickness to ensure that the heat storage associated with the annual cycle of observed sea surface temperature is correctly modelled. The atmospheric carbon dioxide concentration is set at 300 ppm (present concentration around 335 ppm), and 1,200 ppm by volume respectively (hereafter, these experiments are referred to as $1 \times CO_2$ and $4 \times CO_2$ experiments). By using this model the seasonal variation of zonal mean surface air temperature is investigated. In low latitudes, the warming due to the quadrupling of the carbon dioxide content in the air is relatively small and depends little on season, whereas in high latitudes, it is generally larger and varies markedly with season particularly in the northern hemisphere. Over the Arctic Ocean and its neighbourhood, the warming is at a maximum in early winter and is small in summer. It is observed that the sea-ice from the $4 \times CO_2$ experiment is everywhere less than the sea-ice from the $1 \times CO_2$ experiment. Therefore the authors suggest that the $1 \times CO_2$ atmosphere is insulated by thicker sea-ice from the influence of underlying seawater and has a more continental climate with a larger seasonal variation of temperature than the $4 \times CO_2$ atmosphere. They consider that although the poleward retreat of highly reflective snow cover and sea-ice is mainly responsible for the relatively large warming in high latitudes, the change of the thermal insulation effect of sea-ice strongly influences the seasonal variation of the warming over the Arctic region. The seasonal variation of the difference in the surface air temperature between the two experiments over the model continents is significantly different from the variation over the model oceans. At high latitudes the zonal mean surface air temperature over the continents is at a maximum in early winter, being influenced by the large warming over the Arctic Ocean. However, there is a secondary centre of relatively large warming around 65°N in April. This results from a large reduction in surface albedo in spring when the insolation acquires a near maximum intensity. Manabe and Stouffer consider that their study shows two interacting mechanisms, each acting to produce its own sensitivity maximum. The maximum warming of the early winter over the Arctic Ocean and

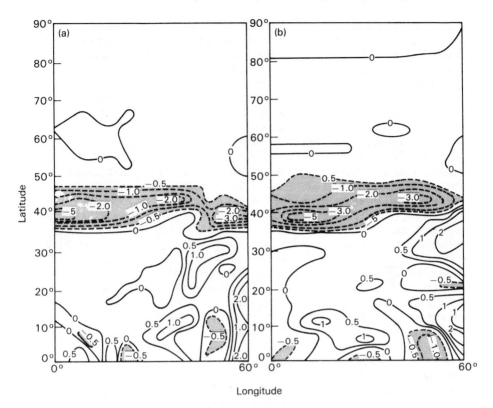

Figure 7.5 Horizontal distribution of the change of soil moisture (cm) over the continent in response to (a) a doubling of CO_2 content, (b) a quadrupling of CO_2 content (after Manabe and Wetherald, 1980).

its neighbourhood is caused by the change in sea-ice thickness, and the relatively large warming over the continents in spring is produced by snow albedo feedback. The area mean change of the annual mean surface air temperature of the model atmosphere which occurs in response to the quadrupling of the carbon dioxide content in the atmosphere is about 4°C. This result suggests that the warming caused by the doubling of the carbon dioxide content would be about 2°C, which is significantly less than the warming which is estimated by the general circulation model of Manabe and Wetherald (1975).

The temperature response from a global doubling of atmospheric carbon dioxide content is likely to be strongest in the surface layer of the polar regions. Here the temperature response could well be 3–4 times that of the global average of 1.5–3°C for a doubling of carbon dioxide, and a value of 7–8°C at 80°N is suggested by Manabe and Wetherald (1980). The question of what would happen to the Arctic Ocean ice pack in the face of such a warming trend has been investigated by Parkinson and Kellogg (1979). The sea-ice model used to perform these experiments is described in detail in Parkinson and Washington (1979). The principal object of the model is to simulate a reasonable yearly cycle of the thickness and extent of the ice over the Arctic Ocean. It was found that with a 5°C increase in surface temperature the ice pack disappeared completely in August and September but reformed in the central Arctic Ocean in the autumn. Also when atmospheric temperature increases of 6–9°C were combined with order-of-magnitude increase in the upward heat flux from the ocean, the ice still reappeared in winter.

Ramanathan *et al.* (1979) have used radiative transfer model calculations to show that the radiative heating of the surface–troposphere system caused by an increase of carbon dioxide undergoes substantial latitudinal and seasonal variations. The increase in zonal seasonal surface temperature for increased atmospheric carbon dioxide, as predicted by their seasonal model, shows little seasonal variability at low latitudes, but at high latitudes there is a pronounced spring/summer enhancement. At 75°N, for example, the carbon dioxide induced enhancement in surface temperature is roughly two times greater in summer than in winter, while at 85°N it is more than three times greater.

Another approach to the problem of climate in a warm, high carbon dioxide world is to use the past as an analogue for the future. Several authors have used this approach and an interesting paper has recently been published by Wigley *et al.* (1980). They compared a composite of the five warmest years in the period 1925–74 with a composite of the five coldest years in this period. For the 65°N to 80°N zone the five warmest years were 1937, 1938, 1943, 1944 and 1953, and the coldest years were 1964, 1965, 1966, 1968 and 1972. The average temperature difference between the warm and cold year groups is 1.6°C for the high latitude zone, and 0.6°C for the northern hemisphere as a whole, and the years themselves reflect the general warmth of the 1930s and 1940s and the subsequent cooling of the northern hemisphere. When the winters are compared (for the high latitude zone), the warm-year group is, on average, 1.8°C warmer than the cold-year group; for the summers the corresponding difference is 0.7°C. The spatial patterns of temperature and precipitation differences between the cold and warm-year groups provide a scenario for (but not a prediction of) the changes in climate which might accompany a carbon dioxide-induced global warming.

A study of the differences between cold and warm-year groups shows that maximum warming occur in high latitudes and in continental interiors, with up to more than five times the hemispheric mean increase in a region extending from Finland across the northernmost parts of Russia and Siberia to about 90°E. This area coincides with the region of greatest natural temperature variability. Some regions show negative differences, for example: Japan, much of India, an area including and adjacent to Turkey, and a region in central Asia. Most of the large differences which occur at high latitudes, although in the same sense for all seasons, are greatest in winter. This is in general agreement with the findings of Manabe and Stouffer (1979). Wigley *et al.* found that an examination of the surface pressure-pattern differences between cold and warm years shows that the warm years have intensified high-latitude (50°N–70°N) westerlies, greater cyclonic activity in the arctic and sub-arctic regions of the eastern hemisphere, and a westward displacement of the Siberian High in the winter's half of the year.

Wigley *et al.* comment that the precipitation patterns generally have greater small-scale spatial variability than temperature patterns. When averaged over the northern hemisphere it was found that the warm-year groups showed a slight (statistically significant) increase in annual precipitation of between 1 and 2 per cent compared with the cold-year group. The most important features in Asia are decreases in precipitation over the central Russian plain and over Japan, and increases in precipitation over India and the Middle East. There is evidence of a more intense monsoon circulation in the warm years.

It has been suggested that a carbon-dioxide induced warming may cause seasonal melting of the Arctic sea-ice. Wigley *et al.* (1980) examined sea-ice extent in their cold and warm-years groups and found there to be notable differences. Two of the warm years (1937 and 1938) were remarkable for a major reduction in the summer ice along the north coast of Asia and increased outflow of ice from the Arctic in the East Greenland current.

Wigley *et al.* (1980) have used the past as an analogue for the future. Flohn (1979) comments that in the geological past one can distinguish several phases with a climate slightly warmer than today and a continental distribution fairly similar to that of the present. Flohn has listed four possible candidates for analogues for a carbon-dioxide induced warm earth. These are the medieval warm period (about 800–1200 AD), the Holocene warm period known as the Atlantic, Hypsithermal or Altithermal (around 4,000 to 8,000 yr BP), the last (Eemian) interglacial (about 120,000 yr BP), and the last period when the Arctic Ocean was believed to be ice free (before 2.5×10^6 yr BP). The most detailed analysis of a past epoch as an analogue for a future warmer world is Kellogg's (1977, 1977–78, 1979) analysis of the Hypsi-thermal. For this period he produces a map showing those regions of the world which were wetter or drier, but his data are a little uncertain.

2.2 Aerosols

Bach (1979) states that the total global aerosol production is at present about 3×10^9 tons yr^{-1} of which roughly 1/10 is of man-made origin. Unlike CO_2 with a resi-dence time of 2–5 years resulting in a uniform atmospheric mixing ratio, aerosols with a mean tropospheric residence time of 9 days show regionally high concentra-tions near urban and industrial agglomerations and near areas with agricultural burning. Aerosol observations are so sparse that it is impossible to draw a global pattern of mean man-made aerosol pollution from them alone.

Kellogg *et al.* (1975) have made crude estimates of man-made aerosol distribu-tions. They found that man-made aerosols generally lie in a belt at mid-latitudes in the northern hemisphere, the major exception being the North Pacific. There is little cross-equatorial flow, and the countries of the southern hemisphere con-tribute far less than the industrialized countries of the northern hemisphere. A large plume originates over Japan and extends southeast in January and northeast in July. The bulk of the Pacific is free from the effects of man-made aerosols. This is confirmed by the observations of Ellis and Pueschel (1971), who, using measure-ments of solar transmittance at Mauna Loa, Hawaii, found no evidence of marked increases in man-made aerosols.

Recent studies seem to indicate that the heat budget is not so much influenced by the scattering properties of aerosols, but rather by the absorption coefficients of the particles (National Academy of Sciences, 1979; Glazier *et al.*, 1976). It also appears that the cooling aspect of aerosols has been exaggerated (Rotty and Mitchell, 1976). Although in fact the thermal impact of aerosols cannot yet be assessed reliably, indications now point rather towards a slight net warming (Mitchell, 1975).

In a recent paper Grassl (1979) explains why it is not yet possible to assess reliably the effects of aerosol particles on the planetary albedo and hence on the tempera-ture. In cloud-free areas a particle increase may either lead to an albedo increase or decrease depending upon the surface albedo and the imaginary part of the refrac-tive index. In cloudy areas one has to distinguish three albedo-changing effects of aerosols. Acting in combination, these will increase the cloud albedo for thin clouds but decrease the cloud albedo for thick clouds. Bach (1979) comments that since we reliably know neither the imaginary parts of the refractive index of aerosols nor the mean optical depth of clouds, it is at present practically impossible to estimate the planetary albedo and temperature changes due to aerosol changes.

2.3 Waste heat

World energy use has increased from 0.1×10^{12} W in 1860 to about 8×10^{12} W in 1975 (Rotty and Mitchell, 1976). Almost all of this heat input has come from energy removed from long-term storage. Bach (1979) comments that if energy consumption continued at a similar growth rate, then by 2050 a little over 400×10^{12} W could be reached. If, on the other hand, by the year 2050 a population of 10×10^9 is assumed together with an energy use of 20 kw per capita, then only 200×10^{12} W, or half the previous estimate would be reached (Kutzback, 1974). Compared to the solar input at the top of the atmosphere these two estimates amount to $0.002\,S$ and $0.001\,S$, respectively. Bach (1979) states that using a planetary albedo of 0.284, one obtains changes in the earth's equilibrium temperature of $0.15°C$ and $0.07°C$ for the above scenarios, which are hardly significant.

Valuable insight can be gained by comparing heat emissions at different scales, because potential climatic alterations depend upon both the size of the area and the density of the energy fluxes from that area. Figure 7.6 (after Bach, 1979) compares a variety of man-induced energy flux densities at different area and time-scales with the global net surface radiation of about 100 W m^{-2}, and the available potential energy of about 2.4 W m^{-2}, the latter being the amount of energy used in all weather processes. Energy sources on the mesoscale and below are intense enough to produce local climate changes. This is seen in the well known local heat islands associated with many cities. At the extreme end of the scale, industrial complexes and especially fossil and nuclear power plants are excessive heat islands. These so-called heat islands are small (a few degrees) thermal 'bumps' on the lowest few hundred metres of the planetary boundary layer of the atmosphere.

Regional and global effects of heat emissions can be assessed with the help of atmospheric model experiments. Using the three-dimensional general circulation model of the National Center for Atmospheric Research in his first experiment, Washington (1971) tested the response of the model atmosphere for an extremely large addition of waste heat (about 250 times the present global heat production) distributed over all non-ocean areas. Surface temperature increases of $8-10°C$ over Canada and Asia and $1-2°C$ over Africa were obtained. Clearly this first experiment had a number of major deficiencies. In a more realistic experiment in which the heat input of 200×10^{12} W (about 25 times the present value) was distributed according to population density it was found that the thermal effects upon the atmosphere were within the noise level of the model's natural variability (Washington, 1972). Testing for regional effects in January, the present energy flux density of 90 W m^{-2} for Manhattan was assumed to exist in the year 2000 from the Atlantic Seaboard to the Great Lakes and Forida in the USA (Llewellyn and Washington, 1977). The main conclusion from this experiment is that temperature may increase by $12°C$ in the vicinity of the anomalous heating but that the heating effect is restricted to the boundary layer (a depth of 3 km). These experiments have been extended by Washington and Chervin (1978). The results from their later experiments indicate that a megalopolis over the eastern USA with an energy flux density of 90 W m^{-2} would lead to significant temperature changes not only over the region itself but also immediately upstream and downstream from it both in winter and summer. Chervin (1980) found in similar computations for both January and July statistically significant changes over the eastern USA with respect to mean temperature, vertical velocity, precipitation and soil moisture.

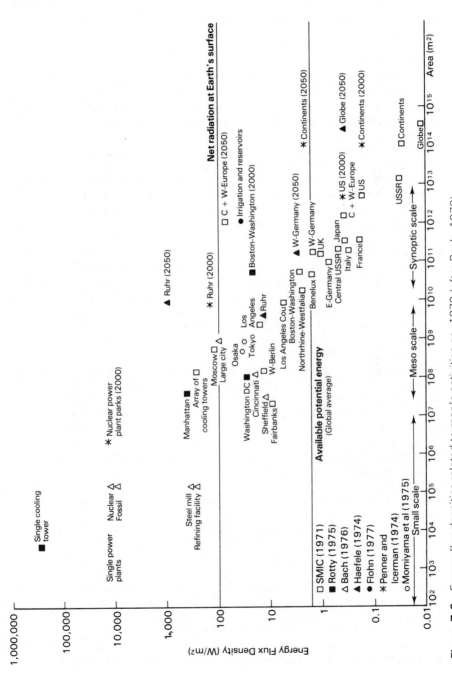

Figure 7.6 Energy flux densities related to man's activities ca. 1970 (after Bach, 1979).

2.4 Urban climates

The effects of aerosols and waste heat are most marked in urban areas, it is therefore necessary to consider urban climates. The climatic modifications caused by human settlements can be of the same order as those produced by small mountains or lakes. Clearly, the size of the city, its type of activity, and its location in the macroclimatic zones will have a decisive influence on the degree of differentiation from the surrounding environment.

According to Landsberg (1981b) some of the causes of distinctive urban climates are as follows:

(1) Replacement of the natural surface by buildings, often tall and densely assembled, causing increased roughness. This can reduce surface wind speeds.

(2) Replacement of natural soil by impermeable roads and roofs, combined with drainage systems, reduces evaporation and humidity and leads to faster run-off.

(3) Roads and building materials have physical constants substantially different from natural soil. Generally, many have lower albedos and greater heat conductivity and capacity. This alters the radiation balance and has consequences for air temperature. In most instances, it leads to heat storage from the solar radiation received. Also, in effect a new primary radiative surface is centred in densely built-up areas at roof level, and this leads to considerable alteration of the lapse rates in the lowest layer.

(4) Heat is added by human activities, and this can be a substantial part of the local energy balance. This, together with the effects described under (1) and (2) above leads to an increase in temperature above that of the surroundings (Figure 7.7). In turn, this can lead to convective rising of air causing cloudiness and promoting precipitation.

(5) Addition of foreign substances, such as water vapour, fumes and gases from combustion and industrial processes has obvious affects. Because of the large numbers of hygroscopic nuclei released, visibilities are reduced, radiation is intercepted, fogs are formed, and precipitation is probably increased.

A gradual consensus is developing about the influence of city size on the magnitude of the temperature difference city–country. One of the difficulties is that the spectacular growth of the large cities in the temperate latitudes of the northern hemisphere in the twentieth century has been paralleled by a slow general warming of the climate until about 1940. To this are added the uncertainties of station changes and exposures. The magnitude of the urban–rural temperature difference was first related by Summers (1964) to meteorological parameters in the surface layers. His simple model is expressed as:

$$\Delta T_{(u-r)} = \frac{2xQ(\partial\theta/\partial z)}{\bar{u}\rho C_p} \qquad (7.1)$$

where $\Delta T_{(u-r)}$ = urban–rural temperature difference

x = wind fetch from periphery to the centre of urban area radius of city area.

Q = heat input of city

$\bar{u}$ = mean wind speed

ρ = air density

C_p = specific heat of air at constant pressure

$\frac{\partial\theta}{\partial z}$ = lapse rate of potential temperature in rural area.

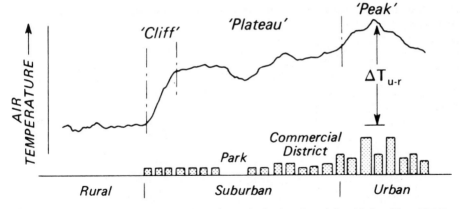

Figure 7.7   Generalized cross-section of a typical urban heat island (after Oke, 1978).

Many of the values are generally only approximately known, hence the checking of the above equation is difficult.

Oke (1973) developed from observations in a number of towns of varying size a relation for the maximum value of $^{-}\Delta T_{(u - r)}$:

$$\Delta T_{(u - r)MAX} \sim \log \text{Pop.} \qquad (7.2)$$

where 'Pop' is the population number. The greatest differences develop in the evening 2–3 hours after sunset, when winds are calm. The slopes of the logarithmic population line is different for European and American towns, as shown in Figure 7.8. Landsberg (1981b) considers that the reasons for this difference are not entirely clear but may be sought principally in the different construction and density of housing and the different modes of energy production and use.

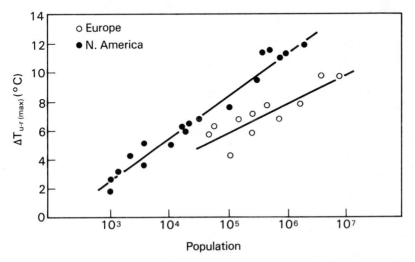

Figure 7.8   Relation between maximum observed heat island intensity ($T_{u-r(max)}$) and population (P) for North American and European settlements (after Oke, 1978).

3 Climatic controls on energy use and the effects of climatic variation

Energy use has been found to depend partly upon the variations in weather and climate in the region concerned. This can be illustrated for electricity generation by monthly New Zealand electricity generation and associated weighted temperature departures, April–September, 1961–70. According to Maunder (1974) the analysis indicates that 25 per cent of the variance in the random oscillation of New Zealand monthly electricity generation was associated with the departure of the monthly New Zealand temperature indices (weighted according to the distribution of the human population) from the 1931–60 norms. Maunder comments that information about the weather can play a very important role in the decision-making processes associated with management of 'weather-sensitive enterprises'.

According to Barnett (1972), electricity demand may be forecast using a model containing the following elements:

Electricity demand = basic level + weather demand
+ day of week correction
+ random component.

Barnett quotes the following values to illustrate the sensitivity of electricity demand to weather changes in Britain during the winter period:

(a) Temperature – a fall in temperature of 1° will increase the demand by about 1.8 per cent.
(b) Wind – a change in wind speed from 4 knots to 9 knots will increase the demand by about 1 per cent assuming temperature does not change;
(c) Cloud – a change in lighting conditions from a clear to a half-obscure sky on a dry day would cause an increase in demand of about 1.4 per cent, assuming there are two cloud layers present. On a wet day the response will be doubled.

It should be noted that the above sensitivities are average levels for a typical winter weekday. Sensitivity to weather also has a diurnal as well as a seasonal pattern. The above values may not apply in other countries and particularly not in tropical countries. Nevertheless increases in temperature in tropical urban areas will cause increased use of electricity for cooling, and thus an inverse of the temperature relationship noted above.

Some indication of the natural variability of temperature can be gained by studying maps of the interannual variabilities of monthly mean air temperature. Craddock (1964) considers that the interannual variability is greater in the northern winter than in the northern summer, greater in the interior of the continents than it is in the oceanic and maritime areas, and very low in all months in the tropics. The regions of highest variability are generally grouped round the Arctic Circle, and southwards there is a general tendency for variability to decrease with latitude. In January the axis of greatest variability exists over the west coast of Hudson Bay, from Greenland to Novaja Zesula in central Siberia, and over the northern Rocky Mountains, Alaska and northeast Siberia.

3.1 Long-term climatic changes

Temperature trends may be economically important, and thus have to be considered. They are best approached by considering the historical variations in climate, since future trends may be similar to those in the immediate past.

Pittock *et al.* (1978) suggest that an inspection of the available data indicates that the 'Little Ice Age' appears to have begun in the tenth century in East Asia and temperatures there reached a minimum in the twelfth century. The cooling apparently spread westward reaching European Russia in the mid-fourteenth century and central Europe in the mid-fifteenth century. Pittock *et al.* comment that such displacements and longitudinal relationships are suggestive of adjustments in the tropospheric wave structure.

In the first half of the twentieth century, the world was enjoying a full recovery from the Little Ice Age of around 1700. Northern latitudes warmed by ~ 0.8°C between the 1880s and 1940 (see Figure 7.9), then cooled ~ 0.5°C between 1940 and 1970. Figure 7.9 also shows that low latitudes warmed 0.3°C between 1880 and 1930, with little change thereafter. Southern latitudes warmed 0.4°C in the past century, while the global mean temperature increased ~ 0.5°C between 1885 and 1940, with slight cooling thereafter. Mitchell (1977) suggests that in the first half of the twentieth century there are indications that the circumpolar westerlies contracted towards the poles, and that, in the northern hemisphere at least, the amplitude of the planetary waves underwent a decrease. A general warming of the earth occurred, which was most pronounced in the Atlantic sector of the sub-arctic. A rapid worldwide retreat of mountain glaciers and a poleward extension of the ranges of many flora and fauna took place. Mitchell further comments that there is considerable evidence that, between the 1940s and about 1970, the climatic changes of the earlier part of this century had tended to undergo a reversal. Temperatures had mostly fallen, especially in the Arctic and the Atlantic sub-Arctic, where sea-ice has been increasing. The circulation of the northern hemisphere appears to have shifted in a manner suggestive of an increasing amplitude of the planetary waves and greater extremes of weather conditions in many parts of the world. Unfortunately, the situation in the southern hemisphere has not been so well documented. These events dramatize the fact that climatic variability is to be expected no less on time-scales of months and years than on time-scales of centuries and millennia.

For the period 1942–72, zonally averaged temperature changes have been computed by Williams and van Loon (1976). Table 7.1 lists the average change in 1942–72 between 15° and 80°N in each season, the values being weighted by area. The values in Table 7.1 combined give an annual temperature change for the period 1942–72 of − 0.26°C. Further analysis shows that the temperature changes were large at high latitudes but relatively small in the tropics.

Table 7.1 Average temperature change 15–80° N during the period 1942–72 (*after Williams and van Loon, 1976*)

Winter	− 0.21°C
Summer	− 0.19°C
Spring	− 0.36°C
Autumn	− 0.29°C
Year	− 0.26°C

Isopleths of the slope of the regression line of winter mean temperature for 1942–1972 have been calculated by van Loon and Williams (1976a). They show that the largest drop of temperatures was in the polar region, but also demonstrate that there was a net rise in the latitudes between 42°N and 54°N and in the tropics as far south as the analysis goes. The average trend from 1942–1972 between 15°N and 75°N was − 0.21°C, and the negative sign is obviously the influence of the change north of 54°N. Indeed, van Loon and Williams (1976a) comment that the

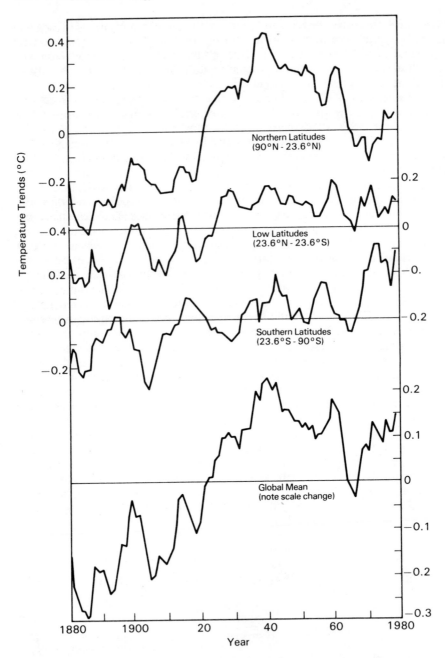

Figure 7.9 Observed surface air temperature trends for three latitude bands and the entire globe. Temperature scales for low latitudes and global mean are on the right (after Hansen *et al.*, 1981). Copyright 1981 by the American Association for the Advancement of Science.

sign of the overall winter temperature trend in the northern hemisphere to a large extent is determined by the direction of the change in higher latitudes. Summer mean temperature changes were also analysed and are positive south of 30°N as far south (15°N) as the analysis goes, but negative at middle and high latitudes. The biggest differences between winter and summer were over the polar cap where the negative changes were 4–7 times larger in winter than in summer, and in 45–50°N where the zonal mean winter temperature rose but the summer temperature fell. The average change in 1942–72 between 15° and 80°N in summer was – 0.19°C, weighted by area. The change was influenced mainly by the large negative values in the middle latitudes, whereas the average change in winter was dominated by the trend in the Arctic.

According to Hansen *et al.* (1981) a remarkable conclusion from Figure 7.9 is that global temperature is almost as high today as it was in 1940. The common misconception that the world is cooling is based on northern hemisphere experience to 1970.

3.2 Scenario of future climatic changes

Man has upset the global carbon cycle by burning fossil fuels and, probably, by deforestation and changing land use. The net result of these activities has been to increase the CO_2 content of both the atmosphere and the oceans. As discussed previously, it is generally believed that if fossil fuel use continues to increase at about the rate of increase experienced over the past century then atmospheric CO_2 levels will reach roughly twice today's level by the middle or end of the next century. The enhanced greenhouse effect could well cause a significant increase in global mean annual surface temperature, and numerical model experiments suggest this could amount to 2 or 3°C. Some of the numerical models suggest that the surface warming due to increased carbon dioxide should be detectable about now at high latitudes. It is not (Madden and Ramanathan, 1980), possibly because the predicted warming is being delayed more than a decade by ocean thermal inertia, or because there is a compensating cooling due to other factors, such as volcanic activity. Thus in the short term the present temperature trends may well continue. In the longer term the increasing energy use noted earlier may well take place against a background of climatic changes induced by increasing atmospheric carbon dioxide though this now appears unlikely before about 1990 or the end of this century.

References

BACH, W. 1976: Global air pollution and climatic change. *Review of Geophysics and Space Physics*, 14, 429–74.
— 1979: Short-term climatic alterations caused by human activities: status and outlook. *Progress in Physical Geography*, 3, 55–83.
BARNETT, C. V. 1972: Weather and the short-term forecasting of electricity demand, 2. In Taylor, J. A. (ed.) *Weather Forecasting for Agriculture and Industry*. Newton Abbott: David & Charles, 209–23.
BROWN, N. E. and HOWE, J. W. 1978: Solar energy for Tanzania. *Science*, 199, 651–7.
BRYSON, R. A. 1974: A perspective on climatic change. *Science*, 184, 753–60.
BUDYKO, M. I. 1974: *Climate and Life*. New York: Academic Press.
CHERVIN, R. M. 1980: Computer simulation studies of the regional and global climatic impacts of waste heat emission. In Bach, W., Pankrath, J. and Williams, J. (ed.) *Interactions of Energy and Climate*. Dordrecht: Reidel, 399–415.

CLARK, W. C. 1982: *Carbon Dioxide Review* 1982. Oxford: Oxford University Press.

COOKE, E. 1971. The flow of energy in an industrial society. In *Energy and Power*. New York: W. H. Freeman & Co., 134–47.

CRADDOCK, J. M. 1964: Interannual variability of monthly mean air temperature over the northern hemisphere. *Meteorological Office Scientific Paper 20*, London.

DELMAS, R. J. J., ASCENCIO, M. and LEGRAND, M. 1980: Polar ice evidence that atmospheric CO_2 20,000 years BP was 50% of present. *Nature* 284, 155–7.

ELLIS, H. T. and PUESCHEL, R. F. 1971: Absence of air pollution trends at Mauna Loa. *Science*, 172, 845–6.

FLOHN, H. 1977: Man-induced input of latent and sensible heat and possible climatic variation. In *Proceedings of the Dahlem Workshop on Global Chemical Cycles and their alterations by Man*. Berlin: Abakon Verlagsgesellschaft.

— 1979: Can climate history repeat itself? Possible climatic warming and the case of paleoclimatic warm phases. In Bach, W., Pankrath, J. and Kellogg, W. W. (eds.) *Man's Impact on Climate*. Amsterdam: Elsevier, 15–28.

GLAZIER, J., MONTEITH, J. L. and UNSWORTH, M. H. 1976: Effects of aerosols on the local heat budget of the lower atmosphere. *Quarterly Journal of the Royal Meteorological Society*, 102, 95–102.

GRASSL, H. 1979: Possible changes of planetary albedo due to aerosol particles. In Bach, W., Pankrath, J. and Kellogg, W. W. (eds) *Man's Impact on Climate*. Amsterdam: Elsevier, 229–41.

HAEFELE, W. 1974: A systems approach to energy. *American Scientist*, 62, 438–47.

HANSEN, J., JOHNSON, D., LACIS, A., LEBEDEFF, S., LEE, P., RIND, D. and RUSSELL, G. 1981: Climate impact of increasing atmospheric carbon dioxide. *Science*, 213, 957–66.

IIASA ENERGY SYSTEMS PROGRAM, 1981: *Energy in a Finite World: a Global Systems Analysis*. Cambridge, Mass: Ballinger.

JÄGER, J. 1983: *Climate and Energy Systems*. Chichester: Wiley.

KELLOGG, W. W. 1977: Effects of human activities on global climate. *World Meteorological Organisation Technical Note* No. 156. Geneva: World Meteorological Organization.

— 1977–78: Effects of human activities on global climate. *World Meteorological Organisation Bulletin*. Part 1, 26, 229–40 (1977); Part 2, 27, 3–10 (1978).

— 1978: Global influences of mankind on the climate. In Gribbin, J. (ed.) *Climatic Change*. Cambridge: Cambridge University Press, 205–7.

— 1979: Influences of mankind on climate. *Annual Review of Earth and Planetary Science*, 7, 63–92.

KELLOGG, W. W., COAKLEY, J. A. J. R. and GRAMS, G. W. 1975: *Effects of Anthropogenic Aerosols on the Global Climate*. In Proceedings of the WMQ/IAMAP symposium on long-term climatic fluctuations. Geneva: World Meteorological Organization, 323–30.

KELLOGG, W. W. and SCHWARE, R. 1981: *Climate Change and Society*. Boulder: Westview Press.

KUTZBACH, J. E. 1974: *Possible Impact of Man's Energy Generation on Climate*. Paper prepared for the International Conference on the Physical Basis of Climate and Climate Modelling, Stockholm.

LA MARCHE, V. C., Jr 1974: Paleoclimatic inferences from long tree-ring records. *Science*, 183, 1043–8.

LAMB, H. H. 1965: The early Medieval warm epoch and its sequel *Palaeogeography, Palaeoclimatology, Palaeoecology*, 1, 13–37.

— 1977: *Climate: Present, Past and Future*. London: Methuen.

LANDSBERG, H. E. 1981a: *The Urban Climate*. New York: Academic Press.

— 1981b: City climate. In Landsberg, H. E. (ed.) *General Climatology* 3. Amsterdam: Elsevier, 299–334.

LANDSBERG, H. H. and PERRY, H. 1977: Energy consumption in 2075. In *Energy and Climate*. Washington DC: National Academy of Sciences, 27–31.

LLEWELLYN, R. A. and WASHINGTON, W. M. 1977: Regional and global aspects. In *Energy and Climate*. Washington DC: National Academy of Sciences. 106–18.

LOVINS, B. 1980: Economically efficient energy futures. In Bach, W., Pankrath, J. and

Williams J. (eds.). *Interactions of energy and climate*. Dordrecht: Reidel, 1–31.

MACHTA, L., HANSON, K., and KEELING, C. D. 1977: Atmospheric carbon dioxide and some interpretations. In Andersen, N. R. and Malakoff, A. (eds). *The Fate of Forsil Fuel CO₂ in the Oceans*. New York: Plenum Press, 131–44.

MADDEN, R. and RAMANATHAN, V. 1980: Detecting climate change due to increasing carbon dioxide. *Science*, 209, 763–8.

MANABE, S. and WETHERALD, R. T. 1975: The effects of doubling the CO_2 concentration on the climate of a general circulation model. *Journal of Atmospheric Sciences*, 32, 31–5.

MANABE, S. and STOUFFER, R. J. 1979: A CO_2-climate sensitivity study with a mathematical model of the global climate. *Nature*, 282, 491–3.

MANABE, S. and WETHERALD, R. T. 1980: On the distribution of climatic change resulting from an increase in CO_2 content of the atmosphere. *Journal of Atmospheric Sciences*, 37, 99–118.

MAUNDER, W. J. 1974: National econoclimatic models. In Taylor, J. A. (ed.) *Climatic Resources and Economic Activities*. Newton Abbott: David & Charles, 237–57.

MITCHELL, J. M., Jr 1975: A reassessment of atmospheric pollution as a cause of long-term changes of global temperature. In *The Changing Global Environment*. Boston: Reidel, 149–73.

— 1977: The changing climate. In *Energy and Climate*. Washington DC: National Academy of Science, 51–8.

MOMIYAMA-SAKAMOTO, M., TAKEUCHI, J. and KATAYAMA, K. 1975: Signs seen in Japan of deseasonality in human mortality. *Meteorology qnd Geophysics*, 26, 30–3.

NATIONAL ACADEMY OF SCIENCES 1975: *Understanding Climatic change*. Washington, DC.

OESCHGER, H., SIEGENTHALER, U. and HEIMANN, M. 1980: The carbon cycle and its perturbation by man. In Bach, W., Pankrath, J. and Williams, J. (eds.) *Interactions of Energy and Climate*. Dordrecht: Reidel, 107–27.

OKE, T. R. 1973: City size and the urban heat island. *Atmospheric Environment*, 7, 769–79.

— 1978: *Boundary Layer Climates*. London: Methuen.

PARKINSON, C. L. and KELLOGG, W. W. 1979: Arctic sea-ice decay simulated for a CO_2-induced temperature rise. *Climatic Change*, 2, 149–62.

PARKINSON, C. L. and WASHINGTON, W. M. 1979: A large-scale numerical model of sea-ice. *Journal of Geophysical Research*, 84, 311–37.

PENNER, S. S. and ICERMAN, L. 1974: *Demands, resources, impact, technology and policy I*. Reading: Addison-Wesley.

PERRY, H. and LANDSBERG, H. H. 1977: Projected world energy consumption. In *Energy and Climate*. Washington DC: National Academy of Sciences, 35–50.

PITTOCK, A. B., FRAKES, L. A., JENSSEN, D., PETERSON, J. A. and ZILLMAN, J. W. 1978: *Climatic Change and Variability*. Cambridge: Cambridge University Press.

RAMANATHAN, V., LIAN, M. S. and CESS, R. D. 1979: Increased atmospheric CO_2: zonal and seasonal estimates of the effect on the radiation energy balance and surface temperature. *Journal of Geophysical Research*, 84, 4949–58.

ROTTY, R. M. 1975: *Energy and Climate. Oak Ridge: Institute for Energy Analyses*. IEA Research Memorandum IEA (M)–76–4.

ROTTY, R. M. and MARLAND, G. 1980: Constraints on fossil fuel use. In Bach, W., Pankrath, J. and Williams, J. (eds.) *Interactions of Energy and Climate*. Dordrecht: Reidel, 191–212.

ROTTY, R. M. and MITCHELL, J. M., Jr. 1976: *Man's Energy and the World's Climate*. AICHE Symposium Series Volume on Air Pollution and Clean Energy.

ROTTY, R. M. and WEINBERG, A. M. 1977: How long is coal's future. *Climatic Change I*, 45–57.

SMIC REPORT 1971: Inadvertent climate modification. *Report of the Study of Man's Impact on Climate*. Cambridge, Mass: MIT Press.

SUMMERS, P. W. 1964: *An Urban Ventilation Model Applied to Montreal*. PhD Thesis, McGill University.

VAN LOON, H. and WILLIAMS, J. 1976a: The connection between trends of mean temperature and circulation at the surface: Part 1, winter. *Monthly Weather Review*, 104, 365–80.

— 1976b: The connection between trends of mean temperature and circulation at the surface: Part II, summer. *Monthly Weather Review*, 104, 1003–11.

WASHINGTON, W. W. 1971: On the possible uses of global atmospheric models for the study of air and thermal pollution. In Matthews, W. H., Kellogg, W. W. and Robinson, G. D. (eds.) *Man's Impact on the Climate*. Cambridge, Mass: MIT Press 265–76.

— 1972: Numerical climate-change experiments: the effects of man's production of thermal energy. *Journal of Applied Meteorology*, 11, 768–72.

WASHINGTON, W. W. and CHERVIN, R. M. 1978: Regional climatic effects of large-scale thermal pollution: simulating studies with the NCAR General Circulation Model. *Journal of Applied Meteorology*, 18, 3–16.

WATTS, I. E. M. 1969: Climates of China and Korea. In Arakawa, H. (ed.) *Climates of Northern and Eastern Asia*. Amsterdam: Elsevier, 1–117.

WEN-HSIUNG, S. 1974: Changes in China's climate. *Bulletin of American Meteorological Society*, 55, 9–24.

WIGLEY, T. M. L., JONES, P. D. and KELLEY, P. M. 1980: Scenarios for a warm, high-CO_2 world. *Nature*, 283, 17–21.

WILLIAMS, J. and VAN LOON, H. 1976: The connection between trends of mean temperature and circulation at the surface: Part III, Spring and Autumn. *Monthly Weather Review*, 104, 1591–6.

WORLD CLIMATE PROGRAM, 1981: *On the Assessment of the Role of CO_2 on Climate Variations and their Impact*. Geneva: World Meteorological Organization.

WOODWELL, G. M., WHITTAKER, R. H., REINERS, W. A., LIKENS, G. E., DELWICHE, C. C. and BOTKIN, D. B. 1978: The biota and the world carbon budget. *Science*, 199, 141–6.

YAMAMOTO, T. 1972: On the nature of the climatic change in Japan since the 'Little Ice Age' around 1800 AD. *Japanese Progress in Climatology*, 97–110.

8 Climate and Food

Man relies on relatively few crops for the greater part of his food supply. Now climate is the major factor determining the world distribution of crop species, since it is the climate of a locality which restricts the choice of species which can be grown. Other factors, particularly economic and social considerations and soil type, then determine which of the suitable species a farmer grows. According to Bunting *et al*. (1982) the main supplies of dietary energy in most parts of the world come from two groups of plants, the cereals and the root and tuber crops (cassava, potatoes, sweet potatoes, yams). In 1978 approximately 757 million million ha of cereals were harvested, and this was about 74 per cent of the harvested area of all crops. Wheat, rice and maize alone provided 49 per cent of the harvested area. The harvested area of root and tuber crops was 50 million ha or 5 per cent of the total harvested area. Bunting *et al*. (1982) have calculated (Figure 8.1) for each country the ratio of the harvested area of root and tuber crops to that of cereals. The ratio was greater than 100 per cent in two areas only – equatorial Africa and Papua New Guinea. In these areas the climate is generally continuously wet and humid, which hinders the setting of grain in cereals. In most tropical countries the ratio is between 1 per cent and 10 per cent, in North America, Australia and the Middle East, it is less than 1 per cent.

Bunting *et al*. (1982) show in Figure 8.2 the first-ranking cereal in each country. It is seen that wheat is the principal cereal of Europe, USSR and Oceania, it occupies second place in North and Central America, South America and Asia and fourth place in Africa. The distribution of rice is much more restricted since Asia produces 95 per cent of the total output. Sorghum is produced in Asia, North and Central America and Africa, and 93 per cent of the world production of bulrush millet is in Asia and Africa. The predominant root and tuber crops in each country are shown in Figure 8.3. It appears that important as evolutionary and other factors are on crop distribution, adaptation to climate is plainly the most significant determinant of the current distribution of crops.

1 Climate and plant growth

According to Bunting *et al*. (1982) for a crop to be grown successfully the following conditions must be satisfied:

1 The species must be able to germinate and grow at the prevailing temperatures. Extreme cold or warm temperatures must not destroy the crop.
2 There must be a supply of water available to the plant to meet the demand for transpiration and maintain turgidity during the period of growth. The water may come from crop-season rainfall, storage in the soil or irrigation.

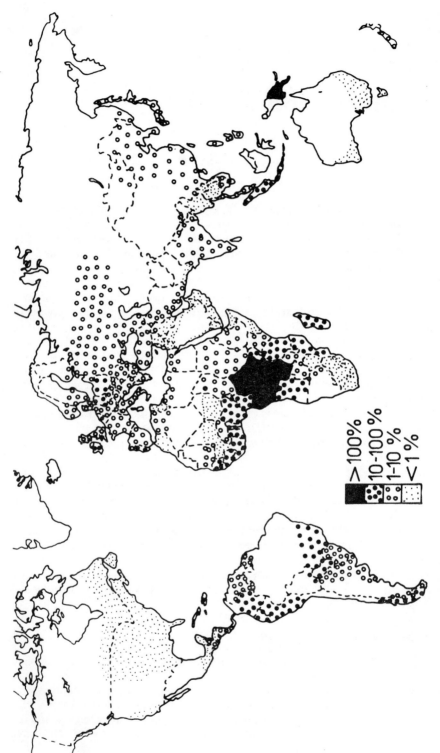

Figure 8.1 The area of root and tuber crops as a percentage of the area of cereal crops for each country in 1978 (after Bunting *et al.*, 1982).

Legend:
- >100%
- 10-100 %
- 1-10 %
- <1%

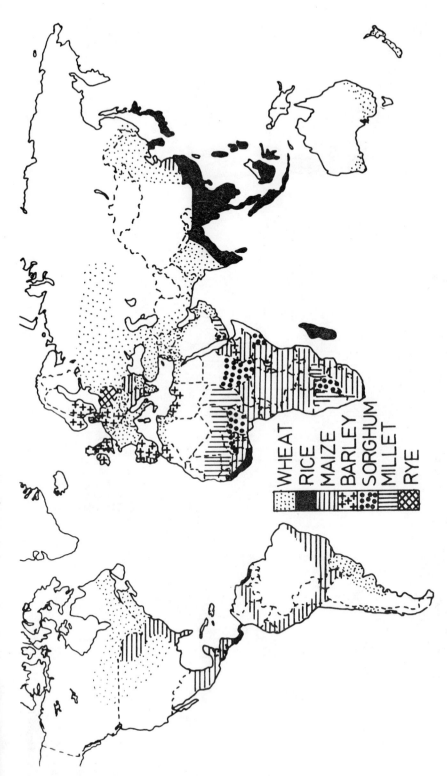

Figure 8.2 The predominant cereal crop, by area occupied, for each country in 1978 (after Bunting *et al.*, 1982).

WHEAT
RICE
MAIZE
BARLEY
SORGHUM
MILLET
RYE

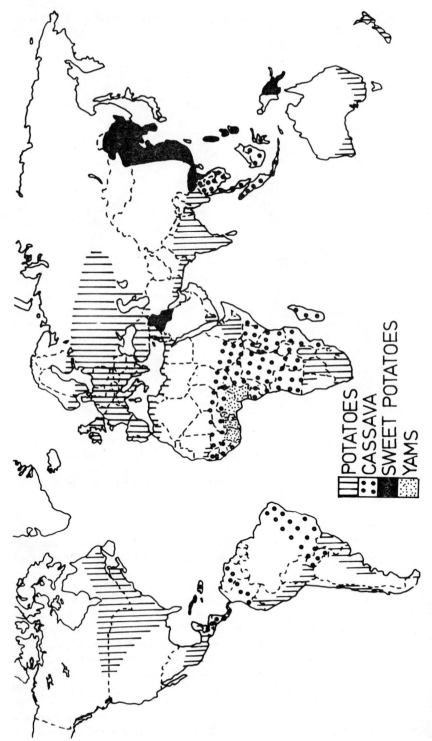

POTATOES

CASSAVA

SWEET POTATOES

YAMS

Figure 8.3 The predominant root and tuber crop, by area occupied, for each country in 1978 (after Bunting *et al.*, 1982).

3 The development pattern of the crop over time (which depends mainly on temperature and photoperiod), must be such that its growth cycle can fit into the constraints imposed by 1 and 2.
4 The climate must not favour the vigorous development of serious diseases.

1.1 Temperature

Temperature is one of the primary factors affecting plant growth. Thus there are limits of temperature beyond which plants cannot survive, and these may be called the lethal temperatures. There are also temperature limits beyond which plant growth ceases, but between these limits there exists a narrow range of temperatures over which plant growth is most favoured. These maximum, optimum and minimum temperatures are generally referred to as 'cardinal temperatures'. The cardinal temperature varies with plant species, and within different stages of growth for a given plant. Generally for cool-season crops such as wheat, oats, barley, rye, etc., the cardinal (minimum, optimum, and maximum) temperatures are $0°$–$5°C$, $25°$–$31°C$, and 31—$37°C$, respectively (Yao, 1981).

When the development of a crop from germination to maturity is limited physiologically rather than by bad weather, the length of the growth period is simply the sum of time spent in each developmental phase. According to Monteith and Scott (1982), much of the evidence from the laboratory and from the field demonstrates that the reciprocal of time in each phase (a measure of development rate) is a linear function of temperature below an optimum T_0. By extrapolating the linear relation between a measured rate of development r and a corresponding temperature T, it is possible to determine a base temperature T at which development would stop. When a rate of development is expressed as the reciprocal of the time for a complete developmental phase, i.e. $r = 1/t$, the change of temperature per unit change of rate or $(\partial R/\partial T)^{-1}$ is the 'accumulated temperature' required for the completion of the phase. In calculating this requirement, it is essential to use the appropriate base temperature. Monteith (1981) comments that an arbitrary value of $6°C$, which has no special physiological significance, slipped into the literature over a century ago and has been widely used by climatologists unaware of its origins and defects. Angus *et al.* (1980) have recently shown that a wide range of crops grown in Australia have base temperatures distributed in two main groups with modal values of 3 and $11°C$ but no species between 5 and $7°C$.

1.2 Soil moisture

Plant growth is directly related to soil moisture because most agricultural plants absorb their water from the soil. Cell division and cell expansion are the first processes affected during a period of drought, but with increasing stress, photosynthesis is restricted partly as a result of stomatal closure and partly because of increased mesophyll resistance. For practical purposes shortage of water is usually expressed as a soil water deficit – the amount of water needed to restore the soil to field capacity expressed in the same units as rainfall. Penman (1971) analysed the results of irrigation experiments at Woburn and at Rothamsted in terms of a 'limiting deficit' (D_1) – the amount of soil-water lost from the profile up to the point where growth effectively stops. Monteith and Scott (1982) comment that the limiting deficit must increase during the growing season as roots penetrate to greater depths. In temperate climates with a transpiration rate of 2 to 4 mm day^{-1}, the penetration by the root system of an arable crop, once it is established, will usually keep pace with the demand for water until the roots are impeded mechanically or stop growing at an appropriate point in the developmental timetable. In either case, Monteith and Scott consider that it is appropriate to quote a single limiting deficit describing the response to water for a particular crop on a particular site. In regions where the transpiration rate is 4 to 8 mm day^{-1}, the concept of a single limiting deficit will be less useful, particularly in light soil when the downward penetration of the root system cannot keep pace with the demand for water.

Monteith (i.e. see 1981 paper) generalized the Penman equation by specifying two resistances to vapour transfer: from evaporating surfaces to the air within a canopy (r_s) and through the air (r_a) to a reference height. This equation is applicable to all crops but its use requires that r_s be measured or related to simple attributes of the soil and crop in the same way that r_a can be estimated from crop height and windspeed. Russell (1980) has related the seasonal trend of r_s to crop development and soil water status.

Russell (1980) has investigated the relationship between r_s and soil water deficit for barley and pastures. He found for barley that r_s remains constant at about 30 s m^{-1} till the deficit exceeds 30 mm after which r_s increases steadily to 210 s m^{-1} at 100 mm. The pasture results were more variable, until the deficit reached 40 mm, r_s remained at about 40 s m^{-1} but as the soil dried still further, r_s increased rapidly to 180 s m^{-1} at a soil moisture deficit of 100 mm.

Russell (1980) has also investigated the relationship between soil-water deficit and the ratio of actual to potential evaporation rate. This ratio should reach unity only if the foliage remained wet because a dry canopy has a finite value of r_s. Russell found that for barley the ratio remained near 0.6 until the deficit exceeds 30 mm and then declined linearly to 0.17 at a soil-moisture deficit of 100 mm. For pasture the ratio declined gently from about 0.77 when there was no deficit to 0.74 at 50 mm and then more sharply to 0.45 at 100 mm. Monteith (1965) quotes a range of 0.62–0.76 for field crops adequately supplied with water.

As explained in earlier chapters, it is the difference between rainfall and evaporation plus run-off which determines the rate of change of soil-water content. Observations show that in all parts of the world where crops are grown, the variability of rainfall in time and space is much greater than the corresponding variability in evaporation. In most of Britain for example, crop evaporation during the summer proceeds at a mean ratio of 2.5 to 3 mm day^{-1}, while annual potential evapotranspiration lies between 450 and 500 mm with little year-to-year variation. In contrast, the amount of rain during any month can change by an order of magnitude from one year to the next. The rainfall in any month may exceed the rate of potential evapotranspiration so that the soil is always wet, or it may be confined to occasional light showers equivalent to the water lost by evaporation in a few days.

2 Climate and grain production

The influence of climate on food production is well illustrated by considering one important grain crop – wheat.

Wheat is an annual grass belonging to *T.monococcum*, and varieties are classified as winter, intermediate, and spring. Winter wheat is generally sown in the autumn, because when it is sown in the spring it usually remains prostrate on the ground throughout the growing season, producing no seed. Spring wheat is sown in the spring, but it can be sown in the autumn, while the difference between spring and intermediate wheat varieties is that the intermediate wheat varieties can not be sown in late spring in order to complete their life cycle whereas the spring wheat varieties can.

The main areas of wheat production in those countries in which the harvested area of wheat is more than 5 per cent of all arable land are shown in Figure 8.4. Wheat is grown extensively throughout the world, because it can adapt to a wide range of climate. The most extensive wheat growing areas in the world are in the north and south temperate zones. The approximate sowing and harvesting times are shown in Figure 8.4 and according to Bunting *et al.* (1982) show three patterns of wheat production. The first is winter wheat which overwinters in the vegetation stage and is harvested in the following summer. This is the pattern of production in northwest Europe and the southern Great Plains of the USA. The second pattern is spring wheat, which is often grown where the winters are too severe for winter wheat, e.g. Canada. The third pattern of production is common in the Mediterranean area. Wheat is sown in the autumn and harvested in the spring, and the plants grow continuously throughout this period. The most favourable climatic conditions for growing wheat were described by Nuttonson (1955): a cool, moderately moist, growing season during which the basal leaves become well developed and tillering proceeds freely, merging gradually into a warm, bright, and preferably dry harvest period.

Bunting *et al.* (1982) found that monthly mean temperatures at sowing mostly range from 8 to 16°C for winter wheat, 6 to 12°C for spring wheat and 12 to 20°C for Mediterranean wheat. Harvesting normally occurs at warmer temperatures than sowing; 16 to 24°C for winter wheat, 10 to 24°C for spring wheat and 18 to 26°C for Mediterranean wheat. In most wheat climates the mean temperature of the growing season lies between 7 and 17°C and little wheat is grown where mean temperatures exceed 18°C.

Yao (1981) comments that wheat is not grown extensively in the humid regions due primarily to the prevalence of diseases and leaching of nutrients from the soil. Bunting *et al.* (1982) found that most wheat is grown in climates in which annual rainfall lies between 200 and 800 mm. Wheat is grown in drier areas but this requires irrigation (as in India and Egypt) or a fallow system (as in parts of Australia). Little wheat is grown where rainfall exceeds 1,000 mm.

The importance of weather, or the longer phase of weather, i.e. 'climate', in influencing plant growth and final yield is unquestioned (Yao, 1981). Given that soil fertility is more or less constant, and varieties and other technological improvement change is in a steady upward direction, year to year variations in yield are a function of fluctuations in weather. Thus year-to-year change in the global food supply is dependent to a large extent on the year-to-year variability of climate.

Sakamoto *et al.* (1980) define a favourable (unfavourable) year as a yield departure of at least + 10 per cent (– 10 per cent) from the yield trend line. Normal yield is defined by a yield trend line obtained by applying a linear least-square fit for a specified sub-period of the total record. This sub-period, in turn, is determined by inspection of the plotted series, supported by known technological advances within the specified country. For example, in India, the introduction of high yielding plants in the 1960s was responsible for the upward surge in the yield series beginning about 1966. Sakamoto *et al.* (1980) illustrate this procedure by the Australian wheat yield series shown in Figure 8.5. They fitted a piece-wise linear trend to the data series, shown in the top of the figure; the residuals from the trend are plotted in the lower portion. Any point greater than + 10 per cent (less than – 10 per cent) is considered a favourable (unfavourable) crop year. The frequency of runs including one, two or three consecutive years of crop yield greater than + or – 10 per cent was then determined.

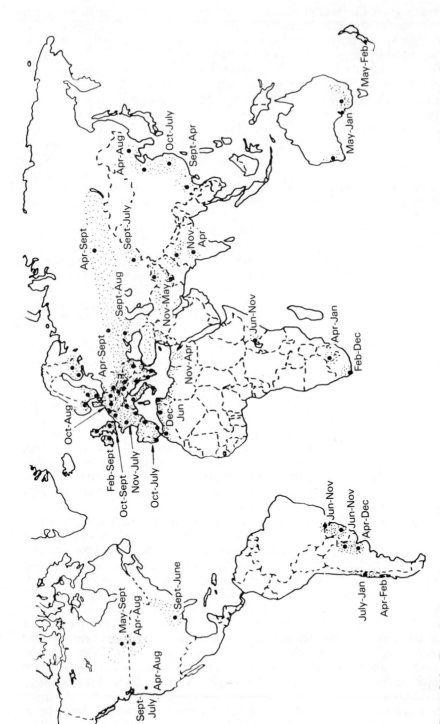

Figure 8.4 The wheat-growing areas of countries with more than 5 per cent of their arable land under wheat. The months of sowing and harvesting are shown (after Bunting *et al.*, 1982).

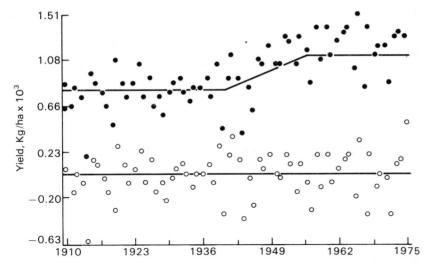

Figure 8.5 Wheat yield data series for Australia, showing the fitted piecewise linear trend (top), and the residuals from this trend, plotted by (o) in the lower portion of the figure (after Sakamoto *et al.*, 1980).

In Table 8.1, Sakamoto *et al.* (1980) summarizes the chance of one, two or three favourable or unfavourable wheat yield years in seven countries. Several interesting items emerge from this table. Canada has the highest probability of having two consecutive favourable (22 per cent) as well as unfavourable (17 per cent) years. Furthermore, the chance of having three successive unfavourable (favourable) years is 13 (11) per cent. Thus Sakamoto *et al.* (1980) consider that of all the major wheat growing countries, Canada has a greatest chance of crop failures or successes due to the fluctuations of weather. Except for Canada and the spring wheat areas of the Soviet Union, all the other countries have less than a 10 per cent chance of experiencing two consecutive poor yield years. The probability of wheat yield departures being above or below normal for any given year is much higher in the USSR than in the United States.

Investigations of climate and crop relationships have a long history, but in earlier studies most of the analyses were limited to correlations between monthly temperature or precipitation or both. With the advent of high-speed computers it became possible to carry out analysis of weather and crop yields involving multiple variables. For example, Thompson (1962, 1963, 1969) studied corn, wheat, soybean, and sorghum in the Corn Belt and in the Great Plains of the USA. Eight climatic variables (pre-season precipitation, and monthly precipitation and temperature during most of the growing season), their quadratic terms, and some of the interaction between temperature and precipitation were used. A linear trend for technological improvement such as fertilizer application was used in the analysis. For wheat Thompson found that the weather variables and the time trend factor accounted for 82 per cent or more of the yield variation in each state (Kansas, Nebraska, Oklahoma, and North and South Dakota).

Recently Steyaert *et al.* (1978) discussed a new approach to crop yield modelling. They use principal components of large-scale atmospheric general circulation fea-

Table 8.1   Per cent probability of one, two or three successive favourable/unfavourable (per cent of normal) wheat yield years in selected countries (1910–1975)* (*after Sakamoto et al., 1980*)

Country	Favourable Years ($\geqslant +10\%$) 1	2**	3**	Unfavourable Years ($\leqslant -10\%$) 1	2**	3**
Argentina	31	9 (10)	2 (3)	25	6 (6)	3 (2)
Australia	35	6 (12)	0 (4)	23	8 (5)	3 (1)
Canada	38	22 (14)	11 (5)	33	17 (11)	13 (4)
India	24	8 (6)	2 (1)	17	8 (3)	5 (0.5)
United States	20	8 (4)	3 (1)	17	9 (3)	5 (0.5)
United States (1950–1975)	12	0 (1)	0 (0.2)	19	8 (4)	0 (1)
USSR (all wheat) (1886–1930; 1950–1975) minus 1917–1921	33	13 (11)	5 (4)	27	6 (7)	2 (2)
USSR winter wheat (1945–1977)	25	6 (5)	0 (1)	28	6 (5)	3 (1)
USSR spring wheat Kazahk (1955–1974)	37	11 (14)	0 (5)	31	6 (10)	0 (3)
USSR (1950–1975)	23	4 (5)	0 (1)	23	0 (5)	0 (1)
USSR (spring wheat)† (1900–1974)	33	11 (10)	4 (4)	44	13 (19)	4 (9)
USSR (winter wheat)† (1900–1975)	38	14 (14)	5 (5)	32	9 (10)	3 (3)

† Simulated after models developed by Steyaert *et al.* (1978).
* Except where noted.
** Number in parenthesis is the probability assuming events independent from year-to-year and probability for a single year is estimated correctly by the sample.

tures as predictors for wheat or rice yield. According to Sakamoto *et al.* (1979) the usefulness of atmospheric pattern recognition can be shown by the wheat yield modelling being done at the United States Center for Environmental Assessment Services (CEAS). For example, principal component analysis of monthly sea-level pressure as predictors of yield in a linear regression was applied to large production areas in the Soviet Union. Sea-level pressure was used because it depicts large-scale circulation features, which in turn determine the weather and climate for an area.

Sakamoto *et al.* (1979) comment that certain weather patterns over the Soviet Union are associated with a good or poor crop year. Soviet scientists have observed from agro-climatological studies that when precipitation is sparse in the eastern part of their steppe area, it is abundant in the western part, and conversely. This observation has been used to justify the development of 50,000,000 hectares of grassland in Kazakhstan for wheat growing since the 1950s. The grassland is separated from the agriculturally productive Ukraine and European Russia by about one-half wavelength of the stationary waves in the upper atmosphere. According to Sakamoto *et al.* the significance of this observation is that if the circulation is anticyclonic and dry over the Ukraine, it is likely to be cyclonic and wet over northern Kazakhstan. In 1972, while the winter wheat areas of the Ukraine, the Volga area, and European Russia suffered from reduced yields as a result of reduced moisture and winterkill, northern Kazakhstan recorded its highest yield, owing to sufficient soil moisture and weather that was cooler than normal.

3 Drought

One of man's worst natural enemies is drought. Its beginning is subtle, its progress is insidious and its effects can be devastating (Hounam *et al.*, 1975). Since rainfall is still the main source of fresh water for agricultural, domestic and industrial use, drought can have an impact ranging from slight personal inconvenience to endangered nationhood.

A study of drought requires an objective definition, but according to Hounam *et al.* no universally acceptable definition has so far been developed. It can be assumed that the basic cause of drought is inadequate precipitation, but the exact meaning of the word 'inadequate' needs careful definition. The distribution of average annual precipitation does not by itself give an indication of drought incidence or intensity although obviously it has a marked control over normal land use. The use of such a distribution may lead to areas of low average rainfall being identified as drought areas. Aridity is usually defined in terms of low average rainfall and, ignoring the possibility of climatic change, is a permanent climatic feature of a region. Hounam *et al.* suggest that drought is a temporary feature in the sense that, considered in the context of variability, it is experienced only when rainfall deviates appreciably below normal. Aridity is, by definition, restricted to regions of low rainfall, and usually of high temperature, whereas drought is possible in virtually any rainfall or temperature regime. Activities in arid zones are poorly geared to meet the 'permanence' of aridity but a drought situation results in at least some interruption of normal activities in all zones. Man's use (or over-use) of the rainfall resource in semi-arid zones in Africa has produced recurrent drought.

Drought should be regarded as a hydrologically extreme state in exactly the same manner that flooding is regarded as a state of extreme river flow. In many ways drought is the hydrological opposite to flooding. Dry spells with large soil-moisture deficits are normal at most lowland middle-latitude sites in late spring and summer. Such dry spells should not be referred to as drought unless they reach an unusual intensity or are abnormally long for the particular sites. The analogy with river flow is close, since a river flood is only said to occur when the rate of run-off is unusually high, and this particular state will vary from river to river. Thus drought should be defined in terms of prolonged and abnormal moisture deficiency.

One major reason for the lack of a universally agreed definition of drought is that concepts of drought are related to particular water uses. For instance, British agriculture is chiefly concerned with adequate summer rainfall to offset the evaporation—transpiration which occurs in that season and gives rise to large soil-moisture deficits. On the other hand, British water supply interests lie much more with winter rain which provides run-off for reservoirs and percolation to re-charge aquifers. On a world scale, ideas on agricultural drought vary depending on the exact type of agricultural activity and the normal local climate.

Agricultural drought may be expressed in terms of the degree to which growing plants have been adversely affected by an abnormal soil moisture deficiency. The deficiency may result either from an unusually small moisture supply or an unusually large moisture demand. While a deficiency in the water supply for live-stock may be regarded as a facet of agricultural drought, it is really a different problem – because it does not depend primarily on soil moisture.

Since every crop has its own drought-sensitive periods, a proper analysis of agricultural drought should cover each crop separately. Tabony (1977) has considered grassland drought in Britain, where grass covers about 70 per cent of the agricultural land. Grass growth becomes restricted when the availability of water from the upper layers of the soil becomes restricted. Tabony therefore suggests that a possible measure of 'grassland' drought D_a is given by:

$$D_a = E_p - E_g \qquad (8.1)$$

where E_p is the potential evapotranspiration and E_g is the transpiration effective for growth, Tabony obtained E_g from E_p by using 25.4 mm (1 inch) root constant.

In contrast, hydrologists in Britain are interested in adequate winter rain to fill reservoirs and re-charge aquifers. The hydrologically effective rainfall (R_e) for such purposes is the water surplus remaining after evaporation has taken place and any soil moisture deficit has been removed. Therefore, Tabony considers that R_e is a possible measure of 'hydrological' drought.

3.1 Definitions of agricultural drought

Drought is caused mainly by prolonged insufficient rainfall. For example, Barger and Thom (1949) defined drought in terms of rainfall deficit. Any period of n weeks during the corn-growing season in Iowa, where $1 \leqslant n \leqslant 16$, which has a rainfall total smaller than the minimum amount determined to be just sufficient for that duration experiences a rainfall deficit. The largest deficiency for the season, regardless of the length of period involved, is taken to be the measure of drought intensity and resulting effect on corn yield for that year.

To understand the variability of water it is necessary to examine the statistical properties of frequency distributions of rainfall. Figure 8.6 represents a 'normal'

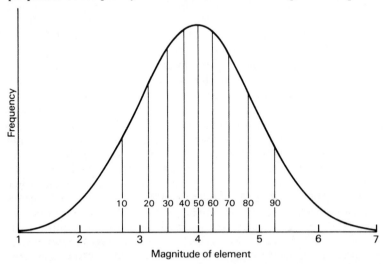

Figure 8.6   Frequency distribution of magnitude of elements in a normal distribution. Scale of abscissa representing magnitude of elements has been selected in arbitrary fashion except that standard deviation is one unit. Frequency of occurrence of magnitude of elements within a given range is proportional to the area under the curve and between vertical lines representing that range. Figures on vertical lines indicate percentage of area of curve to the left of the line (after Gibbs, 1981).

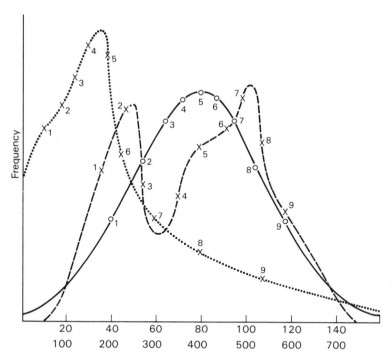

Rainfall – mm (upper scale Melbourne, lower scale Darwin)

Figure 8.7   Frequency distributions of monthly rainfall compared with a normal distribution. Figures on curves are decile numbers.
Continous line –   Normal distribution
Dotted line   Melbourne 37° 49 S 114° 58E
                           January 109 years
Dashed line   Darwin 12° 24 S 130° 48E
                           January 86 years
(after Gibbs, 1981).

(Gaussian) frequency distribution with the abscissa giving the magnitude of the elements of the distribution and the frequency of occurrence of magnitudes within a given range being proportional to the area of the curve subtended by the curve within the limits of that range. Gibbs (1981) states that unfortunately the frequency distribution of totals of most daily rainfalls, many monthly rainfalls and some annual rainfalls are far from normal (Gaussian) and the use of arithmetic means and standard deviations is quite inappropriate for many purposes including the study of drought. Figure 8.7 gives two examples of the departure of monthly rainfall frequency distributions from normal. In this figure Gibbs (1981) shows the frequency distributions of January rainfall for Darwin and Melbourne compared with a normal distribution. The area of these curves is equal. It is seen that the frequency distribution of Darwin January rainfall exhibits the bimodal characteristic commonly found in tropical cyclone areas while the Melbourne distribution shows a strongly skewed character with a long 'tail' indicating the occasional month of heavy rainfall. Gibbs (1981) remarks that complications arise when frequency distributions of rainfall amounts contain a number of zero values, as in the case with

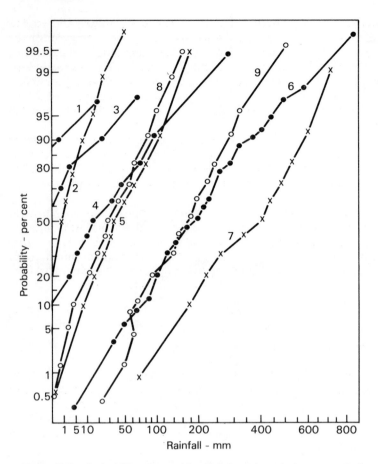

Figure 8.8 Probability of monthly rainfall total not being exceeded.

1. Mardie	21° 12 S	115° 57 E	October	74	years
2. Beijing	39° 48 N	116° 28 E	January	114	''
3. Halls Creek	18° 14 S	127° 40 E	July	74	''
4. Alice Springs	23° 36 S	133° 36 E	January	91	''
5. Melbourne	37° 49 S	144° 58 E	January	·109	''
6. Beijing	39° 48 N	116° 28 E	July	113	''
7. Darwin	12° 24 S	130° 48 E	January	86	''
8. Shanghai	31° 10 N	116° 28 E	December	100	''
9. Shanghai	31° 10 N	116° 28 E	June	100	''

(after Gibbs, 1981).

all hourly, most daily, many monthly and some annual totals. Such a distribution cannot conveniently be represented by curves such as those in Figures 8.6 and 8.7.

A useful device is to use a diagram in which the ordinate is percentage cumulative frequency and the abscissa the magnitude of the element. A curve plotted on such a diagram shows the percentage of occasions a particular value is exceeded. If the sample of observations is sufficiently large the curve may be used to deduce the probability of occurrence. In Figure 8.8 Gibbs (1981) produces curves showing monthly rainfall probabilities for a number of locations in Australia and China.

Gibbs suggested that the square roots of hourly, daily and monthly rainfall tend to be normally distributed, and in Figure 8.8 has used a square root scale for the abscissa. The scale of the ordinate is such that if the magnitude of the elements represented on the abscissa is normally distributed the curve will be a straight line. In practice most plots approximate to straight lines and to the range of probability of rainfall indicated by the plots. Gibbs also remarks on the near coincidence of the monthly rainfall distributions of Melbourne January rainfalls and Shanghai December rainfalls.

After examining a study of the history of drought in Australia, Gibbs and Maher (1967) found good agreement between perceived drought and the occurrence of annual rainfall in the first decile range. This suggests that in marginal lands, such as Australia man uses available water so that he experiences a deficency (drought) one year in ten.

Mooley and Parthasarathy (1983) have used a similar approach to study drought in India. They assume that water-dependent economic activities adjust themselves to the water normally available and the variability of that water. It is therefore natural to express the rainfall of any season in a way which takes into account normal seasonal rainfall and its variability. Such a criterion is the standard deviate, t_i, given by

$$t_i = (X_i - \overline{X})/S \tag{8.2}$$

where X_i is the seasonal rainfall in year *i*.
    X is the long-term average rainfall.
    S is the standard deviation of the rainfall series.

The criteria adopted for indentifying a season of drought or alternatively of flood are standard deviates -1.282 or $+1.282$, respectively. These correspond to the 10 per cent points under the normal distribution, i.e. in 10 per cent of the years droughts (or floods) will be experienced. Thus Mooley and Parthasarathy found that in 1877, 1899, 1918 and 1972, more than 40 per cent of the total area of India experienced drought.

Rainfall studies do not take into account soil moisture, which in many cases is very important, especially during the early growing season. The water balances and vegetation covers of the continental interiors vary widely, and obviously a water shortage in one area would pass unnoticed in another. Palmer (1965) considers that in spite of the differences which exist, people in humid climates seem to mean much the same thing when they refer to drought as do the people in a semi-arid region; viz., that the moisture shortage has seriously affected the established economy of their region. He therefore defines drought as a prolonged and abnormal moisture deficiency, and suggests that the problem, be approached by the use of hydrologic accounting and the use of indices of moisture anomaly.

Palmer summarizes his procedure for drought determination as follows:

1 undertake a detailed water-balance study for each month of a long series of years;
2 summarize the results to obtain coefficients which are dependent on the climate of the area being analysed;
3 re-analyse the series using the derived coefficients to determine the amount of moisture required for 'normal' weather during each month;
4 convert the departures to indices of moisture anomaly;
5 analyse the index series for the start and finish of drought periods and also the severity.

In dry climates it is usual for the actual evapotranspiration to fall below the potential values, and this can be expressed in the form

$$\alpha = \frac{\overline{E}_a}{\overline{E}_t}$$

where α is the coefficient of evapotranspiration, and $\overline{E}_a$ and $\overline{E}_t$ are the monthly averages of the actual and potential evapotranspiration respectively.

Similarly, coefficients of recharge (β), run-off (γ), and loss (δ) can be derived:

$$\beta = \frac{\overline{R}}{\overline{PR}}$$

where $\overline{R}$ is the average monthly recharge and $\overline{PR}$ is the average monthly potential recharge defined as the amount of moisture required to bring the soil to field capacity;

$$\gamma = \frac{\overline{RO}}{\overline{PRO}}$$

where $\overline{RO}$ and $\overline{PRO}$ are the actual and potential run-off values, and potential run-off is the difference between the field capacity of the soil and the potential recharge (equal to the available moisture at the start of the month).

$$\delta = \frac{\overline{L}}{\overline{PL}}$$

where the actual and potential values are as above, potential loss being defined as the amount of moisture that could be lost from the soil provided the precipitation during the period was zero, and the potential evaporation and initial soil-moisture conditions were as observed.

Palmer considers that these coefficients, when used in conjunction with the potential value for a particular month, enable an estimate to be made of the value that is 'climatically appropriate for existing conditions' (CAFEC). It is therefore possible to calculate for any particular month the CAFEC values (denoted by a circumflex) for evapotranspiration, recharge, run-off, and loss:

$$\hat{E}_d = \alpha E_t, \qquad\qquad \hat{RO} = \gamma PRO,$$
$$\hat{R} = \beta PR, \qquad\qquad \hat{L} = \delta PL.$$

The CAFEC precipitation is given by

$$\hat{P} = \hat{E}_a + \hat{R} + \hat{RO} - \hat{L}, \tag{8.3}$$

and the difference (d) between this and the actual precipitation (P) for each month

$$d = P - \hat{P} \tag{8.4}$$

provides a meaningful measure of the departure of the precipitation from normal. It is apparent that the significance of the departure will depend on the locality and the seasons. So Palmer took two areas, one in central Iowa and the other in western Kansas, assumed that the driest year in each would have similar consequences for the inhabitants, and compared the values of d. On this assumption, it is possible to obtain a weighting factor, K, which can be applied to the values of d to make them of equal significance:

$$K = \overline{d}_{KAN}/\overline{d}_{IOWA} \tag{8.5}$$

Further research led Palmer to believe that the climatic characteristic K can be estimated for each calendar month as follows:

$$K = (\bar{E}_t + \bar{R})/(\bar{P} + \bar{L}) \tag{8.6}$$

It is now possible to define a moisture anomaly index, z, such that

$$z = dK$$

Each value of z expresses on a monthly basis the departure of the moisture climate from the average for the month.

Both the US Department of Agriculture and the US Weather Bureau have used the terms mild, moderate, severe and extreme to describe droughts. The severity of a drought is a function of both the value of the index z, and also of the duration of the spell, because clearly the situation will tend to deteriorate with time. Drought severity X is therefore given by the expression

$$X_i = \sum_{t=1}^{i} z_t/(0.309t + 2.691) \tag{8.7}$$

where t is the time in months. Palmer has assigned values of X to various categories of drought, which are listed in Table 8.2; these values may be used to determine the start and finish of a drought and its severity at any time during its occurrence.

Table 8.2 Classes for wet and dry periods (*after Palmer 1965*)

X	Class
$\geqslant 4.00$	Extremely wet
3.00 to 3.99	Very wet
2.00 to 2.99	Moderately wet
1.00 to 1.99	Slightly wet
0.50 to 0.99	Incipient wet spell
0.49 to -0.49	Near normal
-0.50 to -0.99	Incipient drought
-1.00 to -1.99	Mild drought
-2.00 to -2.99	Moderate drought
-3.00 to -3.99	Severe drought
$\leqslant -4.00$	Extreme drought

Analysis by Palmer of a period of 76 years in western Kansas shows that drought occurred in 37 per cent of the months, wet spells in 37 per cent and near-normal conditions in 12 per cent. This indicates that in the interior grasslands the weather tends towards extremes and so-called average conditions are rare. He further found that in western Kansas the drought was mild during 11 per cent of the months, moderate in 11 per cent, severe in 8 per cent, and extreme in 6 per cent. The drought during the 1930s was the longest and most serious so far recorded in western Kansas, for between August 1932 and October 1940 there were 38 months of extreme drought. These years were further characterized by unusually strong winds which created terrible duststorms and combined with the drought to produce really disastrous conditions. Other serious droughts occurred in 1894, 1913, and from 1952 to 1956.

Wigley and Atkinson (1973) have used values of soil moisture to define agricultural drought. Long homogeneous series of precipitation and evapotranspiration data for Kew (51°28′N, 0°19′W) going back to 1698 have been constructed and discussed by Wales-Smith (1917, 1973a, 1973b); these are the longest such series available anywhere in the world. Wigley and Atkinson have constructed soil moisture deficit values back to 1698 using the Kew data. They averaged the soil-moisture deficit over the whole growing season and used a composite of short- and long-rooted vegetation results in order to provide a general agricultural drought index. Using this index, the 14 most severe agricultural droughts over the interval 1698–1976 are: 1705, 1731, 1762, 1781, 1844, 1893, 1901, 1921, 1934, 1938, 1944, 1965, 1974, 1976. The worst years are, in order, 1976, 1934, 1944, 1893, 1938, 1921, 1974. 1976 shows up as the worst agricultural drought since at least 1698, although only fractionally more severe than 1934. However it is probably more realistic to state that 1976 was among the three worst agricultural droughts during the 279 years period of record.

Tabony (1977) has listed the 10 most severe grassland droughts at Kew from 1871 to 1975, obtained from his index D_a. These are in order of magnitude: 1959, 1921, 1893, 1972, 1975, 1949, 1911, 1938, 1899, 1933. He comments that calculations of D_a had not been made at the time of publishing his paper, but that the summer of 1976 is likely to compare with those of 1959 and 1921.

3.2 Teleconnections
The term teleconnections refers to the statistically or empirically determined coupling of large-scale abnormalities of the atmospheric circulation in time and space (Fleer, 1981). Such links were first discovered by Walker (1924) who found linear correlations between sea-level pressure in different parts of the world. Fleer (1981) has undertaken a global survey of rainfall teleconnections. He computed more than 1,000 cross-spectra using several reference stations. Figure 8.9 shows the regional distribution of phase relationships at the period of 5 years between Nauru and 289 selected stations in the zone 30°S to 35°N. Positive correlation (i.e. in-phase) and negative correlation (i.e. out-of-phase) are indicated in this figure. Several interesting relationships are found. Between the equatorial Pacific and Indonesia there exists a negative correlation. The equatorial African rainfall is in-phase with the equatorial Pacific rainfall, while the equatorial South American rainfall and the majority of West African stations are out-of-phase with the equatorial Pacific. There is an internally consistent region on the Indian peninsula, which is out-of-phase with Nauru. Fleer also found that anomalies in the equatorial Pacific are in-phase with those in southern Japan, the Near East and the subtropics of North America, but out-of-phase with those in most of the northern part of the African continent. Such investigations are important because they give an insight into the relationships between wet and dry spells in different parts of the tropics. A physical interpretation of such coincidence of rainfall anomalies in different parts of the tropics can be given in terms of the zonally extended, thermally driven Walker circulation.

3.3 Drought and increasing atmospheric CO_2
The broad hydrological consequences of increasing atmospheric CO_2 were discussed in Chapter 7. Manabe *et al.* (1981) have investigated the hydrologic changes of climate in response to an increase of CO_2 concentration in the atmosphere. The results from numerical experiments with three climate models are analysed and compared with each other. All three models consist of an atmospheric general cir-

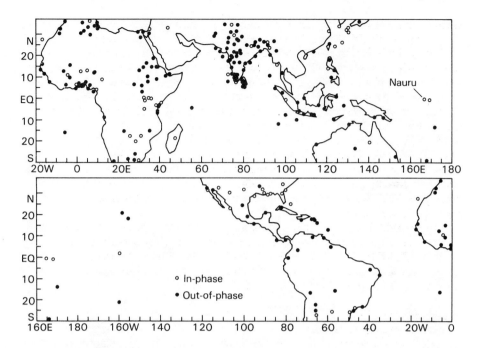

Figure 8.9 Regional distribution of coherent relationships between rainfall anomalies; reference station Nauru (after Fleer, 1981).

culation model and a simple mixed layer ocean with a horizontally uniform heat capacity. The first model has a limited computational domain and simple geography with a flat land surface. The second model has a global computational domain with realistic geography. The third model is identical to the second except that it has a higher computational resolution.

Manabe *et al.* use the first numerical model to obtain zonal mean distributions over the simple continental surface. Figure 8.10 illustrates the latitude – time distribution of the difference in zonal mean soil moisture over the continent between the $4 \times CO_2$ – and the standard experiment. According to Figure 8.10 the difference of the zonal mean soil moisture in high latitudes has a large positive value throughout most of the year except for summer. Figure 8.10 also indicates two distinct zones of reduced soil wetness in middle and high latitudes during the summer season. Because of this increased dryness in middle latitudes, the subtropical dry zone extends polewards by about 5° of latitude during summer in response to the quadrupling of the atmospheric CO_2 concentration.

The soil-moisture distributions are much more complicated in the case of the global general circulation models. Figure 8.11 shows the geographical distribution of soil moisture difference for the spring (March–April–May) season. Qualitatively, the same general pattern exists for both model distributions in the northern hemisphere. For example, there is a general increase of wetness in high latitudes and a reduction of wetness in middle latitudes over both the Asian and North American continents. The corresponding summer distributions are shown in Figure 8.12. Here both models show a general reduction of soil moisture over both middle and high latitudes for Asia and North America. The results from the three

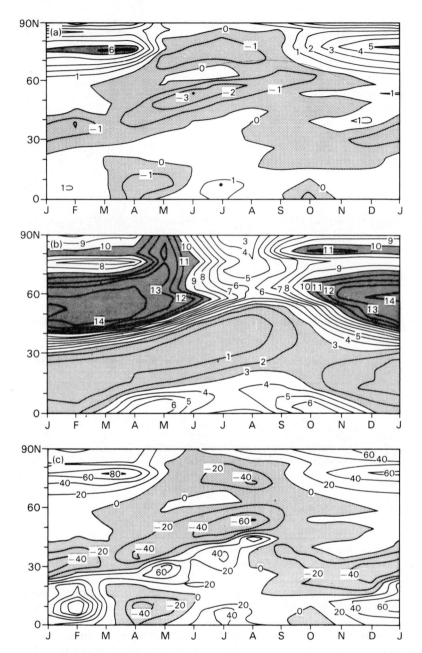

Figure 8.10   The latitudinal and seasonal variation of the zonal mean (a) difference of soil moisture (cm) between the 4 × CO_2 and the standard experiments of Manabe *et al.* (1981), (b) soil moisture (cm) for standard experiment, and (c) percentage change of soil moisture from the standard to the 4 × CO_2-experiment.

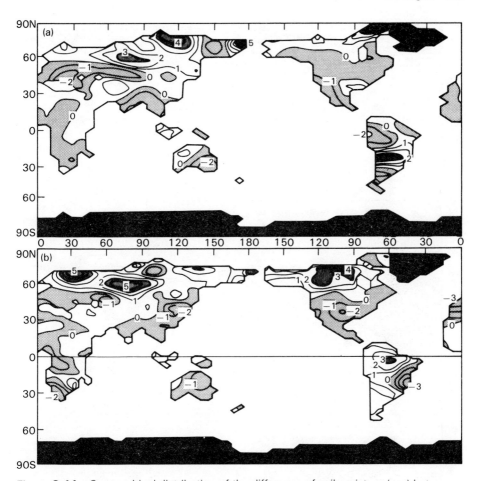

Figure 8.11 Geographical distribution of the difference of soil moisture (cm) between the 4 × CO_2 and the standard experiments of Manabe *et al.* (1981) for the (March–April–May) period, (a) Second model, (b) Third model.

models suggest that this overall increase of aridity during the summer season in middle and high latitudes is a common and significant feature of all the numerical experiments.

According to Manabe *et al.* there are a number of reasons for the increased summer aridity in middle and high latitudes. The snowmelt season in the 4 × CO_2 experiments ends earlier than the corresponding season in the standard experiment. Thus, the warm season of rapid soil moisture depletion begins earlier due to enhanced evaporation resulting in less soil moisture during summer in the 4 × CO_2 experiment. This mechanism is mainly responsible for the CO_2 induced dryness during summer in high latitudes. In middle latitudes, a smaller snowmelt rate and stronger evaporation are also responsible for the faster reduction of soil moisture in late spring in the 4 × CO_2 experiments as compared with the standard experiments. This smaller rate of snowmelt in late spring results not only from the earlier occurrence of the snowmelt season, but also from the smaller total snow accumula-

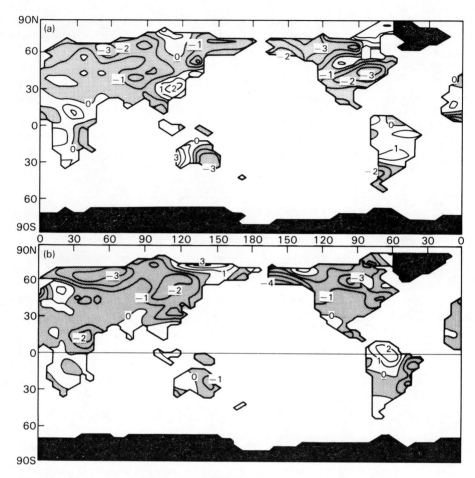

Figure 8.12 Geographical distribution of the difference of soil moisture (cm) between the $4 \times CO_2$ and the standard experiments of Manabe *et al.* (1981) for the (June–July–August) period, (a) Second model, (b) Third model.

tion in middle latitudes where the surface air temperature is near the freezing point during winter. In addition, there is another important factor which maintains the CO_2 induced dryness of soil moisture during summer in middle latitudes. As the seasons progress from winter to summer, the middle latitude rainfall of a given model shifts poleward in both the $4 \times CO_2$ and the standard experiments. Generally, the middle latitude rainbelt in the $4 \times CO_2$ experiments is placed poleward of the corresponding rainbelt in the standard experiment partly due to the penetration of warm moisture-rich air into higher latitudes. This implies that, in middle latitudes, the spring maximum of rainfall rate in the $4 \times CO_2$ experiment arrives earlier than the corresponding maximum in the standard experiment. Thus, the spring to summer reduction of rainfall rate begins earlier and helps to maintain drier soil during summer in the $4 \times CO_2$ experiment.

It is a little early to state the exact implications to agriculture of the increasing atmospheric CO_2 content. From the point of view of agricultural meteorology the summer reductions in soil moisture content must be important. This must be particularly so since many of the soil-moisture reductions are predicted to take place in or near major middle-latitude agricultural regions. Increasing atmospheric CO_2 may imply an increasing frequency of summer drought in many middle-latitude agricultural areas. The situation in the tropics is less clear, with some areas showing increases in soil moisture with high atmospheric CO_2 content. Nevertheless, some parts of the tropical world may become more liable to drought.

There are other effects which may be important but which are not completely within the scope of this book. Increasing atmospheric CO_2 may increase plant growth and also allow plants to grow more readily in dry areas. The associated higher temperature may also improve plant growth. These factors may more than compensate for reduction in plant growth due to lack of soil moisture.

References

ANGUS, J. F., CUNNINGHAM, R. B., MONCUR, M. W. and MACKENZIE, D. H. 1980: Phasic development in field crops. I. Thermal response in the seedling phase. *Field Crops Research* 3, 365-78.

BARGER, G. L. and THOM, H. C. S. 1949: Evaluation of drought hazard. *Agronomy Journal*, 41, 519-26.

BUNTING, A. H., DENNETT, M. D., ELSTON, J. and SPEED, C. B. 1982: Climate and crop distribution. In Blaxter, K. and Fowden, L. (eds.), *Food, Nutrition and climate*. London: Applied Science Publishers, 43-78.

FLEER, H. E. 1981: Teleconnections of rainfall anomalies in the tropics and subtropics. In Lighthill, J. and Pearce, R. P. (eds.) *Monsoon dynamics*. Cambridge: Cambridge University Press, 5-18.

GIBBS, W.J. 1981: The nature of drought and strategies to reduce its effects. In *Proceedings of the Technical Conference on Climate – Asia and Western Pacific*. Geneva: World Meteorological Organization.

GIBBS, W. J. and MAHER, J. V. 1967: Rainfall deciles as drought indicators. *Bulletin 48*. Australian Bureau of Meteorology.

HOUNAM, C. E., BURGOS, J. J., KALIK, M. S., PALMER, W. C. and RODDA, J. 1975: *Drought and Agriculture*. Technical Note 38. Geneva: World Meteorological Organization.

MANABE, S., WETHERALD, R. T. and STOUFFER, R. J. 1981: Summer dryness due to an increase of atmospheric CO_2 concentration. *Climatic Change*, 3, 347-86.

MONTEITH, J. L. 1965: Evaporation and environment. *Symposium Society Experimental Biology* 29, 205-34.

—— 1981: Evaporation and surface temperature. *Quarterly Journal of the Royal Meteorological Society*, 107, 1-27.

—— 1981: Climatic variation and the growth of crops. *Quarterly Journal of the Royal Meteorological Society*, 107, 749-74.

MONTEITH, J. L. and SCOTT, R. K. 1982: Weather and yield variation of crops. In Blaxter, K. and Fowden, C. (eds.), *Food, Nutrition and Climate*. London: Applied Science Publishers, 127-53.

MORLEY, D. A. and PARTHASARATHY, B. 1983: Droughts and floods over India in summer monsoon seasons 1871-1980. In Street-Perrott, A., Beran, M. and Ratcliffe, R. (eds.), *Variations in the Global Water Budget*. Dordrecht: Reidel, 239-52.

NUTTONSON, M. Y. 1955: *Wheat-Climate Relationships and the Use of Phenology in Ascertaining the Thermal and Photo-Thermal Requirements of Wheat*. Washington, DC: American Institute of Crop Ecology.

PALMER, W. C. 1965: Meteorological drought. *Research Paper No. 45*. Washington, DC: Weather Bureau, US Department of Commerce.

PENMAN, H. L. 1971: *Irrigation at Woburn*, VII. Report of Rothamsted Experimental Station for 1970. Part 2.

RUSSELL, G. 1980: Crop evaporation, surface resistance and soil water status. *Agricultural Meteorology*, 21, 213–26.

SAKAMOTO, C., LEDUC, S., STROMMEN, N. and STEYAERT, L. 1980: Climate and global grain yield variability. *Climatic Change* 2, 349–61.

SAKAMOTO, C., STROMMEN, N. and YAO, A. 1979: Assessment with agro-climatological information. *Climatic Change* 2, 7–20.

STEYAERT, L. T., LE DUC, S. K. and MCQUIGG, J. D. 1978: Atmospheric pressure and yield modeling. *Agricultural Meteorology* 19, 23–4.

TABORY, R. C. 1977: Drought classifications and a study of droughts at Kew. *Meteorological Magazine* 106, 1–10.

THOMPSON, L. M. 1962: Evaluation of weather factors in the production of wheat. *Journal of Soil Water Conservation* 17, 149–56.

—— 1963: *Weather and Technology in the Production of Corn and Soybean*. CAED Report 17, Ames, Iowa: Center for Agricultural and Economic Development, Iowa State University.

—— 1969: Weather and technology in the production of wheat in the United State. *Journal of Soil Water Conservation* 23, 219–24.

WALES-SMITH, B. G. 1971: Monthly and annual totals of rainfall representative of Kew, Surrey, for 1697–1970. *Meteorological Magazine* 100, 345–62.

—— 1973a: An analysis of monthly rainfall totals representative of Kew, Surrey, from 1697–1970. *Meteorological Magazine* 102, 157–71.

—— 1973b: Evaporation in the London area from 1698–1970. *Meteorological Magazine* 102, 281–91.

WALKER, G. T. 1924: Correlation in seasonal variations of weather, IX: A further study of world weather (world weather II). *Memoirs of the India Meteorological Department* 54, 79–87.

WIGLEY, T. M. L. and ATKINSON, T. C. 1977: Dry years in southeast England since 1698. *Nature* 265, 431–4.

YAO, A. Y. M. 1981: Agricultural climatology. In Landsberg, H. E. (ed.), *General Climatology*, 3, Amsterdam: Elsevier, 189–298.

Index

ablation 89–90, 93, 94
absorption coefficients, of aerosols 249
abyssal layers 37
accumulated temperature 265
adiabatic warming 144
advection 26, 28, 44, 63, 84, 123, 144, 177, 182
aerodynamic evaporation 171
aerodynamic resistance 179–82, 195, 205–7, 226
aerosols 249–50
agricultural drought 271, 272–8
agroclimatology 131–2
Ahmed, S.B. and Lockwood, J.G. 160, 162
air masses 11
albedo 5, 7, 30–1, 38, 143–5, 153, 170–1, 209, 245
    and aerosols 249–50
    and evolution of atmosphere 33
    definition of 158
    diurnal variation 159
    in Arctic 79
    of crops 161, 162–4
    of deserts 125–6
    of forests 190–1
    of grasslands 156, 157–64
    of urban areas 252
    planetary 8
    seasonal 162–4
albedo feedback mechanism 143, 145, 246–7
Alice Springs, Australia 127, 130, 131
Amazon 223, 229–31
Ångström ratio 153
anoxic conditions 98
Antarctic
    cloudiness 76
    glaciation 98–102
    ice-sheet 73, 74–5, 101, 102–3
    net radiation 81, 85, 86
anticyclones 17, 54, 55
Arabian Sea 38
Arctic 24, 76, 79
Arctic Ocean 73, 74
arid climates 122–4
aridity 124–5, 271
    severity of 135
astronomical theory of climatic change 110–17
atmosphere, evolution of 32
atmosphere — ocean interface 37–43
    interactions 62–8
atmospheric circulation 3, 4, 13–15
    see also jet streams, winds
atmospheric counter radiation 157
atmospheric moisture balance model 123–4
atmospheric turbulence 165–9

baroclinic disturbances 147
baroclinic instability 11
Benguela current 137
Ben Nevis 90, 92
Berlage, H.P. 14–15, 65
Bjerknes, J. 14, 65, 66–7
blocking pattern 23, 55
boreal coniferous forest 203
boundary layer, above forest canopy 194
Bowen ratio 153, 170–1, 174–6
Bunting, A.H. et al. 261, 267
buoyancy flux 99–100

canopy 3
    drainage 180–1, 204
    storage 195–7, 204–7

carbon dioxide   241−8, 278−82
cardinal temperatures   265
cascading system   2
Center for Environmental Assessment Services, US (CEAS)   270
cereals   261, 262, 263
Chad, Lake   132, 138−9
Charney, J.G.   46, 126
circulation
   atmospheric   4, 13−15
   see also jet streams, winds
   oceanic   35−7
   see also ocean currents
CLIMAP project   106, 108
climate
   and energy use   254−7
   as energy source   241
'climatic point'   89
climatic systems
   components of   3−4
   processes within   3
   properties of   3
closed lakes   138−42
cloud cover   157, 190
cloudiness   64, 192
   and aerosols   249
   and albedo   158, 160−1
   in polar regions   75−6
   in Sahara desert   127
clouds   5, 8, 14, 27−8, 46, 47, 48, 51, 157
cold-water regions, of oceans   14
computer simulation of evolution of earth's atmosphere   32−3
condensation, of water vapour   48
conditional instability of the second kind   46−7
coniferous forests, and transpiration   176
coniferous stands in temperate deciduous areas   202
continental arid climate   124
continental ice-sheets   73, 103, 105, 113
continentality   56, 58
convection   169
convective clouds   5, 48
   see also cumulonimbus, cumulus
convergence zones   46−7

see also intertropical convergence zone
coreless winter phenomenon, Antarctica   86−7
Coriolis force   8, 11, 18, 165, 168
Cretaceous   97−8
crop coefficient   183
crop evaporation   266
crops
   albedo of   161, 164
   growth   261, 265
   see also root crops, wheat
crop yield modelling   269−70
cumulonimbus clouds   27−8, 48, 51
cumulus clouds   46, 47, 48, 148
cut off cold pools   128
cyclones, mid-latitudes   17, 55−6

Dalton equation   170
day-length   75
Deacon, E.L.   127, 168, 226
deep-water circulation, in Cretaceous   98
deforestation   137, 145, 223, 229−31, 257
desertification   143−8
deserts
   climate   122−5
   global   136−8
'development theorem'   54
diabatic heating   28−9, 30
diurnal sensitivity, of electricity demand   254
diurnal variation
   of albedo   159
   of radiation balance   156, 173
diurnal temperature range   126−7, 193
downwelling   36, 63
drought   134, 135−6, 271−83
   and atmospheric carbon dioxide   278−83
   definition of   271, 272−8
   severity of   277
dry-land farming   129−35 *passim*
dust   241

easterly waves   48−51
eccentricity   110−14, 142

eddy flux 25−6
eddy viscosity 168
Eemian interglacial 105
Ekman Drift 36−7, 45, 168
electricity demand, and weather 254
El Niño 63, 65, 67−8
energy, and atmospheric circulation 24−5, 25−8
energy balance 170
energy fluxes 25
  density 238−41
  man-induced 250−1
energy use
  and climate 254−7
  in future 241−3
Eocene 98−9, 136−7
equatorial oceans 44−5
equatorial rain forest, interception loss 198
equatorial trough 14, 46−7
equatorial wave disturbance 48−9
evaporation 125, 127
  in tropical forests 223−4, 226, 227, 229
  over oceans 41
evapotranspiration 123, 124, 130−5, 143, 169−76, 276
  see also potential evapotranspiration
evapotranspiration climatonomy 209
evapotranspiration rates, predicted 224−5
evolution of atmosphere, model 32
experimental studies, forests 221−3
extinction coefficients 191−2

fallow system 267
Ferrell cell 12
Fleer, H.E. 65, 68, 278
Flohn, H. 102, 103, 106, 112, 113−14, 209
flow of water, to evaporating surface 172
fluxes, advective and eddy 24, 25−6
  see also energy fluxes, latent heat flux, sensible heat flux
foliar transpiration 177
forecasting, extra tropical summer rainfall 148
forest clearance 137, 209−23

forests
  and interception loss 198−203
  microclimates of 190−5
  types 189
Frakes, L.A. 95, 97, 136
frictional drag 35
friction layer 165−6, 168
future climatic changes 257
future energy use 241−3

Gash, J.H.C. *et al.* 207, 224
general circulation model 64, 96, 106, 142, 246
geostrophic wind 18, 46, 165, 168, 224−6
glacial climate 74
glaciation level 94−5
glaciations
  and Milankovitch theory 103, 110−17
  and orbital variation 112−14
  Antarctic 98−102
  Pre-Cambrian 95−7
  Quaternary 103−7
  Tertiary 97−102
glaciers 73, 74
global deserts 136−8
global radiation 5, 6, 75−6, 154, 170
global temperature changes in recent past 255−7
global wind 35
grain production 266−71
  see also crops, wheat
grasslands
  albedo 156, 157−64
  microclimates 154−169
'Greenhouse effect' 31, 154, 244
Greenland ice core 105−7
Greenland ice sheet 87, 101, 102
growth
  of crops 261, 265
  of plants 261−6
growth periods 131
gyres 35, 96

Hadley cells, 8, 11, 12, 14, 26−30, 103, 122, 145−6
Halley, E. 15
harmonic analysis 58

Hart, M.H.   30, 31, 32−3, 96
Hastenrath, S.   44, 47
heat balance   153
heat budget
    oceanic   38, 44
    of polar regions   102, 103
heating coefficient   157
heat islands   250
heat transfer   166
high pressure belts   45
'hot tower convection hypothesis'
    27−30
humidity   41, 53, 124, 126, 157, 179,
    193, 224
hurricanes *see* tropical cyclones
hydrological cycle   4, 123
hydrological models   211, 278−82
hydrology
    of grasslands   176−7
    of forests   195−208

ice, seasonal formation of   112−3
ice-albedo feedback   145
icebergs   59
ice calving   116−17, 142
ice-sheets   87
    Antarctic   73−4, 102−3
    continental   73, 103, 105, 113
    Greenland   101, 102
    numerical modelling of   114−17
    physics of   88−9
IIASA Energy Systems Programme
    244
index cycles   22−3
index of continentality   58
Indian monsoon
    and global deserts   136−8
    rainfall   65, 68
Indonesia   14
infra red radiation   154
inland basins   125
insolation   110, 111−14
    see also global radiation
interception, of rainfall
    by vegetation   195
    in forests   3, 223−31
interception loss   180−2,   196−7,
    212−6, 221, 223, 224−9
    and wind speed   227

experimental catchments   222
in boreal coniferous forest   203
in coniferous stands in temperate
    deciduous area   202
in equatorial rain forest   198
in mediterranean forest   200
in temperate deciduous forest
    200−1
in tropical forest   199
vertical distribution of   214, 215
intertropical convergence zone (ITCZ)
    47, 136
inversions   86, 169
    see also trade wind inversion
irrigation   261, 267
isostatic equilibrium   88, 115−16
isotopic temperatures   98, 102

jet streams   12, 16, 17, 96−7, 127−8,
    142

Kellogg, W.W.   242, 249
Kelvin waves   37
Kennett, J.P.   99, 100, 101
Kew, drought statistics   278
kinetic energy   25, 35, 103
kinetic properties of climatic systems
    3
King, K.M.   169, 170, 172
Kukla, G.J.   59, 103, 111−12

lake-level fluctuations   139−41
laminar boundary layer   165, 166
Landsberg, H.   129, 192, 238, 252
lapse rates, polar temperatures   87
latent heat   25, 26
latent heat flux   25−30, 35, 38, 41, 43,
    46, 125−6, 138, 143, 169−76,
    179−80, 226, 245−6
lateral advection   44
leaf-area index   177−9, 191, 215, 216,
    224
leaf temperatures   175
leaves, resistances   177
Lettau, H.H.   168, 209, 229
'Little Ice Age'   254−5
long-wave radiation   156−7
low-pressure systems   48, 52

Manley, G.   90, 92, 94, 95
MANTA multilayer crop model 209–21, 224–5
marginal seas   99, 100
maritime arid climate   124
mass balance, Antarctic ice-sheet 74–5
mass budget   88
mean-sea-level pressure-field index   65
mediterranean cyclones   128
mediterranean forest   200
meltwater release   142
meridional circulation   11–12
Milankovitch theory   103, 110–17
Miocene   101
models
    astronomical theory of ice-age   114
    atmospheric moisture balance 123–4
    crop yield   269–70
    evolution of atmosphere   32
    general circulation   64, 96, 106, 142, 246
    monsoon circulation   136
    multilayer crop   209–21, 224–5
    numerical modelling of ice-sheets 114–17
    run-off   211, 213–14, 218–19
    soil moisture   182–5
    statistical dynamic climate   145
moisture advection   123
moisture transport   246
momentum   35
monsoons
    African   142–3
    and mountain topography   136
    and orbital variations   142–3
    Indian   65, 68, 136–8
    monsoon circulation model   136
    South Asian   15–17, 136
Monteith, J.L.   156, 157, 162, 168, 177, 265, 266

Namias, J. and Cayan, D.R.   62, 64
net radiation *see* radiation balance
Nimbus III data   125
non-radiative processes   84, 85
numerical modelling   64
    of ice-sheets   114–17

oasis effects   173–4
obliquity   110–14, 142
ocean currents   35–6, 38, 41
oceanic bottom water   99–100
oceanic climates   58
oceanic energy transfer   25
oceanic heat budget   38, 44
oceanic sedimentation, in Cenozoic 137
oceanic thermal anomalies (OTA) 63–4
oceans
    absorption of solar radiation   38–9
    carbon dioxide content   242
    circulation patterns   35–7
    effect in atmosphere   64
    polar   59–61, 86
    precipitation over   56
    salinity, early Tertiary   98, 100
    surface temperature anomalies 62–3
    tropical   44–5
Oerlemans, J.   86
Oligocene   101, 137
omega-block   55
orbital variations   110–12, 142–3
    and glaciation   103, 112–14
orographic influence   56
oxygen isotope temperature estimates 98, 102

pack-ice   59
Palaeocene   98, 137
palaeogeography   98, 100, 101
    18,000BP continents   108–9
Penman-Monteith equation   171, 179–80, 195
perennial ice   73
permafrost   73
planetary albedo   8
plant growth   261–6
Pleistocene, pluvial period   138
'point of recurvature'   52
pollutants   241
pollution   241–53
polar night   75
polar climates   75–87
polar oceans   59–61, 86–7
polar regions
    atmospheric circulation   24

global radiation   75–6
see also Antarctic, Arctic
Pollard, D.   114
potential energy   25
potential evapotranspiration   130–1, 132, 135, 169, 172, 182, 266
potential recharge   276
potential vorticity   46, 51
precession   110–14, 142
precipitation   139, 142–3, 147, 196, 197, 199–203, 209
    over polar oceans   87
    over temperate oceans   56–8
    see also rainfall, snow
pressure, sea-level   68
pressure correlations   14
pressure force   13, 15
psychrosphere   100
pycnocline   74

Quaternary   101
    glaciations   103–7

radiation
    absorption by oceans   38–9
    and forest microclimates   191–3
    and temperature   125–7
    Antarctic Studies   85
    global   5, 6, 75–6, 77–8, 79–81, 154, 170
    infra-red, absorption of   154
    long-wave   8, 38, 40, 81, 156–7
    solar   75–86, 94, 157–8
radiational index of dryness   153–4
radiation balance   8, 9–10, 38, 84–5, 125–6, 144, 153, 170–1, 193
    Antarctic   86
    Arctic surface   82–3
    diurnal variations of   156
    in grasslands   154–7
    in forests   192–3
    in oases   173
rainfall   14, 52, 64, 273, 274
    condensation type   148
    cumulus convection type   148
    desert   124, 127–30
    forecasting   148
    tropical   47–8, 52–3
    see also precipitation

rainfall deficit   272
rainfall teleconnections   68
resistances   177–9, 266
    see also aerodynamic resistance
reference crop evapotranspiration   182–3
root crops   261, 262, 264
Rossby number   102–3
Rossby waves   17
rotation of earth   8, 12, 36
    and Pre-Cambrian Ice Age   96–7
roughness   252
roughness length   167–9, 195
run-off   137, 139, 224, 246, 276
    and deforestation   223, 228–31
    in tropical forests   223–31
    in urban climates   252–3
    models   211, 213–14, 218–19
    rates   220, 221, 223
Rutter, A.J. *et al.*   204, 205, 224, 227
Rutter Model   197, 204–5

Sahara   8, 127, 131–5, 138
Sahel, W. Africa   134–5
Sakamoto, C. *et al.*   267, 269, 270
salinity of oceans, early Tertiary   98, 100
sea-ice   59, 73, 74, 100, 112–13, 116–17, 142, 246–9
seasonal variations
    in albedo   162–4
    in carbon dioxide   242–3
    of electricity demand   254
sea surface temperature (SST)   14, 47, 66–7
    anomalies   62–3, 68
    polar oceans   86–7
Sellers, P.J.   168, 215, 227
Sellers, P.J. and Lockwood, J.G.   207, 210, 224, 228
sensible heat   126, 155
sensible heat flux   25–30, 35, 38, 41–2, 126, 138, 143, 166, 174–6, 179–80
Shukla, J. and Mintz, Y.   147–8, 177, 209, 229, 231
Shuttleworth, W.J.   182, 183, 184, 224
slantwise convection   11
snow   73

on mountains 89–95
seasonal 73
snow-albedo feedback 246–7
snow-balance in Britain, present day 90–5
snow cover 87, 90–1, 94, 95, 102
snowmelt 89, 281–2
    see also ablation
soil moisture 4, 130–1, 132–5, 143, 153–4, 176–7, 215–18, 228–9 246–7, 250, 275, 277–8, 279–81
    and drought model 279–81, 283
    and plant growth 265–6
    deficiency 271
    in deserts 146–8
    models 182–5
soil water availability 130–1, 172
soil water potentials 172
soil temperature 143–4
solar declination 8
solar luminosity 30
solar power 241
solar radiation 75–86, 94, 157–8
    see also global radiation
solar refraction 75
southern oscillation 14–15, 65–8
spiral layer 165, 166, 168–9
statistical dynamic climate model 145
Stefan-Boltzmann law 156
stem area index 224
stemflow 196, 197, 198–203, 205
stomatal control (conifers) 189
stomatal resistance 178–9
storms, tropical 51–3
    see also thunderstorms
stratus clouds 5
subtropical anticyclones 45, 103
subtropical jet streams 12, 16, 17, 127–8
sunshine duration 157, 211
surface layer 165, 166
surface pressure — patterns 248
surface resistance 196, 178–82
surface temperatures, in Eocene 99–100
    in Cretaceous 98
    polar 86–7
    see also sea-surface temperatures (SST)

surface windspeed 252
systems 1–3

teleconnections 68, 135–6, 278–9
temperate deciduous forest 200–1
temperate oceanic weather systems 54–8
temperature
    and plant growth 265
    in forest microclimates 193, 194
    interannual changes 66–7
    lapse rates, polar 87
    of clouds 8
    of oceans 37
Tertiary glaciations 97–102
thermal patterns 18–22
thermal properties of climatic systems 3
thermal steering effect 54
thermal vorticity 54
thermal wind 18, 54, 55
thermocline 37, 45
thickness lines 18–20
thickness patterns 55
throughfall 180–1, 196, 197, 198–203, 204
thunderstorms 52, 56, 125–6
Tibetan high 17
topographical factors and aridity 124–5
trade winds 12, 13–14, 45–6, 47
trade wind inversion 13–14, 46, 51, 123
transpiration 169, 212–15, 224–9
    see also evapotranspiration
tropical cyclones 51–2, 53
tropical depression 51
tropical forest 199
tropical oceanic weather systems 45–53
tropical oceans 44–5
tropical rainfall 47–8
tropical storms 51–3
tropical synoptic disturbances 48
turbulent plume 99
turbulent transfer 35, 169–70
typhoons *see* tropical cyclones

upwelling 36–7, 38, 44, 59, 63, 65–7, 137

urban climates 252–3
urban heat island 253

vapour density 193, 194
vapour pressure deficit 227
vegetation
    and desert climate 143–6
    and water balance 153–4
vertical advection 63
vertical exchange 195
vertical motion 54
vertical wind shear 166
volcanic eruptions 141, 142
vorticity 44, 46
    thermal 54
vorticity equation 51
Vowinckel, E. and Orvig, S. 76, 79, 84, 86

Walker circulation 14–15, 65
Walker, G.T. 65, 68, 146, 147, 278
warm-temperate evergreen broadleaf forest 192
waste heat 250–1
water balance 131, 153, 180–8, 275–7
    and evapotranspiration 130–5

and vegetation 153–4
of closed lake 138–9
water power 241
water vapour 53, 157
    density 41
    flux 230–1
water yield 137, 153
wave patterns 22–3
westerlies 103
    see also trade winds
wheat 263, 266–7, 268
    yield 269, 270
windblown snow 92
wind power 241
winds 10–11
    and tropical cyclones 52
    thermal 54, 55
    upper 16, 17
    westerlies 4, 22–3, 76
wind speed 13, 169, 226, 227
wind stress curl 44, 45
wind systems 4
World Climate Programme 244
Würm glaciation 105–9, 137–8

zenith angle 161
zero-plane displacement 168, 195
zonal index 22–3, 55, 127